FOOD CHEMISTRY

FOOD CHEMISTRY

Editor

Thiemo Dahl

scitus academics

Food Chemistry

Edited by **Thiemo Dahl**

This Book is Printed on Acid-Free Paper. ♾

ISBN: 9781681178912

Library of Congress Control Number: 2017948919

SCITUS Academics LLC,

616, Corporate Way, Suite 2, 4766,

Valley Cottage, NY 10989

www.scitusacademics.com

Notice

Printed in United States of America.

Preface

Food chemistry is the study of chemical processes and interactions between all biological and non-biological components of foods. The enduring objectives of research in food chemistry are to understand relationships between the structure and functional properties of food molecules and to advance the nutritional, safety and organoleptic aspects of food. It is similar to biochemistry in its main components such as carbohydrates, lipids, and protein, but it also includes areas such as water, vitamins, minerals, enzymes, food additives, flavors, and colors. This discipline also encompasses how products change under certain food processing techniques and ways either to enhance or to prevent them from happening. An example of enhancing a process would be to encourage fermentation of dairy products with microorganisms that convert lactose to lactic acid; an example of preventing a process would be stopping the browning on the surface of freshly cut red delicious applies using lemon juice or other acidulated water. Instruments that are popular in the vicinity of chemistry are employed in food chemistry. Flavors, preservatives, emulsifiers, thickeners, stabilizers, sweeteners, colors are some of the materials that are produced from food chemistry. And a consideration of the development of these materials from their crude source through research, development, production, regulation and commerce; would telltale of how expansive is the importance of food chemistry.

'Food Chemistry' deals with the advancement of the chemistry and biochemistry of foods or the analytical methods or approaches used concerning to major and minor components of food, their nutritional, physiological, sensory, flavor and microbiological aspects; bioactive constituents of foods, including antioxidants, phytochemicals, and botanicals. It also provides a review of traditional and non-traditional food preservation approaches and ingredients used as food additives, contaminants, and other agro-chemicals, together with their metabolism, toxicology and food fate. The objective of this book is to provide knowledge appropriate for students, university researchers, and in general, for anyone wishing to obtain knowledge of food processing and to improve the food product quality.

Table of Contents

CHAPTER 1

The Contribution of Alliaceous and Cruciferous Vegetables to Dietary Sulphur Intake

Joanne F. Doleman [a] , Katrijn Grisar [a] , Lena Van Liedekerke [a] , Shikha Saha[a]

[a], Mark Roe b , Henri S. Tapp c , Richard F. Mithen a,⇑

[a] Food & Health Programme, Institute of Food Research, Norwich, UK

[b] Food Databanks, Institute of Food Research, Norwich NR4 7UA, UK

[c] Analytical Sciences Unit, Institute of Food Research, Norwich NR4 7UA, UK

ABSTRACT

Despite its importance in many areas of human metabolism, there are no recommended daily intake guide lines for sulphur. It is generally assumed that most dietary sulphur originates from intake of methionine and cysteine. We estimated sulphur intake from food diaries, and validated the results with the use of a duplicate diet analyses. Sulphur intake estimations were highly correlated with that obtain through an elemental analysis of duplicate diets, with a mean ± sd daily intakes of 956 ± 327.9 mg estimated from diet diary analyses and 935 ± 329.9 mg estimated by a duplicate diet analyses. Sulphur intake from alliaceous and cruciferous vegetables contributed up to 42% of total sulphur intake. Daily intake estimation comparisons through diet diary analyses and duplicate diet for other elements showed good agreement, except for sodium and zinc, in which analyses of 24 h diet dairies overestimated intake by 35% and 52%, respectively.

Keywords: Sulphur; Sulphur amino acids (SAA); Sulphate; Diet diary; Duplicate diet

1. INTRODUCTION

It is generally considered that most dietary sulphur is derived from the sulphur-containing amino acids (SAA), methionine and cysteine, with other forms of organic sulphur, such as the specialised metabolites that accumulate in alliaceous and cruciferous vegetables contributing a very small percentage to total sulphur intake. While there are recommended daily allowance (RDA) for SAA, based upon the requirements to maintain nitrogen balance, there is not a RDA or an 'adequate intake estimation' for sulphur, despite its importance in many aspects of metabolism (Nimni, Han, & Cordoba, 2007). This lack of recommended intakes is reflected in current methods of dietary analyses that do not provide an estimate of sulphur intake, and do not attempt to identified important sources of sulphur in the diet other than SAA.

The likely major sources of sulphur in the diet other than SAA are inorganic sulphate in drinking water and foods, and the sulphur-containing specialised metabolites that are characteristic of *Allium* vegetables, such as onions, leeks and garlic, and cruciferous vegetables, such as cabbages, kales and broccoli. The levels of sulphate in tap water varies around the world, with low concentrations measured in tap water from the Netherlands (below 260 μ/L) and higher levels in central Canada (up to 22 mmol/L). A water quality survey of British tap water suppliers measured a mean sulphate concentration of 620 μmol/L, and a maximum of 2458 μmol/L (Powell, Bailey, & Jolly, 1987). Sulphate ingestion has been of interest due to the potential importance of sulphate reducing bacteria (SRB) in the colon, and their association with ulcerative colitis, irritable bowel disease, and Crohn's disease (Salonen, de Vos, & Palva, 2010), and inorganic sulphur is required for synthesis of 3′-phosphoadenosine-5′-phosphosulfate required for detoxification of endogenous metabolites and xenobiotics. The consumption of both alliaceous and cruciferous vegetables have been associated with health benefits (McNaughton and Marks, 2003 ; Nicastro *et al.*, 2015), including a reduced risk of developing various cancers including breast, prostate, lung, pancreatic and gastrointestinal (Herr & Buchler, 2010), cardio vascular disease (Zhang et al., 2011), reduced risk of type-2 diabetes (Kurotani et al., 2013), and protection from neurodegenerative disease (Giacoppo et al., 2015). For cruciferous vegetables, these health benefits are associated with the consumption of glucosinolates, the sulphur-rich compounds that accumulate in cruciferous vegetables, and their metabolic derivatives (Armah et al., 2015; Maheo et al., 1997; Yamagishi &

Matsui, 2016). Likewise, for alliaceous vegetables, these putative health benefits are attributed to S-2-propenyl-l-cysteine S-oxide and other related cysteine S-oxides (Block, 1992) which have been shown to inhibit cell proliferation and induce cell death of cancerous cells (Borlinghaus, Albrecht, Gruhlke, Nwachukwu, & Slusarenko, 2014).

In this paper, we provide a method for evaluating sulphur intake based upon food diary analyses and validate this with the use of a duplicate diary analyses. Subsequently, we provide an estimation of the proportion of dietary sulphur that comes from non-SAA sources. The study was extended further to look at daily intakes for other commonly consumed minerals including sodium, calcium, iron, zinc, magnesium, phosphorous and potassium and the validity of current intake estimations from diet diary analysis.

2. MATERIALS & METHODS

2.1. Analysis of commonly consumed foods for sulphur content

Thirty-two commonly consumed foods were purchased for sulphur analysis including fruits and vegetables, meat, fish, pasta, rice and eggs. Fruits and vegetables were analysed raw and included spring onions, red onions, brown onions, onions, leek, garlic, parsnip, potato, sweet potato, green pepper, courgette, carrot, tomato, sugar snap peas, dwarf beans, Braeburn apple, lettuce, wild rocket, radish, savoy cabbage, sweetheart cabbage, curly leaf kale, broccoli and cauliflower (frozen). Cooked ham, plain flour, white bread and minced beef (20% fat) were analysed as purchased. Chicken breast fillets and boneless cod fillets were pan fried without fat prior to analysis. Long grain rice, spaghetti pasta and medium free range eggs were boiled prior to analysis. All samples were weighed, freeze dried and weighed again prior to grinding to a fine powder. Freeze dried samples were sent to Eurofins UK (Eurofins Food Testing UK Ltd, i54 Business Park, Valiant Way, Wolverhampton, WV9 5 GB) for analysis of amino acids and minerals, in which the sulphur, sodium, potassium, calcium, magnesium, phosphorus, iron and zinc content of each sample was measured. Eighteen of the vegetable samples were also analysed in-house using established methods for glucosinolate content (Gasper et al., 2005).

For analysis of sulphate, 25 mg PVPP (polyvinylpolypirrolidone; washed and ground) was soaked in 1 ml H_2O, overnight at 4 °C. 20–30 mg ground, frozen sample was added and mixed and shaken for 1 h at 4 °C. The sample was incubated for 15 min at 95 °C, centrifuged 15 min at 4 °C and supernatant filtered by PVDF syringe filter into an HPLC vial for analysis by LC-MS with a Agilent 1100 HPLC system and a mass spectroscopy detector (Agilent Technologies, Waldbronn, Germany). Samples were eluted at 0.5 ml/min with a gradient mobile phase. Mobile Phase A was 200 mM ammonium acetate buffer at pH 4.0 and B was 60% acetonitrile. The gradient started at 5% solution A increasing over 15 min to 90% A and finally re-equilibrated to 5% A for 5 min. Sulphate was monitored using mass spectrometry in selective ion monitor mode ($m/z = 97$) in negative polarity with electrospray ionization. The quantification was performed using matrix match calibration curve.

2.2. Sulphur database construction

An on-line literature search was conducted to identify sources of published food composition data for the sulphur content in foods. One key paper, two databases and five book supplements were identified that contained sulphur content information and these were imported into our own sulphur database containing 1222 different foods from the different reference sources (FoodStandardsAgency, 2008; Gnagnarella and Salvini, 2008; Holland *et al.*, 1988; Holland *et al.*, 1991; Holland *et al.*, 1992; Holland *et al.*, 1989; Masters and McCance, 1939 ; PublicHealthEngland, 2015). To aid navigation through the database, the foods were grouped into the following eleven categories: vegetables, fruits and nuts, cereals, meat, fish, dairy, beverages, condiments, fats, cooked dishes and miscellaneous foods.

2.3. Computational analysis software

DietPlan6 (http://www.foresoft.co.uk) is a computer programme for the nutrition analysis of recipes, meals, menus and personal weighed-intake food diaries and contains approximately 5500 types of UK foods and allows other national food data tables to be imported and more foods to be added by the user. The database also includes Dietary Reference Values (Ashwell, 1991) Food Portion Sizes (MAFF, Ministry of Agriculture, & F. a. F, 1993) and Food Labelling Data (EU Regulation 1169/2011). For each of the different foods contained within dietary intervention diaries used in this study a new sulphur database code was developed that incorporated all the existing DietPlan6 nutritional information supplemented with the sulphur content from the newly developed database.

2.4. Validation of sulphur database

The consistency of the collected sulphur data was compared between the different data sources. Few differences were identified and where differences arose we chose the sulphur data to use in the database on a case by case basis. Where data was available from more than one source, the value used was a combination of the most up to date and/or the most frequently quoted data in the source documents. Recipes from a previous publication (Masters & McCance, 1939) were compared to recent recipes to ensure they were still produced using the same kinds of ingredients to maintain validity of the historic sulphur data prior to inclusion in the database.

2.5. Diet diaries

Diet diaries from the BASH human dietary intervention study (rec ref 12/EE/0313 clinical trial Number NCT01929564), which was carried out in collaboration between the Institute of Food Research and University of Reading, were used for dietary analysis. Detailed instructions describing how to complete the diet diary were provided, including photographs of food portion sizes and drinking vessels and their volumes, as well as details of how to record both processed meals and homemade recipes. Participants were instructed to record the details of all meals, snacks and drinks consumed, estimating or measuring the quantity of each item and the time of consumption. Participants were encouraged to be as accurate as possible when completing the dairies and were provided with an example completed diary.

In total, 41 one-day diet diaries from 32 different individuals were coded for the computational analysis of calories and nutrients and sulphur content in DietPlan6 software using the sulphur database codes developed specifically for this analysis. Dishes, recipes and commercial food were broken down into their individual ingredients. Daily sulphur intake was estimated and used to identify, low, moderate and high consumers of sulphur.

2.6. Duplicate diet analysis

Eighteen diet diaries encompassing low, moderate and higher consumers of sulphur, as identified from the initial 41 diet diaries, were selected. The recorded food items were purchased from local supermarkets and stored as appropriate, until required. Using the exact cooking methods and quantities consumed, the total food and drink intake for the 24-h period was recreated and blended to produce a homogeneous sample to represent each diet diary. Diet samples were stored at −20 °C prior to inorganic and moisture content analysis by Eurofins Food Testing Laboratory (Wolverhampton, UK).

2.7. Statistical analysis

Statistical analyses were performed using Minitab version 17.2.1, and Matlab (version 8.5), with the Statistics and Machine Learning Toolbox (version 10.0). Differences between duplicate diet analysis and DietPlan6 estimations were compared using paired t-tests and presented using Bland Altman plots (Bland & Altman, 1986). Linear trends in the differences were assessed from the significance of the Pearson correlation between the sum and difference of the two estimates.

3. RESULTS AND DISCUSSION

3.1. Sulphur partitioning in commonly consumed foods

Thirty-two commonly consumed foods were analysed for total sulphur content and the proportion of sulphur from SAA, and a subset of 18 vegetables were also analysed for the contribution of sulphur from glucosinolates and sulphate (Table 1). As expected the proportion of SAA was very high in fish, chicken and minced beef (74–97%). SAA are also the predominant source of sulphur compounds in rice, pasta and eggs. In alliaceous vegetables, a low proportion of the total sulphur in the vegetables was in the form of SAA (e.g. garlic 10.5%, red onions 10.5%), whereas for starchy vegetables the proportion of total sulphur in the form of SAA was much higher (e.g. sweet potato 80.8%, parsnip 62.7%). In wild rocket, radish, broccoli, sweetheart cabbage and curly leaf kale a proportion of the total sulphur content was from glucosinolates with the contribution ranging from 7.4% in curly leaf kale to 59.3% in radish. Sulphate contribution ranged from 1.0% (garlic) to 61.7% (carrot) of the total sulphur.

3.2. Sulphur database construction and estimating dietary sulphur intakes

The sulphur intake determined from analysis of each 24-h food diary analysed ranged from 390 mg to 1414 mg total sulphur consumed in a 24-h period, and enabled the identification of low (<750 mg/day), moderate (between 750 and 1250 mg/day) and high sulphur intake (>1250 mg/day) diets across the 41 diet dairies analysed.

Table 1. Sulphur content in µmoles/g dry weight of 32 commonly consumed foods, portioned to show the proportion derived from the sulphur amino acids methionine and cysteine and other sulphur (including sulphate). The sulphur partitioning of sulphur containing metabolites of 18 commonly consumed vegetables including tryptophan and methionine derived glucosinolates, sulphur amino acids cysteine and methionine, sulphate and other sulphur is also included. (– indicates not measured).

	Sulphur amino acids Cysteine & Methionine µmoles/g dry weight (% of total sulphur)		Other sulphur including sulphate µmoles/g dry weight (% of total sulphur)		Sulphur partitioning of sulphur containing metabolites in 18 commonly consumed vegetables			
					Sulphate µmoles/g dry weight (%)	Met derived glucosinolate (µmoles/g dry weight)	Trp derived glucosinolate (µmoles/g dry weight)	Other sulphur excluding glucosinolates (µmoles/g dry weight)
Spring onion	23.6	(26.5)	65.3	(73.5)	17.1 (19.2)	0.0	0.0	48.2
Red onion	12.7	(10.5)	108.6	(89.5)	–	–	–	–
Brown onion	9.3	(12.0)	67.7	(88.0)	6.6 (8.6)	–	–	–
Leek	27.5	(22.7)	93.5	(77.3)	16.2 (13.4)	0.0	0.0	77.3
Garlic	26.5	(10.5)	225.5	(89.5)	2.4 (1.0)	0.0	0.0	223.4
Parsnip	12.8	(62.7)	7.6	(37.3)	10.6 (51.7)	0.0	0.0	0.0
Potato	20.9	(54.4)	17.5	(45.6)	13.0 (34.0)	0.0	0.0	4.5
Sweet potato	23.8	(80.8)	5.7	(19.2)	8.8 (29.8)	0.0	0.0	0.0
Green pepper	19.5	(34.4)	37.2	(65.6)	–	–	–	–
Courgette	43.1	(72.8)	16.1	(27.2)	11.7 (19.7)	0.0	0.0	4.4
Carrot	13.0	(42.1)	17.8	(57.9)	19.0 (61.7)	0.0	0.0	0.0
Tomato	25.1	(59.3)	17.3	(40.7)	13.6 (32.1)	0.0	0.0	3.7
Sugar snap peas	26.3	(44.7)	32.6	(55.3)	19.5 (33.0)	0.0	0.0	13.2
Dwarf beans	32.5	(61.4)	20.5	(38.6)	11.7 (22.1)	0.0	0.0	8.8
Lettuce	26.1	(51.0)	25.1	(49.0)	18.9 (37.0)	0.0	0.0	6.2
Wild rocket	82.4	(14.9)	469.1	(85.1)	319.9 (58.0)	126.6	1.8	149.2
Radish	17.8	(49.4)	18.3	(50.6)	27.0 (74.7)	19.6	33.2	0.0
Savoy cabbage	40.7	(21.6)	148.0	(78.4)	–	–	–	–
Cauliflower	44.3	(23.3)	145.6	(76.7)	–	–	–	–
Broccoli	56.1	(20.4)	218.5	(79.6)	87.7 (31.9)	24.2	10.3	130.8
Sweetheart cabbage	28.5	(28.9)	70.1	(71.1)	26.0 (26.4)	14.6	5.7	44.1
Curly leaf kale	49.4	(16.0)	259.5	(84.0)	196.7 (63.7)	17.9	6.8	62.8
[illegible]	[illegible]	[illegible]	0.7	(16.0)				
Cooked ham	140.2	(80.0)	35.1	(20.0)	–	–	–	–
Minced beef	117.9	(74.7)	39.9	(25.3)	–	–	–	–
Chicken breast	247.6	(97.4)	6.6	(2.6)	–	–	–	–
Cod	288.8	(85.0)	51.1	(15.0)	–	–	–	–
Eggs	232.6	(96.8)	7.8	(3.2)	–	–	–	–
Long grain rice	31.2	(91.9)	2.8	(8.1)	–	–	–	–
Spaghetti	37.8	(88.5)	4.9	(11.5)	–	–	–	–
Plain flour	35.2	(91.8)	3.1	(8.2)	–	–	–	–
White bread	31.8	(45.0)	38.9	(55.0)	–	–	–	–

3.3. Sources of sulphur in the diet

Diet diary analysis enabled us to identify the sources of sulphur in the diet. We categorised the sources into eight groups which were meat, fish, dairy, vegetables, fruits, wheat and starchy foods, dishes (e.g. prepared foods such as stock cubes); and miscellaneous (foods that did not easily fit into any of the other categories e.g. eggs, coffee, wine). All 41 diet diaries were analysed and the specific sources of sulphur in each diary determined. The total sulphur from meat and fish consumption contributed up to 458 mg (38.2%) and 884 mg (62.5%) respectively of total sulphur intake. Wheat and starchy foods contributed up to 594 mg (46.9%); dairy contributed up to 426 mg (35.3%) and vegetables contributed a maximum of 359 mg (37.9%) of the total sulphur intake (Table 2). We were particularly interested in the sulphur intake originating from cruciferous and alliaceous vegetables consumption and as such performed an additional analysis to investigate what proportion of the sulphur intake came from these vegetables. Cruciferous and alliaceous vegetables combined contributed up to 41.7% of the total sulphur intake. The highest contribution from cruciferous vegetables was 39.9% and the maximum contribution from alliaceous vegetables in any one diet diary analysed was 7.3% (Table 2). In 20 of the 41 diet diaries analysed cruciferous and alliaceous vegetable consumption represented more than 50% of the total sulphur consumed from all vegetables.

Table 2. Summary data table showing the mean sulphur intake (mg), % of total sulphur intake, intake range and the standard deviation across 8 food categories used in the analysis of 41 × 24-h diet diaries analysed using DietPlan6 with the sulphur database incorporated. The contribution of sulphur from the consumption of cruciferous and /or alliaceous vegetables is also included.

Food Category	Mean sulphur intake, mg	Mean % of total sulphur intake	Range mg (%)	Standard Deviation
Meat	142.1	15.8	0–458 (0–38.2)	135.5
Fish	93.1	10.4	0–884 (0–62.5)	200.5
Dairy	130.7	14.5	0–426 (0–35.3)	98.6
Vegetables	149.6	16.7	0–359 (0–37.9)	101.4
Fruits	22.2)	2.5	0–189 (0–18.6)	32.7
Wheat and starchy foods	152.8	17.0	0–584 (0–46.9)	119.1
Dishes	66.0	7.3	0–516 (0–62.0)	121.4
Other	141.9	15.8	4–559 (0.5–48.0)	125.9
Total sulphur consumed	898.3		361–1417	296.2
Cruciferous & alliaceous vegetables	67.3	7.5	0–335 (0–41.7)	79.9
Alliaceous vegetables	9.7	1.1	0–41 (0–7.3)	13.2
Cruciferous vegetables	57.6	6.4	0–320 (0–39.9)	78.2

3.4. Validity of the sulphur data in the historic databases

Of the 32 different commonly consumed foods selected for sulphur analyses, including vegetables, fruit, meat, fish, pasta and rice, 28 were present in the sulphur database we constructed (absent were garlic, courgette, sugar snap peas and wild rocket). Much of the data collated in the sulphur database originated from the old components dataset of McCance and Widdowson's 'The Composition of Foods Integrated Dataset' (PublicHealthEngland, 2015), the accuracy of this historical sulphur data was checked against the current sulphur content of various foods and vegetables by performing an in-house sulphur analysis of 32 commonly consumed foods. Samples were prepared and analysed as described in the materials and methods section for total sulphur content. The sulphur content of the analysed foods was compared to that in the sulphur database. In most cases, we could show good agreement between our in-house sulphur data and that in the collated sulphur database (Fig. 1).

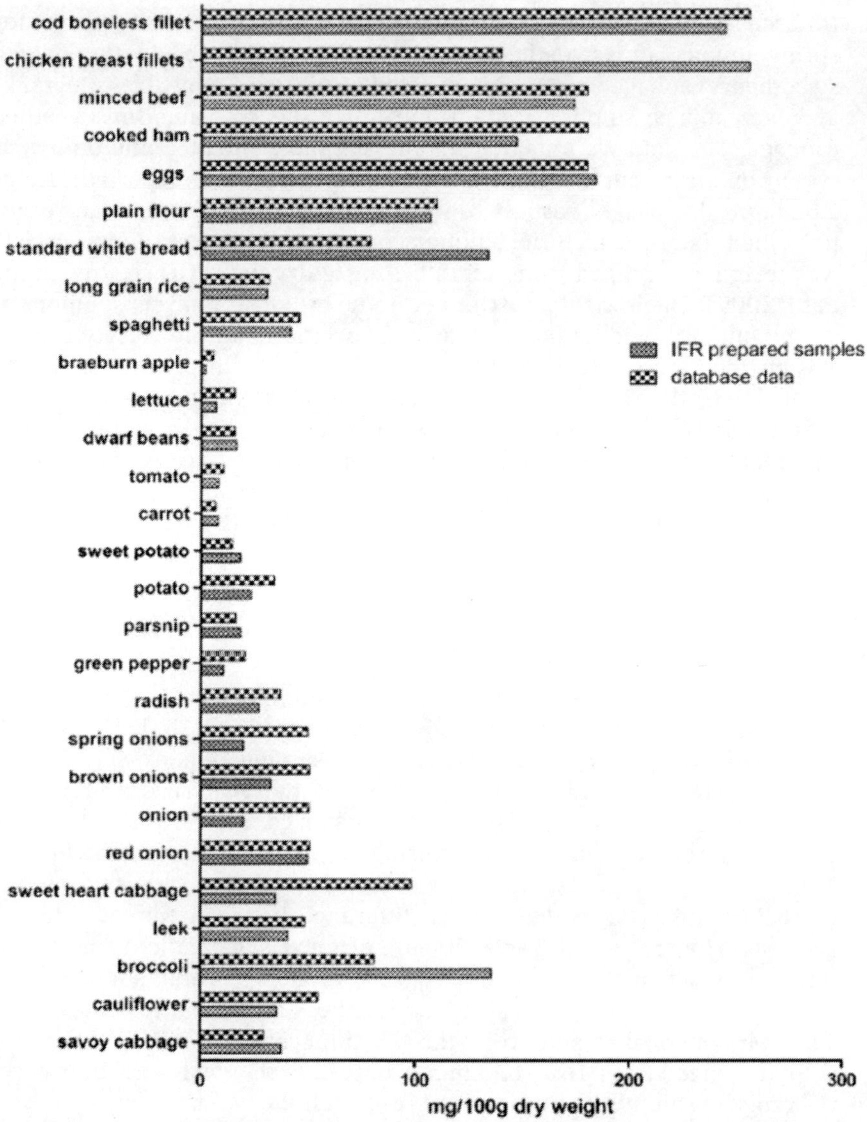

Fig. 1. Sulphur content of 28 commonly consumed foods from the sulphur database versus Eurofins UK analysis of IFR in-house prepared samples. The sulphur content in mg per 100 g fresh weight of each food tested was comparable to that in the sulphur database, although some differences are observed. There does not appear to be any obvious systematic bias in the analysis.

Differences between the two sources of data were evident for onions, spring onions and sweetheart cabbage. The difference in the data for sweetheart cabbage was not unexpected as we were only able to obtain an average cabbage sulphur content and not a specific value for sweetheart cabbage. The data we obtained for the sulphur content of the onions and spring onion by our in-house analysis was lower than expected (1.9 and 2.85 mg/g dry weight respectively), and this was also lower than recently published data in which the sulphur content of 9 different onion varieties was tested and ranged from 3.5 to 5.1 mg/g dry weight (Lee, Yoo, Jifon, & Patil, 2009). The low sulphur content in the in-house analysis of onions and spring onions could be due to seasonal variation and the variety tested in this instance. The historic sulphur data for onions and spring onions, which aligned with the recently published data on the sulphur content of onions, was used in the sulphur database (Lee et al., 2009). Although some differences were observed, there did not appear to be any obvious systematic bias indicating that the sulphur data included in the database was representative of foods currently available to consumers.

3.5. Duplicate diet analysis

Duplicate diet analysis enabled us to assess the validity of the sulphur intake estimates compared to the historical sulphur data used in diet diary analysis. Of the 41 diet diaries analysed a subgroup of 18 diet diaries that encompassed 6 low, 6 moderate and 6 high sulphur intakes were analysed by duplicate diet analysis as described in the materials and methods section. These analytical data were compared to the sulphur intakes calculated from diet diary analysis using the Bland Altman method (Fig. 2). There was a strong positive linear association between the two methods ($r = 0.73$) and little evidence of a difference between the two methods (bias = 20.2 mg; $P = 0.73$; Table 3) or of any systematic trend in the differences against the mean ($P = 0.97$; Table 3). Total sulphur intake estimated by diet diary analysis across the 41 diet diaries ranged from 11.3–44.2 mmol/day and from the 18 duplicate diets the total dietary sulphur ranged from 16.1–48.3 mmol/day, thus showing little difference in the range of sulphur intake estimated between the two methods.

The total daily sulphur intake measured by both methods in this study is comparable to the daily sulphur intake estimations proposed by Florin and Ingenbleek et al. (Florin *et al.*, 1991 ; Ingenbleek and Kimura, 2013). In other studies a moderately high total sulphur intake has been reported to be 38.2 mmol/day (high 42.2 mmol/day & very high 57.6 mmol/day) (Curno, Magee, Edmond, & Cummings, 2008). Using these figures a portion of the diet diaries in this study falls into the moderately high and high total sulphur intake category, however none of the diet diaries analysed could be classified as very high sulphur intake.

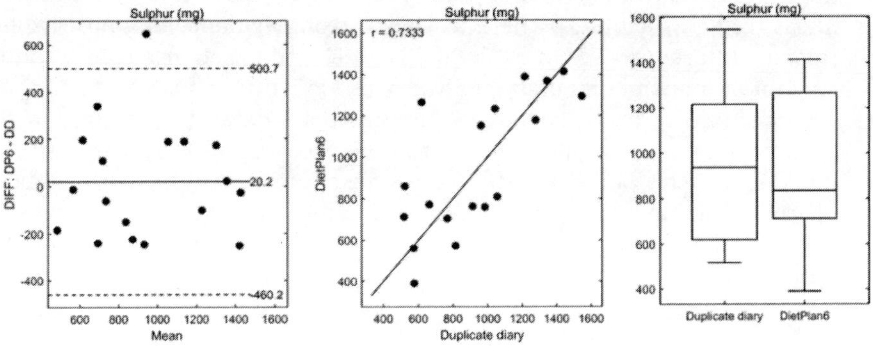

Fig. 2. Sulphur intake estimates of 18 × 24-h dietary intake diaries using DietPlan6 and duplicate diet analysis. *Left panel,* Bland and Altman plot (Bland & Altman, 1986) of the differences, DietPlan6 – duplicate diary, against the mean of the two estimates. Also shown is the bias given as the mean difference and the 95% limits of agreement (mean difference ± 2 × standard deviation). *Central panel,* scatterplot of DietPlan6 estimates against duplicate diary values. Also shown is the Pearson correlation between the two methods and the line of unity. *Right panel,* boxplots summarising the distribution of the two estimates.

Table 3. Summary statistics for a selection of constituents estimated during the diet diary analysis by DietPlan6 and by duplicate diet analysis.

Constituent	*n*	DietPlan6		Duplicate diet		*DIFF: DP6-DD*			Trend
		Mean	±SD	Mean	±SD	Mean	±SD	*P diff*	*P trend*
Sulphur (mg)	18	956	±327.9	935	329.9	20.2	240.2	0.7251	0.9726
Sodium (mg)	18	2500	±1752	1764	1295	736	612	**0.0001**	**0.0003**
Calcium (mg)	18	890	±378.7	955	400.4	−65	299.8	0.3707	0.7576
Iron (mg)	18	14.33	±5.446	13.47	6.696	0.86	4.355	0.4129	0.2173
Magnesium (mg)	18	322.6	±96.65	339.2	98.44	−16.6	47.92	0.1610	0.8795
Phosphorus (mg)	18	1485	±460.1	1349	458.6	136	290.8	0.0625	0.9835
Potassium (mg)	18	3466	±1237	3785	1149	−319	758	0.0918	0.6297
Zinc (mg)	18	11.07	±4.724	8.36	3.110	2.71	2.791	**0.0007**	**0.0079**

n, sample size; *SD*, standard deviation; *DIFF: DP6-DD*, paired difference in the estimates by DietPlan6 and duplicated diary analysis; *P diff*, significance of difference based on paired *t*-test; *P trend*, significance of linear trend based on the correlation between the sum and difference of the two estimates; mg, milligram.

3.6. Analysis of sodium in the diet using duplicate diet and diet diary analyses

As both the duplicate diet analysis and diet diary analysis provided information on other minerals we also investigated the agreement between the two methods for dietary sodium (Fig. 3) and various other minerals

(Table 3). For sodium, there was a strong positive linear association between the two methods ($r = 0.963$). However, DietPlan6 showed a tendency to produce proportionately higher estimates as compared to duplicate diets (ratio of standard deviations, DP6/DD = 1.35, significance of linear trend, $P = 0.0003$). There was also strong evidence of a positive bias (mean difference = 736 mg, $P = 0.0001$). The dataset used for sodium content of foods in diet diary analysis was McCance and Widdowson's 'The Composition of Foods' integrated dataset from 2008 (FoodStandardsAgency, 2008). In recent years' links between sodium intake and health have led the food industry to lower the salt content of processed foods and it was therefore possible that the sodium data used within the diet diary analysis could underestimate current intake.

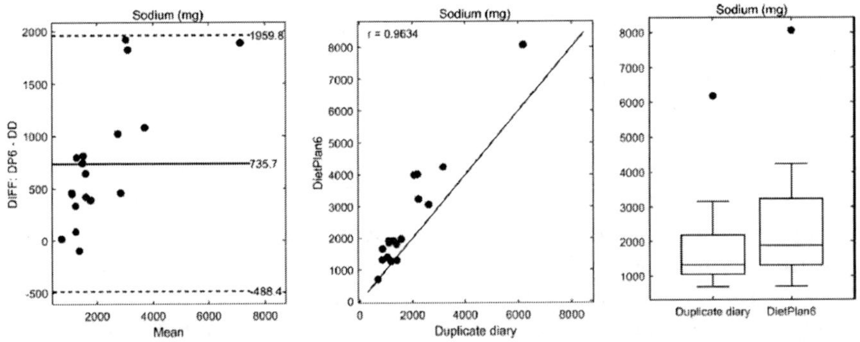

Fig. 3. Sodium intake estimates of 18 × 24-h dietary intake diaries using DietPlan6 and duplicate diet analysis. *Left panel*, Bland and Altman plot (Bland & Altman, 1986) of the differences, DietPlan6 – duplicate diary, against the mean of the two estimates. Also shown is the bias given as the mean difference and the 95% limits of agreement (mean difference ± 2 × standard deviation). *Central panel*, scatterplot of DietPlan6 estimates against duplicate diary values. Also shown is the Pearson correlation between the two methods and the line of unity. *Right panel*, boxplots summarising the distribution of the two estimates.

In March 2015, an updated food composition dataset was published by the UK Department of Health (PublicHealthEngland, 2015) and has since been included as an update for DietPlan (DietPlan7). We therefore developed new dietary codes for the foods consumed in the diet diaries that included both the sulphur content as previously used, and the updated sodium values from the 2015 food composition dataset. The sodium intake as estimated by duplicate diet and DietPlan7 analysis was compared as before. Overall the data between the two methods was more comparable than with the older dataset for sodium intake, although there was still strong evidence of a positive bias (mean difference = 311 mg, $P = 0.0041$) and proportionately higher estimates (ratio of standard deviations, DP7/DD = 1.21, significance of linear trend, $P = 0.0015$). Summary plots for the comparison between DietPlan7 and duplicate diaries are provided as Supplementary figures.

High sodium intake has been linked to increased risk of high blood pressure and cardiovascular disease (He & MacGregor, 2015). Although the average intake of sodium in the UK is on a downward trend, it was still 76 per cent above the recommended maximum Reference Nutrient Intake of 2.4 g/day in 2014 and in this study was found to be 2.5 g/day on average by diet diary analysis (Table 3). As such there has been an active approach to the reduction of salt added to foods in recent years (Webster, Trieu, Dunford, & Hawkes, 2014) and it is therefore important to use up-to-date information on sodium content of foods in dietary analysis calculations. The change in salt added to processed foods is undergoing such a rapid change that even the updated databases in Dietplan7 (from 2015 data) may still be overestimating dietary sodium intake, which may explain the disparity between the diet record estimates and duplicate diet analysis.

3.7. Analysis of calcium, iron, magnesium, phosphorous, potassium and zinc in the diet using duplicate diet and diet diary analyses

Calcium, iron, magnesium, phosphorous, potassium and zinc intake were also estimated for the 18 diet diaries and compared to duplicate diet analysis. Table 3 shows there was little evidence of bias for calcium, iron and magnesium ($P > 0.10$), and weak evidence of bias for phosphorous ($P = 0.063$) and potassium ($P = 0.092$). For zinc, there was strong evidence of a positive bias (mean difference = 2.7 mg, $P = 0.0007$), and to overestimate (ratio of standard deviations DP6/DD = 1.52, significance of linear trend $P = 0.008$) by diet diary analysis. Compound figures showing Bland and Altman, scatter and boxplots for these minerals are provided as Supplementary figures.

Caution is also advised when using dietary analysis software for the estimation of other mineral intake such as iron. Many foods that may contain added minerals, e.g. breakfast cereals, food composition data is often the average data for a range of similar products and the actual amount of fortificant consumed can be brand dependant.

4. CONCLUSIONS

As expected, high levels of sulphur were obtained from consumption of meat and fish, mainly in the form of SAA. However, the analysis of sulphur content and partitioning in commonly consumed foods and vegetables undertaken in this study indicates as much as 89.5% of the total sulphur consumed in a typical diet may not be derived from SAA. Significant sulphur intake was also achieved from consumption of wheat and starchy foods, dairy and vegetables. If only SAA intake is monitored the significant contribution to total sulphur intake from these other foods will be disregarded. Alliaceous and cruciferous vegetables are shown to be an important source of sulphur in the diet.

The results of this study suggest there is no requirement for mass re-analysis of the composition of sulphur in foods and that existing data can be used with consumption data for estimation of sulphur intake. We also found evidence that dietary intake analysis overestimated the amount of sodium consumed which may be a consequence of the compositional data not keeping pace with the trend of manufacturing food with a reduced salt content. Total sulphur intake estimated from dietary intake diaries is comparable to that from duplicate diet analysis and can include a breakdown of the sulphur amino acid, sulphate and sulphur containing metabolites consumed.

AUTHORSHIP CONTRIBUTIONS

Joanne F Doleman: formulation of research question, preparation of the sulphur database, design of study, analysis of diet diaries, data analysis and writing of article.

Shikha Saha: analysis of food samples.

Katrijn Grisar: preparation of the sulphur database, analysis of diet diaries.

Lena Van Liedekerke: preparation of the sulphur database, analysis of diet diaries.

Henri S Tapp: data analysis.

Mark Roe: input to data for sulphur database, assistance in diet diary analysis, contribution to structure, content and writing of article.

Richard F Mithen: formulation of research question, design of study, contribution to structure, content and writing of article.

FUNDING

The study was supported by a strategic programme grant to IFR from the UK Biotechnology and Biological Sciences Research Council [BB/J004545/1].

ACKNOWLEDGEMENTS

The authors would like to thank Dr Charlotte N Armah of IFR, Dr Christos Derdemezis and Dr Julie A Lovegrove of University of Reading for the use of the diet diaries from the BASH study.

REFERENCES

1. Armah, C. N., Derdemezis, C., Traka, M. H., Dainty, J. R., Doleman, J. F., Saha, S., ... Mithen, R. F. (2015). Diet rich in high glucoraphanin broccoli reduces plasma LDL cholesterol: Evidence from randomised controlled trials. Molecular Nutrition and Food Research, 59(5), 918–926.
2. Ashwell, M. (1991). 4: The COMA Report on Dietary Reference Values. Nutrition Bulletin, 16(3), 132–135.
3. Bland, J. M., & Altman, D. G. (1986). Statistical methods for assessing agreement between two methods of clinical measurement. Lancet, 1(8476), 307–310.
4. Block, E. (1992). The Organosulfur Chemistry of the Genus Allium – Implications for the Organic-Chemistry of Sulfur. Angewandte Chemie-International Edition in English, 31(9), 1135–1178.
5. Borlinghaus, J., Albrecht, F., Gruhlke, M. C., Nwachukwu, I. D., & Slusarenko, A. J. (2014). Allicin: Chemistry and biological properties. Molecules, 19(8), 12591–12618.
6. Curno, R., Magee, E. A., Edmond, L. M., & Cummings, J. H. (2008). Studies of a urinary biomarker of dietary inorganic sulphur in subjects on diets containing 1–38
7. mmol sulphur/day and of the half-life of ingested 34SO4(2-). European Journal of Clinical Nutrition, 62(9), 1106–1115.
8. Florin, T., Neale, G., Gibson, G. R., Christl, S. U., & Cummings, J. H. (1991). Metabolism of dietary sulphate: absorption and excretion in humans. Gut, 32(7), 766–773.
9. FoodStandardsAgency. (2008). The McCance and Widdowson's Composition of Foods integrated dataset. In Public Health England. Gasper, A. V., Al-Janobi, A., Smith, J. A., Bacon, J. R., Fortun, P., Atherton, C., ... Mithen, R. F. (2005).
10. Glutathione S-transferase M1 polymorphism and metabolism of sulforaphane from standard and high-glucosinolate broccoli. American Journal of Clinical Nutrition, 82(6), 1283–1291.
11. Giacoppo, S., Galuppo, M., Montaut, S., Iori, R., Rollin, P., Bramanti, P., & Mazzon, E. (2015). An overview on neuroprotective effects of isothiocyanates for the treatment of neurodegenerative diseases. Fitoterapia, 106, 12–21.
12. Gnagnarella P, Salvini S, & M., P. (2008). Food Composition Database for Epidemiological Studies in Italy. In vol. Version 1.2015.

13. He, F. J., & MacGregor, G. A. (2015). Salt and sugar: their effects on blood pressure. Pflugers Archiv, 467(3), 577–586.
14. Herr, I., & Buchler, M. W. (2010). Dietary constituents of broccoli and other cruciferous vegetables: implications for prevention and therapy of cancer. Cancer Treatment Reviews, 36(5), 377–383.
15. Holland, B., Unwin, I. D., & Buss, D. H. (1988). Cereals and Cereal Products. (Vol. Third supplement to 4th edition of McCance and Widdowson's The Composition of Foods). Nottingham: The Royal Society of Chemistry.
16. Holland, B., Unwin, I. D., & Buss, D. H. (1989). Milk Products and Eggs (Vol. Fourth supplement to 4th edition of McCance and Widdowson's The Composition of Foods). Cambridge: The Royal Society of Chemistry.
17. Holland, B., Unwin, I. D., & Buss, D. H. (1991). Vegetables, Herbs and Spices (Vol. Fifth supplement to 4th edition of McCance and Widdowson's The Composition of Food). Cambridge: The Royal Society of Chemistry.
18. Holland, B., Unwin, I. D., & Buss, D. H. (1992). Fruit and Nuts. (Vol. First supplement to 5th edition of McCance and Widdowson's The Composition of Foods.). Cambridge: The Royal Society of Chemistry.
19. Ingenbleek, Y., & Kimura, H. (2013). Nutritional essentiality of sulfur in health and disease. Nutrition Reviews, 71(7), 413–432.
20. Kurotani, K., Nanri, A., Goto, A., Mizoue, T., Noda, M., Kato, M., ... Japan Public Health Center-based Prospective Study (2013). Vegetable and fruit intake and risk of type 2 diabetes: Japan Public Health Center-based Prospective Study. British Journal of Nutrition, 109(4), 709–717.
21. Lee, E. J., Yoo, K. S., Jifon, J., & Patil, B. S. (2009). Characterization of shortday onion cultivars of 3 pungency levels with flavor precursor, free amino acid, sulfur, and sugar contents. Journal of Food Science, 74(6), C475–C480.
22. MAFF, Ministry of Agriculture, F. a. F (1993). Food portion sizes (2nd ed.). London: Her Majesty's Stationery Office.
23. Maheo, K., Morel, F., Langouet, S., Kramer, H., Le Ferrec, E., Ketterer, B., & Guillouzo, A. (1997). Inhibition of cytochromes P-450 and induction of glutathione Stransferases by sulforaphane in primary human and rat hepatocytes. Cancer Research, 57(17), 3649–3652.
24. Masters, M., & McCance, R. A. (1939). CLIX. The sulphur content of foods. Biochemistry Journal, 33(8), 1304–1312.
25. McNaughton, S. A., & Marks, G. C. (2003). Development of a food composition database for the estimation of dietary intakes of glucosinolates, the biologically active constituents of cruciferous vegetables. British Journal of Nutrition, 90(3), 687–697.
26. Nicastro, H. L., Ross, S. A., & Milner, J. A. (2015). Garlic and onions: their cancer prevention properties. Cancer Prevention Research (Philadelphia), 8(3), 181–189.
27. Nimni, M. E., Han, B., & Cordoba, F. (2007). Are we getting enough sulfur in our diet? Nutrition and Metabolism (London), 4, 24.
28. Powell, R., Bailey, R. J., & Jolly, P. K. (1987). Trace Elelments in British Tap Water Supplies. In http://dwi.defra.gov.uk/research/completed-research/reports/dwi0007.pdf: Department of the Environment.
29. PublicHealthEngland. (2015). The McCance and Widdowson's Composition of Foods integrated dataset. In Foods standards agency.
30. Salonen, A., de Vos, W. M., & Palva, A. (2010). Gastrointestinal microbiota in irritable bowel syndrome: Present state and perspectives. Microbiology, 156(Pt 11), 3205–3215.
31. Webster, J., Trieu, K., Dunford, E., & Hawkes, C. (2014). Target salt 2025: A global overview of national programs to encourage the food industry to reduce salt in foods. Nutrients, 6(8), 3274–3287.
32. Yamagishi, S. I., & Matsui, T. (2016). Protective role of sulphoraphane against vascular complications in diabetes. Pharmaceutical Biology, 1–11.

33. Zhang, X., Shu, X. O., Xiang, Y. B., Yang, G., Li, H., Gao, J., ... Zheng, W. (2011). Cruciferous vegetable consumption is associated with a reduced risk of total and cardiovascular disease mortality. American Journal of Clinical Nutrition, 94 (1), 240–246.

CHAPTER 2

Milk Protein

Gabriella Pinto[1], Marina Cuollo[1], Sergio Lilla[1], Lina Chianese[1], Francesco Addeo[1] and Simonetta Caira[2]

[1] *Department of Food Science, University of Naples "Federico II", Parco Gussone, Portici (Naples), Italy*

[2] *Food Science Institute of the National Research Council (C.N.R.), Via Roma, Avellino, Italy*

1. INTRODUCTION

Nutraceuticals, a term combining the words "nutrition" and "pharmaceutical", is *a food or food product that provides medical or health benefits including the prevention and treatment of diseases.* A functional food essentially provides a health benefit beyond the basic nutrition, whereas nutraceutical is used to describe an isolated or concentrated molecular extract of bioactive compounds. Milk is a unique food providing a variety of essential nutrients necessary to properly fuel the body. Inactive food proteins can release encrypted bioactive peptides *in vivo* or *in vitro* by digestive enzymatic hydrolysis. Bioseparation protocols offer unique possibilities for a number of application areas, *e.g.,* hydrolyzate-based nutraceutical ingredients for functional foods, dietary supplements and medical foods.

Many ingredients are included in the wide range of nutraceuticals, such as essential amino acids, conjugated fatty acids, vitamins, minerals and polyphenols. They have already been patented and incorporated in functional foods and nutritional beverages. Such components are believed to improve overall health and well-being, reducing the risk of specific diseases or minimizing the effects of other health concerns. However, milk is devoid of flavonoids, the most common group of vegetable polyphenolic compounds, which act as antioxidants and free radical scavengers. In contrast, the ingestion of soy and green tea extract may reduce the risk of developing prostate cancer and may protect against various other types of cancer [1-2]. An interesting patented invention has made available an extended-release form of polyphenols and riboflavin (vitamin B2) coated with methylcellulose [3]. Coating slows down the release of polyphenols in the nutraceutical preparation [3]. Possible applications of coating technology could be extended to all of the bioactive peptides susceptible to digestive enzymes. For example, glutathione can be maintained in human blood at normal levels by supplying it as dry-filled capsules [4]. Nutrients and bioactive compounds may be microencapsulated by using mixtures of proteins or peptides and oils. Encapsulation of ω-3 *fatty acids* (FA) enhances the stability and bioavailability of bioactive food ingredients [5]. By these means, new transparent bioavailable beverages containing ω-3 rich oils, phospholipids and minerals in an oxidatively stable food system were created [6]. Iron or calcium casein phosphopeptides (CPP) were embedded in the chitosan lactate fiber as a protective agent against oil oxidation [6]. A recent patent relates to a nutraceutical composition consisting of a sweetener admixture for food or drink comprising calcium lactate, calcium acetate, vitamin D3 and sucralose (for fortified zero calorie formulation) or sugar (white/brown; for fortified sugar preparation) [7]. Milk proteins playing a physiological role include proteins such as β-lactoglobulin, α-lactalbumin, immunoglobulins, lactoferrin, heat-stable proteose peptones, serum albumin and various acid soluble phosphoglycoproteins. Casein (CN), representing 80% of total milk proteins, consists of four α_{s1}-, α_{s2}-, β-, and κ-CN families in the approximate ratio 38:11:38:13. Research performed in recent years has shown that caseins and whey proteins are rich in encrypted biologically active peptides such as exorphins (casomorphins), CPP and immunopeptides [8]. The peptides are released by enzymes in the form of mature bioactive components or the precursors thereof [9]. They are 3- to 20-residue long peptides released during *in vivo* gastrointestinal digestion. Historically, the opioid peptides were discovered as the result of a systematic search for exogenous substances, namely (i) first discovered in 1979, opioid agonist peptides derived from milk proteins were characterized in 1986; (ii) in 1982, angiotensin-converting enzyme (ACE)-inhibitory peptides were found to be antihypertensive peptides; (iii) then, fibrinogen-like sequences with antithrombotic activity were found; (iv) phagocytic activity and lymphocyte proliferation of numerous immunomodulating peptides were observed; (v) CPP facilitating the absorption of minerals, especially calcium, magnesium and iron were found; and (vi) antimicrobial peptides were discovered [10]. There are many milk peptides that possess multifunctional activities, *i.e.*, they can play two or more hormone-like roles. Bioactive peptides grouped according to their function in human well-being are shown in Figure 1.

Nutraceutical products comprising short bioactive peptides showing *in vitro* or *in vivo* antimicrobial, ACE-inhibitory activity and/or antihypertensive and/or antioxidant activity are being considered for possible use by the pharmaceutical industry. The CN hydrolyzates could serve as food preservatives to reinforce the body's natural defenses or as pharmaceutical products for facilitating the control of blood and/or bacterial infections [12]. Much research has been devoted to increasing mineral transport by phosphorylated groups of peptides [13]. CPP in commercial hydrolysed casein (Tatua Cooperative Dairy Co. Ltd, New Zealand and Arla Foods Ingredients and Sweden) seem to help the absorption of chelated calcium, iron, copper, zinc and manganese in the intestine (Table 1). Thus, CPP-bound amorphous calcium phosphate (ACP) displayed anticariogenic effect when added to dentifrices or oral care products by localizing calcium and phosphate ions at the tooth surface. Similarly, it has been claimed that a chewing gum or other confectionery product containing a combination of CPP-ACP and sodium bicarbonate as active ingredients can provide dental health benefits [14]. In experiments on humans, synthetic CPP-ACP nanocomplexes incorporated in mouth rinses and sugar-free chewing gums have been proven to be potential anticariogenic agents [15] (Table 1).

2. BIOACTIVE PEPTIDES RELEASED *IN VITRO* BY THE HYDROLYSIS OF MILK PROTEINS

Peptides with various bioactivities can be produced according to two different methods: i) *in vitro* fermentation of milk inoculated with starter cultures and ii) *in vitro* digestion of milk proteins by one or more proteolytic enzymes.

(i) The proteolytic system of lactococci is able to degrade milk proteins using cell-wall-bound proteinases by releasing di-, tri-, and/or oligo-peptides and amino acids supporting the growth of bacteria. In addition, lactococcal peptidases released into the curd/cheese consequently to autolysis can further degrade the internalized peptides to amino acids [17]. The higher the exopeptidase activity in cheeses, the greater the age of the cheese [18]. Consistently, yogurt and other cultured dairy drinks have some of the highest counts of cells that actually survived and thus possessed the lowest number of peptides derived from the aminopeptidase activity. In a comprehensive review of literature, no enzyme with carboxypeptidase (CPase) activity has been reported for either lactococci or other LAB [19-20]. The bacterial peptidases have different and partly overlapping specificities (Figure 2).

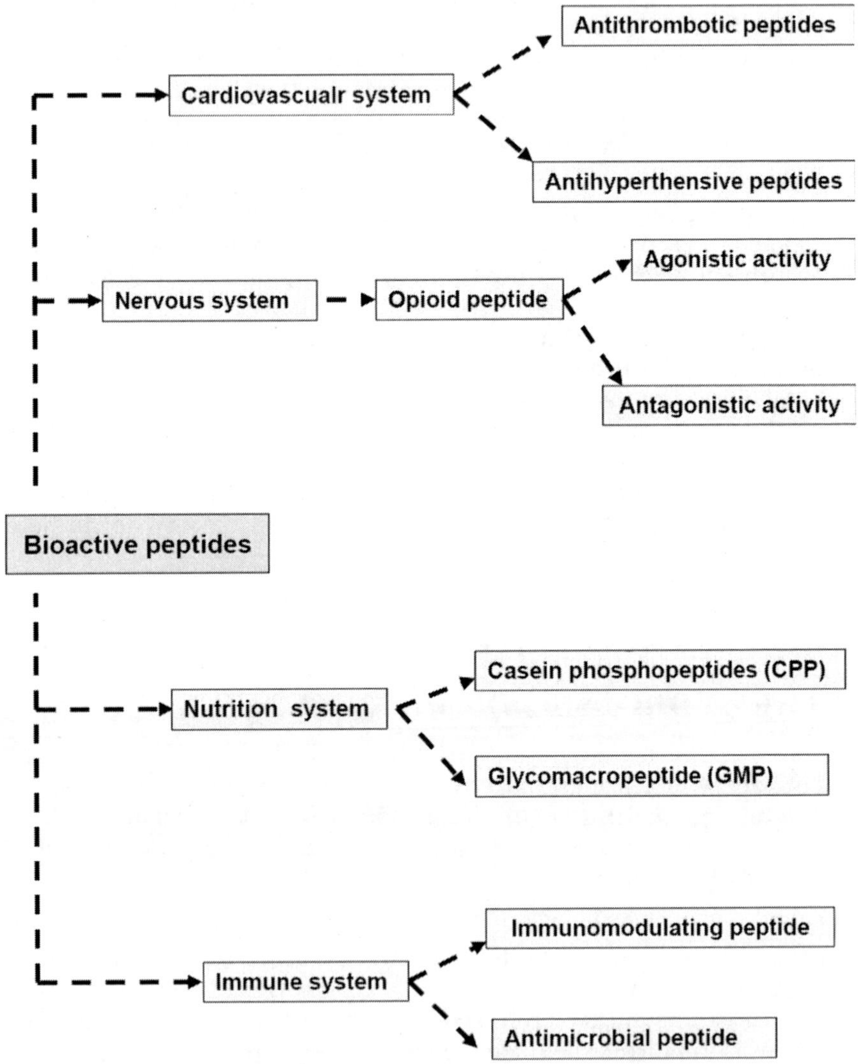

FIGURE 1. Main bioactivity of peptides formed by the enzymatic digestion of milk proteins (source: [11]).

TABLE 1. Commercial dairy products and ingredients with health or function claims based on CPP content (source: modified from [16]).

Brand name	Product type	Claimed functional bioactive peptides	Health/function claims	Manufacturer
HCP102/HCP 105	Hydrolysate ingredient	CPP	Helps mineral absorption	Tatua (New Zealand)
Capolac	Hydrolysate ingredient	CPP	Helps mineral absorption	Arla Foods Ingredients (Sweden)
Recaldent	Chewing gum	ACP-CPP	Anticariogenic	Cadbury Adams (USA)
Recaldent	Toothpaste	ACP-CPP	Anticariogenic	GC Tooth Mousse (USA)

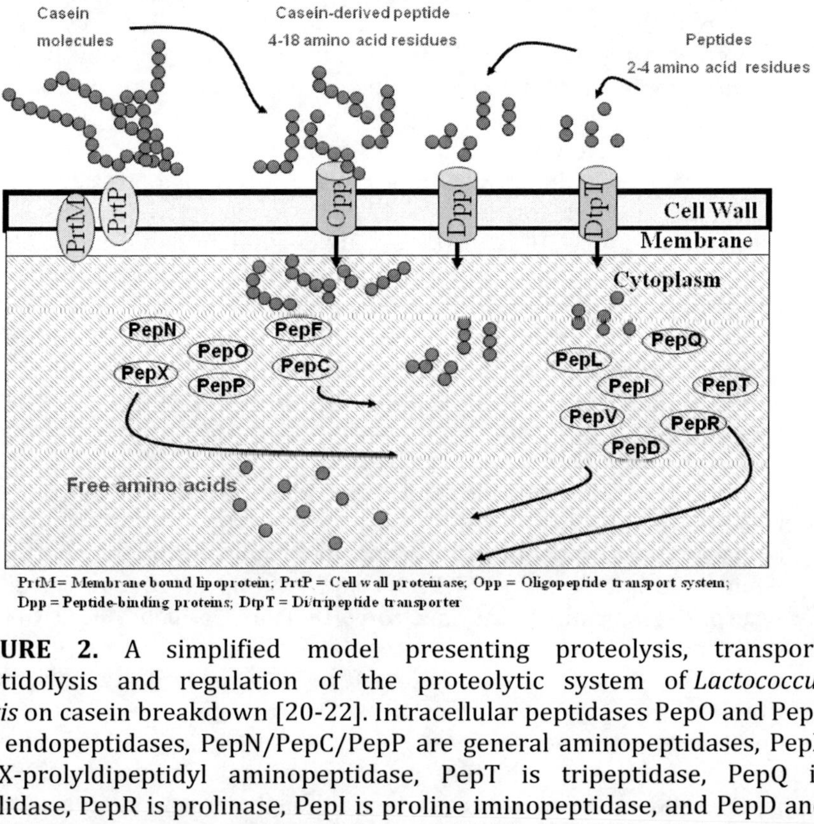

FIGURE 2. A simplified model presenting proteolysis, transport, peptidolysis and regulation of the proteolytic system of *Lactococcus lactis* on casein breakdown [20-22]. Intracellular peptidases PepO and PepF are endopeptidases, PepN/PepC/PepP are general aminopeptidases, PepX is X-prolyldipeptidyl aminopeptidase, PepT is tripeptidase, PepQ is prolidase, PepR is prolinase, PepI is proline iminopeptidase, and PepD and PepV are dipeptidases D and V. The role of PepN, PepC, PepI, PepP and PepA and PepO, PepF, PepX, PepQ, PepR, PepV, PepT and peptidolytic cycles are depicted schematically (various alternative routes of breakdown are possible for most peptides).

Although the intracellular endopeptidases PepN and PepC are unable to hydrolyze casein molecules, the X-prolyl dipeptidyl aminopeptidases (PepX) are active on oligopeptides hydrolyzing the internal bonds of casein-derived peptides. Taken together, these enzymes are able to remove the N-terminal residues from peptides, with the specificity primarily depending on the nature of the N-terminal amino acid [20-21]. Di- and tripeptides generated by endopeptidases, general aminopeptidases and PepX are next subjected to additional cleavage by the tripeptidase, PepT, and dipeptidases, PepV and PepD (Figure 2). Other peptidases with more specific substrate specificities include PepA, which liberates N-terminal acidic residues from 3- to 9-residue long peptides; PepP, which prefers tripeptides carrying proline in the middle position; PepR and PepI, which act on dipeptides containing proline in the penultimate position; PepQ, which cleaves dipeptides carrying proline in the second position; and PepS, which shows preference for peptides containing two to five residues with Arg or aromatic amino acid residues in the N-terminal position [20-21,23] (Figure 2). This proteolytic system is able to support LAB growth to high cell densities (10^9–10^{10} cfu/mL) in milk containing only small amounts of hydrolytic products that are transportable into the cells for assimilation. *Lb. helveticus* and *Lb. delbrueckii* ssp. *bulgaricus* possess cell wall proteinase activity stronger than that of lactococci. The number of intracellular proteins released by *St. thermophilus* is greater than that by *Lb. helveticus* [24]. Proteins released in cheese from a starter-based thermophilic LAB, such as *Lb. helveticus, Lb. delbruecki* subsp. *lactis* and *Streptococcus salivarius* subsp. *thermophilus* and *Propionibacteriumfreudenreichii,* have been identified using 2D-PAGE and mass spectrometry (MS) analysis [24]. Similarly, bioactive peptides have been determined using High Performance Liquid Chromatography (HPLC) and offline Matrix-Assisted Laser Desorption/Ionization Mass Spectrometry-Time-Of-Flight (MALDI-MS-TOF) [25]. These peptides were all generated from CN and released upon proteolysis depending on the bacterial strain [26]. In this manner, proteinases play a primary role, as they are able to generate specific bioactive peptides. Recombinant human α_{s1}-casein digested by trypsin gave rise to several ACE-inhibitory peptides and calcium-binding CPP. These peptides did not form in cheese whey, although they can be formed from CN during fermentation using various commercial dairy starters [27]. The combination of the LAB bacteria and proteolytic enzymes could serve to increase the range of bioactive peptides.

(ii) *In vitro hydrolysis* of CN by pepsin and trypsin could produce many bioactive peptides. Pepsin, an endopeptidase with broad specificity, preferentially cleaves hydrophobic, preferably aromatic, residues. Trypsin specifically hydrolyses peptide bonds just after a lysine or an arginine residue of β-casein $A^1$and A^2 variant (11 and 4, respectively); κ-casein A and B (9 and 5); α_{s2}-CN A (24 and 6); or αs1-CN B (14 and 6). In this manner, tryptic hydrolysates of CN contain uneven peptides of up to 8159 Da and also free amino *acids* [28]. We have demonstrated that CN hydrolysate by pepsin (P) and trypsin (T) in succession does not contain peptides with molecular mass greater than 2431 Da. β-CN resulted extensively hydrolyzed into a high number of oligopeptides by using proteases with well defined cleavage specificity. Controlled partial hydrolysis by proteinases could lead to the formation of partially degraded proteins critical for obtaining new functional products. The choice of digestion enzymes needs to be evaluated carefully because influences the hydrolysate final composition. A high number of peptides with antimicrobial, anti-hypertensive and opiod-like activity has been identified (Table 2), some of which exactly matched those described in the literature for potential bioactivity (Table 2). The potent opioid β-CM7 peptide retained part of its original opioid activity-like food hormone when progressively shortened. The synthetic β-casomorphin derivatives have been shown to be highly specific and potent β-type opioid receptor ligands [29].

We also performed sequential milk protein (powder sample) digestions using various endoproteases facilitating a consistent partial hydrolysis. The high degree of specificity in terms of cleaving peptide bonds exhibited by a cocktail of enzymes (P, T and P432 from Biocatalysts, U.K) yielded a limited number of CPP. After fractionation on an HA column, both non-CPP and CPP were identified as shown in Table 3. Different oxidation rates of Met residues in the same protein resulted in formation of the peptides with different molecular mass. In most cases, a single Met-containing peptide and its oxidized counterpart were identified. However, in many cases, when proteins contained several consecutive endoprotease sensitive bonds, different long peptides containing the same Met-oxidation site(s) were identified. The data in Table 3 suggest that the cocktail enzymes containing amino- and CPases, in addition to P and T, progressively reduce the size of peptides without altering the degree of phosphorylation. Evaluation of protein/peptide quality can take advantage of the tandem MS for the detection of native, partly oxidized and partly dephosphorylated peptides.

TABLE 2. Identity of bioactive peptides found in the PT digest of milk protein powder sample. The bioactivity of the peptide from which they derive and the references are reported.

Identity of bioactive peptides	Bioactivity	References	PT peptides
β-CN (f60-70); (f59-61); (f59-64); (f60-68)	Opiod	[29-30]	(f58-68); (f59-63); (f59-62)
β-CN (f74-76) β-CN (f80-90); (f84-86) β-CN (f108-113) β-CN (f140-143) β-CN (f177-183) β-CN (f169-174) β-CN (f193-198) β-CN (f193-202)	ACE-Inhibitory	[31-33]	(f69-80); (f71-80); (f75-80) (f80-88); (f81-89); (f81-88); (f81-92); (f81-93); (f81-94) (f108-103) (f139-142) (f177-183); (f177-184); (f179-182) (f169-176); (f172-176) (f193-199) (f193-202)
β-CN (f1-25)4P	Mineral carrier, immunomodulatory, cytomodulatory	[34-38]	(f12-17)1P; (f12-25)3P; (f12-25)4P; (f15-25)2P; (f15-25)3P; (f19-25)1P; (f19-25)2P
β-CN (f29-41)1P	Mineral carrier	[13]	(f33-42)1P; (f33-43)1P; (f33-44)1P
β-CN (f84-86) α_{s1}-CN (f23-34) αs_1-CN (f25-27)	ACE-Inhibitory	[31-33]	(f80-87); (f81-88); (f81-89); (f81-92); (f81-93); (f81-94) (f24-32) (f25-31)
α_{s1}-CN (f90-96); (f90-95); (f91-95)	Opiod (agonist)	[39-41]	(f92-95); (f91-95); (f90-95)
α_{s1}-CN (f142-147) α_{s1}-CN (f157-164) α_{s1}-CN (f194-199)	ACE-Inhibitory	[31-33]	(f142-145); (f143-146); (f143-150); (f144-149) (f155-164) (f194-199); (f197-199); (f193-199)
α_{s1}-CN α_{s1}-CN	Mineral carrier	[13]	(f110-119)1P (f41-55)1P; (f68-79)2P
α_{s2}-CN (f174-179) α_{s1}-CN (f189-193)	ACE-Inhibitory	[31-33]	(f174-179) (f189-193)
α_{s2}-CN α_{s2}-CN κ-CN (f33-38)	Mineral carrier Opiod (antagonist)	[13]	(f138-146)1P (f124-137)2P; (f126-136)2P; (f126-137)2P (f58-65)1P

By this means, phosphorylated peptides (2 and 3P/mole) of a precursor of lactophorin (LP) (28 kDa milk glycoprotein), proteose-peptone component 3, and glycosylation-dependent adhesion molecule 1 were detected. In addition, three low molecular weight non-CPP derived from LP were detected.

Native milk proteins used as the substrate for digestion by enzymes did not form CPP. This suggests that denatured LP and other whey proteins could have the tendency to form low molecular-mass peptide aggregates characterized by a poor solubility. For this reason, the use of milk protein and/or any milk substrate powder must be discouraged to eliminate phosphates and salts from the substrate. The enzymatic hydrolysis of the casein implies the use of endoproteases. However, the protein hydrolysate with alcalase is used in infant formulae, dietetic foods, nutraceuticals, ice creams, dressings, fermented products, yogurts, and personal care products. CPP released by alcalase are truncated with respect to those released by trypsin. The identified peptides can be categorized into two groups, one containing multiphosphorylated peptides and the other tri-, di- and mono CPP. Each group contained a number of variously long peptides due to the broad specificity of alcalase cleaving peptide bonds mainly on the carboxyl side of Glu, Met, Leu, Tyr, Lys, and Gln. The exoproteases responsible for the hydrolysis are inactivated by heating for ~10 min to ~85 °C. The *in vitro* sequential use of pepsin, pancreatic proteases and extracts of human intestinal brush border membranes, mimicking the respective gastric, duodenal and jejuneal *in vivo* digestion of CN, exhibited significant bioactive effects. A limited number of CN and whey protein peptides survived the *in vitro* simulated gastro-intestinal digestion. The anionic character seems to confer a marked resistance to multi-phosphorylated CPP hydrolysis by endoprotease. Ten out of 19 CPP contained SerP available for binding minerals, and four of these peptides, α_{s1}-CN (f57-90)5P, α_{s1}-CN (f56-90)5P, α_{s1}-CN (f55-76)5P, β-CN (f1-52)5P, were reported for the first time in the CN digests [42]. Only β-CN (f1-25)4P, 3P and 2P survived the simulated gastrointestinal digest of CN [43].

The ingress of foreign material in general, such as CPP, across the mucosal brush-border into the enterocyte is conditioned by the efficient dephosphorylation of peptides by alkaline phosphatase. This aspect deserves more in-depth investigation.

3. *IN VIVO* DIGESTION OF CASEIN, FORMATION OF CPP AND THEIR PHYSIOLOGICAL IMPORTANCE

Among the biologically active peptides, CPP characterized by SerP and/or ThrP residues account for ~30% of monoesters of hydroxyl amino acids. They mainly occur in the Ser/Thr-Xaa-SerP/Glu/Asp sequence consensus, where Xaa is any amino acid residue but Pro. The three-phosphorylated motif -SerP-SerP-SerP-Glu-Glu- occurs in α_{s1}-CN (f66-70), α_{s2}-CN (f8-12), α_{s2}-CN (f56-60), and β-CN (f17-21). According to the current CN nomenclature, bovine α_{s1}-CN, α_{s2}-CN, β-CN, and κ-CN possess 8-9, 11-13, 4-5, and 1-2 phosphate (P) residues, respectively, and the P number could change according to the casein variant [44]. For example, β-CN D has one SerP residue less than the A counterpart due to the substitution $Lys^{18} \rightarrow Ser^{18}$.

Table 3. Native and partly Met-oxidized CPP isolated from a three enzyme (Pepsin, trypsin and P432) milk protein hydrolyzate.

Parent protein	Molecular mass (Da)		Start	End	Peptide sequence	Peptide modifications
	Expected	Calculated				
α_{s1}-CN	1222.5	1222.5	110	119	(L)EIVPNSAEER(L)	1P
	1517.8	1517.6	68	79	(S)SEEIVPNSVEQK(H)	2P
	1525.7	1525.5	41	53	(L)SKDIGSESTEDQA(M)	2P
	1586.6	1586.5	43	55	(K)DIGSESTEDQAME(D)	2P; Oxidation (Met)
	1672.7	1672.6	41	54	(L)SKDIGSESTEDQAM(E)	2P; Oxidation (Met)
	1785.8	1785.6	41	55	(L)SKDIGSESTEDQAME(D)	2P
	1801.7	1801.6	41	55	(L)SKDIGSESTEDQAME(D)	2P; Oxidation (Met)
	1963.9	1963.8	39	55	(N)ELSKDIGSESTEDQAME(D)	1P; Oxidation (Met)
	1989.9	1989.7	37	52	(K)VNELSKDIGSESTEDQ(A)	3P
	2060.7	2060.7	37	53	(K)VNELSKDIGSESTEDQA(M)	3P
α_{s2}-CN	900.4	900.3	58	65	(S)SEESAEVA(T)	P
	937.4	937.3	141	147	(D)MESTEVF(T)	1P; Oxidation (Met)
	1067.4	1067.3	57	65	(S)SSEESAEVA(T)	2P
	1089.5	1089.4	138	146	(K)TVDMESTEV(F)	1P
	1105.4	1105.4	138	146	(K)TVDMESTEV(F)	1P; Oxidation (Met)
	1252.6	1252.5	138	147	(K)TVDMESTEVF(T)	1P; Oxidation (Met)
	1410.6	1410.5	126	136	(R)EQLSTSEENSK(K)	2P
	1538.7	1538.6	126	137	(R)EQLSTSEENSKK(T)	2P
	1623.7	1623.6	1	13	KNTMEHVSSSEES(I)	2P
	1639.7	1639.5	1	13	KNTMEHVSSSEES(I)	2P; Oxidation (Met)
	1680.8	1680.6	124	136	(L)NREQLSTSEENSK(K)	2P
	1719.7	1719.5	1	13	KNTMEHVSSSEES (I)	3P; Oxidation (Met)
	1808.9	1808.7	124	137	(L)NREQLSTSEENSKK(T)	2P
β-CN	639.3	639.3	12	16	(E)IVESL(S)	1P
	900.5	900.4	19	25	(S)SEESITR(I)	1P
	1067.4	1067.4	18	25	(S)SSEESITR(I)	2P
	1354.6	1354.5	15	25	(E)SLSSSEESITR(I)	2P
	1434.6	1434.5	15	25	(E)SLSSSEESITR(I)	3P
	1447.7	1447.5	33	43	(K)FQSEEQQQTED(E)	1P
	1576.7	1576.6	33	44	(K)FQSEEQQQTEDE(L)	1P
	1689.8	1689.6	33	45	(K)FQSEEQQQTEDEL(Q)	1P
	1775.8	1775.7	12	25	(E)IVESLSSSEESITR(I)	3P
	1855.6	1855.6	12	25	(E)IVESLSSSEESITR(I)	4P
κ-CN	968.4	968.4	145	152	(A)TLEDSPEV(I)	1P
	1734.7	1734.7	147	161	(L)EDSPEVIESPPEINT(V)	1P
Lactophorin	1226.5	1226.51	34	43	(L)SKEPSISRED(L)	1P
	1306.6	1306.5	34	43	(L)SKEPSISRED(L)	2P
	1419.6	1419.56	34	44	(L)SKEPSISREDL(I)	2P
	1499.7	1499.5	34	44	(L)SKEPSISREDL(I)	3P

The *in vivo* digestion of milk proteins takes place mainly in the stomach under the action of pepsins, gastric digestive proteinases that are able to digest ~20% proteins. Afterwards, the pepsin digests pass to the duodenum where peptides are further hydrolyzed by pancreatic enzymes. The digestion is completed by membrane proteases and a variety of peptidases embedded in the brush border of the small intestine and released by the intestinal microflora. These peptidases release an amino acid residue or a dipeptide from the N- and C-terminal side of oligopeptides [45-47]. Phosphatase(s) located in the brush border of the apical membrane of enterocytes, act(s) in removing phosphate groups, thus promoting partial or full peptide dephosphorylation of peptides in different body districts. The phosphorylated sequence is responsible, at the intestinal pH, for binding Ca^{++}, Zn^{++}, and Mg^{++} and for the *in vivo* resistance of the complex to gastrointestinal proteases [48]. Fe complexed to β-CN (f15-25)4P was scarcely hydrolyzed throughout the digestion, suggesting that the coordination of iron ions to CPP inhibits the action of both phosphatase and peptidases [49]. Brush border enzyme alkaline phosphatase activity could improve the absorption of Fe complexed CPP by releasing Fe from peptides. Moreover, Fe complexed to β-CN CPP was absorbed more than Fe complexed to α_{s1}-CN CPP [50]. The differences in protein composition between cow and breast milk could explain some of the differences in the Fe bioavailability of the latter [50]. Iron deficiency, a major worldwide nutritional problem, can be reduced by CPP. Fe complexed CPP prevents the formation of poorly absorbed high molecular weight ferric hydroxides. Zinc absorption can also be enhanced by the formation of Zn complexed to CCP, in particular to β-CN (f1-25)4P [51]. Some portion of the mineral complexed CPP formed in the small intestine was resistant to the digestive and enteric bacteria enzymes and found in the feces of rats fed a casein-based diet [52]. Although literature data regarding intestinal CPP absorption are conflicting, the peptides seem to interact directly with the plasma membrane. One possible mode of CPP action on the transmembrane flux of calcium is that CPP might insert themselves into the plasma membrane and form their own calcium selective channels or act as calcium-carrier peptides rapidly internalized via endocytosis or other processes and eventually provide ionized calcium in the cytosol [53]. Cellular uptake studies of fluorine-18 labeled CPP in human colorectal adenocarcinoma cell line (HT-29) and human head and neck squamous cell carcinoma line (FaDu) cells at 37 °C and 48 °C showed a poor cell penetration because of the poor transport of the phosphopeptides through the cell membrane [54]. The results from *in vivo* studies are still too controversial, as there are many factors affecting Ca availability, such as the various co-present dietary compounds in the intestinal lumen [55]. Despite the vigor of the saturable active transport process by the duodenum, most of the absorption of ingested calcium occurs in the ileum (88% of calcium), jejunum (4%) and duodenum (8%) [56]. An important factor determining the contribution of the ileum to overall calcium absorption is the relatively long transit time of calcium in the segments of the small intestine, accounting for approximately 102 min in the ileum and 6 min in the duodenum [57]. The higher absorption of calcium occurred when inorganic P was added to the Ca-CPP preparation. CPP exhibit a potent ability to form soluble complexes with Ca^{2+} and other trace elements, preventing the formation of Ca-phosphate precipitate in the intestine. CPP could limit the inhibitory effect of phosphate on Ca availability and increase Ca transport across the distal small intestine [55]. All components of the diet reaching the ileum make calcium soluble or keep it in solution within the ileum. Several molecules, particularly CPP, stimulate the passive diffusion of minerals. CPP have been for the first time detected in human ileostomy fluid, confirming their ability to survive gastrointestinal passage into the human distal ileum [58]. CPP released during milk digestion appeared to be stable for up to 8 h in ileostomy contents [58]. The *in vivo* formation of bovine CPP was demonstrated in the small intestinal fluid of minipigs after ingestion of a diet containing casein [59] and in the stomach and duodenum after ingestion of milk or yogurt [60]. The *in vivo* survival of CPP to the prolonged intestinal passage in the distal small intestine is a prerequisite for their function as bioactive substances [58]. CPP are protected from degradation in the gut by the milk matrix, provided that they are ingested as milk constituents and not as isolated CPP. Whole casein or individual casein fractions are used as raw materials to obtain CPP as dietary supplements. Ca could be bound to either SerP or Glu residues [61], suggesting that CPP may enhance the solubility of calcium in the intestinal lumen, thereby increasing the mineral availability for absorption in the small intestine [62,13]. Chemically synthesized CPP, *i.e.*, β-CN (f1-25)4P and αs1-CN (f59-79)5P, carrying the characteristic cluster Ser(P)-Ser(P)-Ser(P)-Glu-Glu, increase the intracellular calcium uptake by the human cultured HT-29 tumor cells [63], Caco-2 cells [64] and osteoblasts [65]. A more pronounced effect has been observed for β-CN-derived peptides than for α_{s1}-CN-counterparts. It has been suggested that CPP promote calcium binding, which would depend on the structural conformation conferred by the two phosphorylated 'acidic motif' and the N-terminal sequence of β-CN [63].

Dental caries are initiated via the demineralization of tooth hard tissue by organic acids directly from the diet or produced from fermentable carbohydrate by dental plaque cariogenic bacteria. CPP can help to replace the minerals that were previously lost consequently to caries [66-67]. Hence, there is a great interest in developing CPP as nutraceutical ingredients for the formulation of functional foods.

4. CPP ENRICHMENT BY DIFFERENT TECHNIQUES

CPP preferably comprise components released by four different casein families, each having a molecular weight greater than 500 Da. Multiply and singly, tryptic CPP can be simultaneously detected using MALDI-TOF, and the location of phosphate groups by a combination of tandem mass spectrometry and computer-assisted database search programs, such as SEQUEST (Trademark, University of Washington, Seattle Wash) [68-69]. Nano-electrospray MS/MS has been used for phosphopeptide sequencing and exact determination of phosphorylation sites [70]. However, mass spectrometric analysis of proteolytic digests of proteins rarely provides full coverage of the phosphorylated sequence, with parts of the sequence often going undetected. In addition, protein phosphorylation is often sub-stoichiometric, such that ionization of CPP present in lower abundance in complex hydrolysates is ordinarily suppressed by strongly ionizable non-phosphorylated peptides. The MALDI-MS desorption/ionization efficiency for phosphopeptides was reported to be an order of magnitude lower than that recorded for the non-phosphorylated counterpart, and ionization became more difficult as the number of phosphate groups increased [71]. Direct analysis of phosphopeptides utilizes two orthogonal MS scanning techniques, both based on the production of phosphopeptide-specific marker ions at m/z 63 and/or 79 in the negative ion mode. These scanning methods combined with the liquid chromatography (LC)-electrospray mass spectrometry (ESI) and nano-electrospray MS/MS allow the selective detection and identification of phosphopeptides even in complex proteolytic digests. Thus, even when the signal of the phosphopeptide is indistinguishable from the background, as in the conventional MS scan, low-abundant and low-stoichiometric phosphorylated peptides can be selectively determined in the presence of a large excess of non-phosphorylated peptides. This strategy is particularly well suited to phosphoproteins that are phosphorylated to varying degrees of stoichiometry at multiple sites [72]. However, the identification and characterization of phosphoproteins would be greatly improved using selective enrichment of CPP prior to MS analysis. An ancient technique for phosphoprotein enrichment consisted of the precipitation of phosphopeptides as insoluble barium salts and recovery by centrifugation, as according to the Manson & Annan method [73]. High-throughput phosphoproteome technologies currently rely on combining pre-separation of proteins, most commonly by high-resolution two-dimensional polyacrylamide gel, in-gel tryptic cleavage of proteins, and subsequent MALDI-TOF or ESI-MS/MS mass spectrometry analysis of peptides [74]. The high resolving power of 2-DE with the sensitive MS requires extensive manual manipulation of samples. Alternative methods are based on chemical derivatization. For example, for β-elimination, a strong base such as NaOH or $Ba(OH)_2$ is used to cleave the phosphoester bonds of phosphoserine and phosphothreonine and form dehydroalanine or dehydroaminobutyric acid, respectively, each able to react with different nucleophiles, such as ethanedithiol (EDT) or dithiothreitol (DTT). This procedure provides a considerably simpler method to enrich CPP. By using cross-linking reagents with affinity tags, such as biotin, interfering non-cross-linked peptides are eliminated, and CPP are highly enriched [75]. Although the chemical derivatization methods are highly selective, they are not widely applied in phosphoproteome studies due to sample loss by the multiple reaction steps and unavoidable side reactions [76]. Immobilized metal-ion affinity chromatography (IMAC, with Fe^{3+}, Ga^{3+}, Ni^{2+} and Zr^{4+} metal ions) and metal oxide affinity chromatography (MOAC, with TiO_2, ZrO_2, Al_2O_3 and Nb_2O_5) have been widely used for the quantitative binding of CPP on resin or adsorbent. Iminodiacetic acid (IDA, a tridentate metal-chelator) or nitrilotriacetic acid (NTA, a quadradentate metal chelator) are often used as IMAC functional matrices reacting with multivalent metal ions to form chelated ions with positive charges useful for the purification of phosphopeptides. Usually Fe^{3+}, Ga^{3+} and Al^{3+} are bound to a chelating support prior to fractionating the complex mixture of peptides before MS analysis [77-78]. Ga^{3+}showed selectivity for CPP higher than Fe^{3+} and Al^{3+} [77,79]. Phosphopeptides bound to IMAC resin are successively recovered in the column effluent by increasing either pH or the phosphate concentration in the buffer [80]. The negatively charged CPP selectively interact with TiO_2microspheres via bidentate binding at the dioxide surface [81-83]. TiO_2-MOAC showed higher specificity than immobilized gallium (Ga^{3+}), immobilized iron (Fe^{3+}), or zirconium dioxide (ZrO_2) affinity chromatography for phosphopeptide enrichment. The main problem associated with the chelating resins is the metal-ion leaching, which leads to CPP loss during the enrichment procedure. The selectivity of these methods was somewhat compromised by the detection of several acidic non-CPP that were also retained by the TiO_2 column [84]. To overcome this drawback, the carboxyl groups are methyl esterified which eliminates the non-specific adsorption of acidic peptides on IMAC [85]. Considerable efforts have been expended to remove acidic non-CPP by washing the resin with 2,5-dihydroxybenzoic acid (DHB) [86] or phthalic acid [87]. It has been found that aliphatic hydroxyl acid modified metal oxide works more efficiently and more specifically than aromatic modifiers such as DHB and phthalic acid in titania and zirconia MOC [88]. However, all affinity techniques developed for the current enrichment strategies of CPP gave reproducible but incomplete results due to *poor binding* of low concentrations of CPP and the insufficient *recovery* of multiple phosphorylated peptides [89]. Recently, a specific hydroxyapatite (HA)-based enrichment procedure has been developed for complex mixtures of phosphoprotein/CPP [90]. Salt such as calcium phosphate, also occurring in bone and tooth tissue in the HA form, with the formula $[Ca_{10}(PO_4)_6(OH)_2]$, has been previously used to enrich bone proteins [91]. The phosphate groups of phosphoproteins interact with crystalline lattice Ca^{2+} [92] more strongly than do the carboxyl groups [93]. Moreover, increasing protein phosphorylation leads to tighter binding of the proteins/CPP to HA [92]. One might conclude that the affinity of the multi-phosphorylated proteins/peptides for HA is significantly higher than that of the same components with lower phosphorylation. Essentially, the HA-based protocol immobilizes on HA microgranules proteins/peptides through their phosphate groups, while the non phosphorylated components are washed out using various buffers. In a previous article, CPP immobilized on HA were progressively eluted, increasing phosphate in the elution buffer, and then identified by off-line MALDI-MS [94]. This procedure was accelerated, and loss during elution was minimized by spotting HA-CN/CPP microgranules onto a MALDI target and analyzing the peptides directly by MALDI-TOF [90]. This method was useful for measuring the phosphorylation level of phosphoproteins/CPP quickly, with less than 2 h elapsing from the fractionation of the protein/CPP to the readout of the MALDI spectra (excluding the trypsinolysis step).

The more important advantages of the procedure are the possibility of (1) detecting phosphorylated proteins/peptides even in complex mixtures, (2) determining phosphorylated sites and those dephosphorylated by phosphatase, (3) attaining information regarding weakly and heavily phosphorylated peptides and (4) adding the HA-CPP complex directly to food, which is enabled by the use of an edible resin such HA [90]. Moreover, use of available commercial CPP preparations by the food industry is difficult for three primary reasons: i) the matrix bound to CPP is often not edible and such products can be hazardous; ii) the preparation of CPP is a long and a laborious procedure that requires cumbersome and expensive manipulation; and iii) CPP have an unpleasant taste even in modest amounts, which disadvantageously limits their direct utilization as a human food ingredient. A novel HA-based method for food grade CPP preparation has been performed on tryptic digests of casein. HA captured all CPP free of non-CPP [90]. There were approximately 32 HA bound CPP, and all non-CPP peptides were eluted [90]. HA-based enrichment procedure has been successfully applied to phosphopeptide recovery from complex biological fluids such as human serum thus providing a great source of potential biomarkers of disease. Four primary phosphopeptides derived from fibrinogen were enriched from human serum (Figure 3a-b, Table 4). A similar set of phosphorylated peptides was previously obtained using a modified IMAC strategy coupled to iterative mass spectrometry-based scanning techniques [95], using the titanium ion-immobilized mesoporous silica particles and MALDI-TOF [96] and cerium ion-chelated magnetic silica microspheres [97].

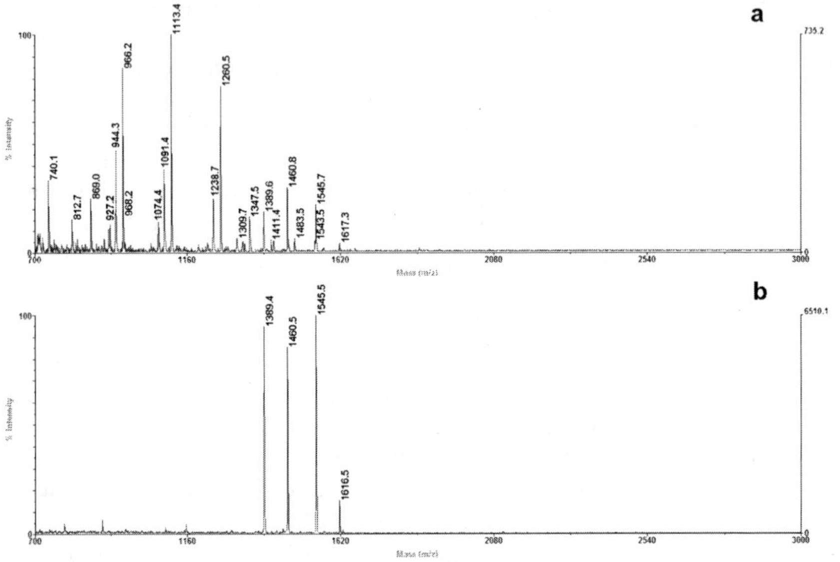

FIGURE 3. MALDI-MS-TOF spectra for the human serum before (a) and after (b) enrichment by HA (insert is the zoomed between 700 and 3000 Da).

Fibrinopeptide A (FPA) (f1-16)1P, (SerP^3), a 16-residue long peptide (1615 Da) (Table 4), is the segment anchored on the thrombin surface [98]. The other three phosphopeptides, (f1-15)1P, (f2-15)1P and (f2-16)1P (Table 4), are hydrolytic products of FPA. The serum level of fibrinogen and its hydrolytic products may reflect the expression and activation of enzymes including kinase, phosphatase, and protease [99]. An altered ratio of FPA (f2-15) and FPA (f1-16) is detected in patients affected by hepatocellular carcinoma; the D^2[pS]GEGDFLAEGGGV^{15} peptide is upregulated, and the A^{1}D[pS]GEGDFLAEGGGVR^{16} peptide is down-regulated greatly. The other two peptides, A^{1}D[pS]GEGDFLAEGGGV^{15} and D^2[pS]GEGDFLAEGGGVR^{16}, varied only slightly between the two groups [96]. The proportions of fibrinogen and their phosphorylation products offer new opportunities for basic research in exploring new frontiers in bio-marker discovery.

TABLE 4. Identification of phosphorylated fibrinogen fragments from human serum immobilized on HA.

Molecular Mass (Da)		Fibrinogen α-chain sequence	Phosphorylation sites
Theoretical	Measured MH^+		
1388.5	1389.4	$D^2SGEGDFLAEGGGV^{15}$	1
1459.5	1460.5	$A^1DSGEGDFLAEGGGV^{15}$	1
1544.6	1545.5	$D^2SGEGDFLAEGGGVR^{16}$	1
1615.6	1616.5	$A^1DSGEGDFLAEGGGVR^{16}$	1

4.1. CPP In Commercial Milk as Specific Indicators of Heated Milks

Because of the lower value, the addition of UHT and milk powder to raw or pasteurized milk is prohibited (EU Directives 92/46 CEE and 94/71 CEE) for cheese milk. The intensity of heat treatment was found to correlate with the furosine content. Glycated proteins and peptides formed during the initial stages of the Maillard reaction are indirectly evaluated through the furosine content [100-101]. The Amadori compound formed upon the reaction of lysine residue with a lactose molecule will prevent the digestive enzymes from reaching the binding sites. Native and lactosylated forms of β-CN (f1-28)4P, (f1-27)4P and α_{s2}-CN (f1-24)4P, although typical of UHT milk and milk powder, are missing in raw, pasteurized milk (71.7 °C for 15 s) and intensely pasteurized milk. The lactosylated peptides that varied with heat treatment characterize UHT milk added in amounts not lower than 10% to raw and pasteurized milk [102]. Milk delactosed with microbial β-galactosidase did not suppress the Maillard reaction; indeed, the furosine concentration increased to 35-400 mg/100 g of protein [103]. As expected, a lactose-reduced UHT milk had β-CN (f1-28)4P glycated mainly by its monosaccharides (Figure 4a).

Therefore, the nonenzymatically glycated CPP derived from the reaction of one molecule of glucose or galactose with a lysine residue (m/z 3641) can be considered to be the signature peptide of lactose-reduced milk (Figure 4).

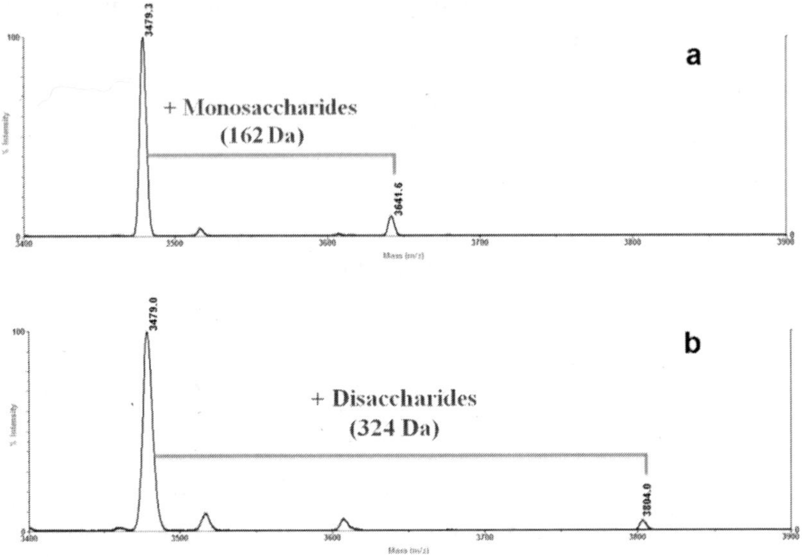

FIGURE 4. Enlarged view of the MALDI spectrum of β-CN (f1-28)4P (MH^+ = 3479 Da) signature glycosylated CPP in lactose-reduced UHT milk (a) and lactosylated CPP in UHT milk (b). The mass differences corresponded to lactose and glucose/galactose residues.

4.2. CPP IN YOGURT

Yogurt is a fermented milk defined as the "food produced by culturing one or more of the optional dairy ingredients (cream, milk, partially skimmed milk, and skim milk) with a characteristic bacteria culture that contains the lactic acid-producing bacteria, *Lactobacillus delbrueckii* subsp. *bulgaricus* and *Streptococcus thermophilus*". A heat treatment of 90 °C for 10 min is considered optimal to obtain a good quality yogurt [104], and the addition of milk powder increases the content of furosine to more than 300 mg/100 g protein [105]. In yogurt, enzymes could give rise to the liberation of a particularly high number of bioactive peptides, among them CPP, which could be partly due to LAB proteolytic activity. Comparison of CPP in raw, pasteurized and intensely heated milks has previously shown that there is a plethora of milk peptides among which a few were glycosylated CPP [102]. Yogurt is prepared from intensely heated milk instead of low-pasteurized drinking milk. The CPP of two preparations were enriched on HA and analyzed by MALDI-TOF (Table 5). The proportion of CPP with molecular masses between 2.5 and 4 kDa was significantly higher in yogurt than in pasteurized milk as shown in Table 5. Only four fragments of CPP-derived peptides produced during yogurt preparation occur in pasteurized milk (Table 5).

TABLE 5. List of HA-enriched CPP identified in commercial samples of yogurt and pasteurized milk. Relative intensity of each peak is reported.

Molecular Mass (Da)		CPP sequence	Yogurt	Pasteurized milk
Measured MH^+	Theoretical		Relative Intensity	
		α_{s1}-CN derived CPP		
1327.5	1326.4	α_{s1}-CN (f44-54)2P	0.3	n.d.
5009.1	5008.7	α_{s1}-CN (f39-79)6P	0.6	n.d.
5089.3	5088.7	α_{s1}-CN (f39-79)7P	1.2	n.d.
6959.0	6958.7	α_{s1}-CN (f35-90)8P	0.6	n.d.
7087.1	7086.9	α_{s1}-CN (f34-90)8P	0.2	n.d.
		α_{s2}-CN derived CPP		
1493.6	1492.7	α_{s2}-CN (f139-150)1P	1.2	n.d.
1617.5	1616.5	α_{s2}-CN (f1-12)3P	2.0	n.d.
1931.7	1930.8	α_{s2}-CN (f1-15)3P	8.6	n.d.
2007.9	2006.7	α_{s2}-CN (f7-21)4P	0.4	n.d.
2356.2	2355.1	α_{s2}-CN (f1-18)4P	20.8	n.d.
2666.7	2665.4	α_{s2}-CN (f51-72)4P	2.3	n.d.
2748.6	2747.5	α_{s2}-CN (f1-21)4P*	3.0	26.9
2876.8	2875.7	α_{s2}-CN (f1-22)4P	1.8	n.d.
3005.2	3004.8	α_{s2}-CN (f2-24)4P	0.4	n.d.
3134.3	3133.0	α_{s2}-CN (f1-24)4P*	24.2	46.0
3382.0	3381.1	α_{s2}-CN (f49-76)4P	8.4	n.d.
3461.9	3461.1	α_{s2}-CN (f49-76)5P	1.4	n.d.
3666.0	3665.9	α_{s2}-CN (f16-45)2P	0.9	n.d.
4166.4	4165.3	α_{s2}-CN (f115-149)3P*	0.6	n.d.
4294.7	4293.5	α_{s2}-CN (f115-150)3P*	0.6	n.d.
		β-CN derived CPP		
960.5	959.9	β-CN (f30-36)1P	0.8	n.d.
1462.5	1461.4	β-CN (f17-27)3P	1.8	n.d.
1511.6	1510.4	β-CN (f17-28)2P	0.6	n.d.
1515.5	1514.4	β-CN (f15-25)4P	1.3	n.d.
1591.4	1590.4	β-CN (f17-28)3P	19.5	n.d.
1628.9	1628.3	β-CN (f15-26)4P	0.7	n.d.
1645.4	1644.3	β-CN (f14-25)4P	1.1	n.d.
1743.8	1742.5	β-CN (f15-27)4P	16.0	n.d.
1791.4	1790.6	β-CN (f15-28)3P	7.9	n.d.
1871.6	1870.6	β-CN (f15-28)4P	100.0	n.d.
1999.8	1998.7	β-CN (f14-28)4P	0.4	n.d.
2240.4	2239.0	β-CN (f8-25)4P	1.6	n.d.
2709.4	2708.5	β-CN (f7-28)4P	4.1	n.d.
2967.4	2966.7	β-CN (f1-24)4P	0.2	n.d.
3080.5	3079.9	β-CN (f4-28)4P	0.6	n.d.
3123.2	3122.9	β-CN (f1-25)4P*	n.d.	5.8
3351.6	3350.2	β-CN (f1-27)4P	3.2	14.8
3479.9	3478.4	β-CN (f1-28)4P*	70.1	100.0
3607.8	3606.6	β-CN (f1-29)4P*	13.2	6.0
3803.8	3802.4	β-CN (f1-28)4P + 1 lactose	0.7	n.d.
3849.3	3848.8	β-CN (f1-31)4P	0.1	n.d.
3978.2	3977.0	β-CN (f1-32)4P	n.d.	0.5
		κ-CN derived CPP		
6229.8	6228.8	κ-CN (f106-163)1P	0.3	n.d.
6788.5	6787.5	κ-CN (f106-169)1P	0.1	n.d.

For example, β-CN (f1-28)4P, a peptide representing 100% intensity (assumed as base peak) of the signals in MALDI spectra, was reduced by approximately 30%; this loss was associated with the transformation of pasteurized milk into yogurt. One can deduce that the original peptide undergoes degradation even considering the higher number of formed CN peptides. In yogurt, β-CN (f1-28)4P, the most common CPP in pasteurized milk, was hydrolyzed into the peptide β-CN (f15-28)4P, which thus becomes the most abundant CPP.

β-CN (f1-29)4P, β-CN (f1-28)4P, β-CN (f1-25)4P, β-CN (f1-24)4P, α_{s2}-CN (f1-24)4P, α_{s2}-CN (f1-21)4P, resulting from the CN hydrolysis by plasmin and enriched on HA, were also found as C-terminally shortened peptides. During fermentation and storage, α_{s2}-CN (f115-150)3P and α_{s2}-CN (f115-149)3P derived by plasmin action did not react further to produce shorter peptides, most likely because of the absence of proteolytic enzymes. In the yogurt fraction recovered by centrifugation, only five multi-phosphorylated α_{s1}-CN and two low-phosphorylated κ-CN, κ-CN (f106-163)1P and κ-CN (f106-169)1P, were identified. Few CPP were less phosphorylated than the native peptides due to the presence of milk phosphatase, which was denatured in all pasteurized cheese-milks. The presence of lactosylated β-CN (f1-28)4P CPP was indicative of yogurt made with high-heat treated or milk fortified with milk protein powder [102]. Proteolysis of milk proteins in model yogurt systems has shown a similar set of primary CPP (Table 5). Therefore, the question is raised how CPP, derived from the enzymatic hydrolysis of yogurt CN are digested and absorbed in adult humans. For this reason, it is important to know the gastrointestinal resistance of CPP if used as a functional ingredient for fruit beverages. In the various stages of human digestion, a large quantity of CPP is produced in the stomach by partial hydrolysis of CN through pepsin action and in the small intestine by trypsin; these peptides are successively refined by endoproteases/exopeptidases. Although analysis of the intestinal contents of milk and yogurt ingestion has revealed the presence of CPP [60], their further resistance to gastrointestinal enzymes is poorly documented. Fragment β-CN (f1-24)4P has been previously identified in the lumen contents of rats after 60 min of digestion as a β-CN (f1-25)4P derived peptide [106]. Moreover, after yogurt ingestion, β-CN (f1-32)4P CPP was released in the human stomach [60] and β-CN (f1-31)4P was found in a yogurt sample. The fragment β-CN (f1-28)4P constitutes a clear example of the multi-functionality of milk-derived peptides because some regions in the primary structure of caseins contain overlapping peptide sequences that exert different biological effects, in this case both mineral binding and immunostimulatory action [107]. Even with the difference in the peptide pattern, it is evident that CPP binding iron (or other metal ions) remains soluble in the digestive tract, where they escape further enzyme digestion [106]. The authors have studied in depth the simulated digestion of CPP from peptide precursors. These studies greatly benefit from the knowledge of enzyme specificities and degradation mechanisms. CPase and chymotryptic activity of pancreatin exhibits broad specificity, cleaving bonds on the carboxyl side of several amino acid residues of CPP. The latter, which are in the mass range 960-7087 Da (Table 5), are good candidates for intestinal absorption and for playing a possible physiological role in mineral bioavailability. However, there are conflicting results on the lack of α_{s1}-CN (f43-52)2P and α_{s2}-CN (f1-19)4P identified by other authors after CN hydrolysis with pancreatin, an enzyme used during the intestinal step of simulated physiological digestion [108]. Generally, the physiological effects of CPP may not always be extended to precursor peptides, although they are structurally similar. There are not enough data concerning the effects of CPP addition to probiotic acid fermented milks. However, probiotic bacteria such as *Lactobacillus acidophilus* and *Bifidobacterium* spp., selected because of their beneficial action, which they may manifest on the health of the consumer, grow slowly in milk because of the lack of the proteolytic activity [109]. For this reason, and also to reduce the fermentation time, probiotic yogurt is manufactured by yogurt bacteria (*Streptococcus thermophilus* and *Lactobacillus delbrueckii* ssp. *bulgaricus*) with the addition of probiotic culture. In parallel, non-digestible food ingredients, *i.e.*, "prebiotics", resist digestion in the small intestine and reach the colon, where they act as a growth factor for *Bifidobacterium* species and are metabolized into short chain fatty acids by a limited number of the microorganisms also comprising the colonic microflora. Prebiotics are principally oligosaccharides (fructo-oligosaccharides, inulins, isomalto-oligosaccharides, lactilol, lactosucrose, lactulose, pyrodextrins, soy oligosaccharides, transgalacto-oligosaccharides, and xylo-oligosaccharides) that stimulate bifidobacteria growth. In probiotic yogurt containing inulin as a prebiotic, the number of bioactive peptides increased, which means that elevation in the proteolytic activity has a synergistic effect with probiotic counts of yogurt cultures. Therefore, the most proteolytic strains of *St. thermophilus* and *Lb. delbrueckii* subsp. *bulgaricus,* spp. enhance the growth of *Lb. acidophilus* and *Bifidobacterium*. In our studies, the overall opiate activity of the bio-yogurt preparation containing *Lb. acidophilus* and *Bifidobacterium* spp. and inulin as a prebiotic was approximately twice that typical of traditional yogurt. In addition, the above yogurt preparation contained a variety of opioid agonistic and antagonistic, immunomodulation, anti-thrombotic, ACE-inhibitor, and anti-microbial activities. Traditional and probiotic yogurt both possess a characteristic soluble fraction composed by peptides exhibiting biological activity, amongst others. This fraction was found to include CPP, β-casomorphins and antithrombotic peptide precursors that did not differ greatly from one another. In a study comparing the proteolytic, amino-, di-, tri- and endopeptidase activity of nine strains of *St. thermophilus*, six strains of *Lb.*

delbrueckii, fourteen strains of *Lb. acidophilus* and thirteen strains of *Bifidobacterium* spp., aminopeptidase activity was detected for all bacterial strains – traditional yogurt strains and probiotic bifidobacteria - both at the extracellular and intracellular levels. High dipeptidase activity was demonstrated by all bacterial strains for *Lb. delbrueckii* ssp. *bulgaricus, Lb. acidophilus,* and *Bifidobacterium* spp., whereas *St. thermophilus* had greater dipeptidase activity at the extracellular level.

4.3. CPP IN A FEW CHEESE VARIETIES

Whole milk contains a variety of endogenous plasmin-mediated CN peptides. In addition, CPP were released following cell lysis and release of intracellular LAB enzymes. This phenomenon was observed especially at the end of ripening in long-ripened cheeses, such as Comté [110], Grana Padano [111], Parmigiano-Reggiano and semi-hard Herrgard cheese [112]. In Grana Padano cheese, 45 CPP were identified, of which 24 originated from β-casein, 16 from α_{s1}-casein and 5 from α_{s2}-casein. These CPP formally derive from three parent peptides, namely β-CN (f7-28)4P, α_{s1}-CN (f61-79)4P and α_{s2}-CN (f7-21)4P [111]. By comparing CPP of Grana Padano and Herrgard cheese, it was clear that CPP were all progressively shortened and dephosphorylated during ripening. CPP were very resistant to enzymatic degradation, especially when SerP residue was at the N-terminal end. The number of CPP identified according to the different procedures was comparable for Grana Padano [111] and Herrgard cheese [112]. In both the cheeses, CPP were progressively shortened and dephosphorylated during ripening, with both cheeses constituting heterogeneous mixtures of peptides phosphorylated at various sites sharing N- and C-terminally truncated CPP. However, some peptides proved very resistant to enzymatic degradation, especially when SerP residue was present at the N-terminal end of CPP [111]. The only SerP residue located at the N-terminus of CPP was subjected to dephosphorylation, exposing the dephosphorylated residue to aminopeptidase action. A heterogeneous CPP pattern also differentiated the cheese samples within a given form because of the phosphatase gradient amongst peripheral and central parts of the Grana Padano cheese form ($3\ 10^5$ *vs.* $3\ 10^2$). This is due to heat sensitivity in the temperature range 57 and 62 °C and the acid pH at which these enzymes are denatured. These data explain the discrepancy in the amount of serine, which varied by as much as 50% of SerP from the periphery towards the center of the cheese form [113]. In contrast, the CPP fraction of Herrgard cheese was more uniform, with the two cheese varieties sharing active plasmin and amino-peptidases from lactic acid bacteria. Because milk pasteurization denatures alkaline phosphatase while it activates plasmin [114], proteolysis in the above cheese is plasmin-dependent. It is therefore likely that CPP of pasteurized milk cheeses are intrinsically more stable than raw milk cheeses [111]. CPP in artisanal PDO ovine Fiore Sardo cheese have been previously reported [115]. Patterns of CPP similar to that observed for bovine cheese indicated that mechanisms of formation and degradation of CPP were similar regardless of the milk species and cheese variety. The dephosphorylation mechanism in Fiore Sardo was different from that found in Grana Padano cheese, most likely because of the use of different rennet types. In PDO Fiore Sardo cheese, no apparent difference in susceptibility to dephosphorylation was found amongst the differently located SerP peptide residues. This resulted in the simultaneous occurrence of partly dephosphorylated peptides, either internally or externally. CPP enrichment by HA, for example of pH 4.6 soluble fractions of hard Parmigiano Reggiano (PR) (30-mo-old), semi-hard, *pasta filata* Provolone del Monaco (PM) (6-mo-old), semi-cooked Asiago d'Allevo (AA) (3-mo-old) and mold-ripened cheese Gorgonzola (GR) (2-mo-old) cheese, has allowed the identification of CPP in high number (Figure 5) which may explain the broad-specificity of the cheese enzymes involved in CN proteolysis. Some CPP were derived from the Lys-X or Arg-X cleavage by plasmin primarily located in the *N*-terminal region of caseins, such as β-CN (f1-28)4P (Lys^{28}-Lys^{29}) or β-CN (f1-29)4P (Lys^{29}-Ile^{30}), α_{s1}-CN (f61-79)5P (Lys^{79}-His^{80}) and α_{s2}-CN (f1-24)4P (Lys^{24}-Asn^{25}). The native plasmin-derived CPP were then further hydrolyzed by cheese aminopeptidases and CPase into shorter peptides.

It is likely that ingested cheese carries a concentrated pH 4.6 soluble CPP fraction and a variable number of CPP according to the cheese variety. Above all, the presence and integrity of plasmin-mediated products of CN is a function of the milk, whether raw or pasteurized. Pasteurization reduces the milk plasmin activity only by ~15 percent, whereas plasmin activity increases during milk storage. UHT does not inactivate the plasmin in milk, and proteolytic activity will continue to damage milk. Heat treatments modify the peptide profile by increasing the content of larger peptides. The CN breakdown occurring during the ripening of PR cheese proceeded more slowly in PM cheese.. This means that eating PR cheese increases the quota of the co-ingested mineral bound CPP. In contrast, GR cheese show a different CPP level such as that of β-CN (f1-27)4P (3350.2 Da), resulting the most abundant CPP, when compared with hard cheeses (Figure 5a and 6); β-CN (f1-27)4P was further hydrolyzed into the shorter β-CN (f7-27)4P (2580.4 Da), β-CN (f10-27)4P (2270.0 Da), β- CN (f12-27)4P (2083.8 Da) and β-CN (f15-27)4 (1742.4 Da). The most abundant CPP in all three cheeses derived from the peptide β-CN (f1-28)4P, but long-ripened PR cheese was dissimilar from the other cheeses in its content of α_{s1}-CN (f62-79)5P (2332.9 Da) (Figure 5d and 6).

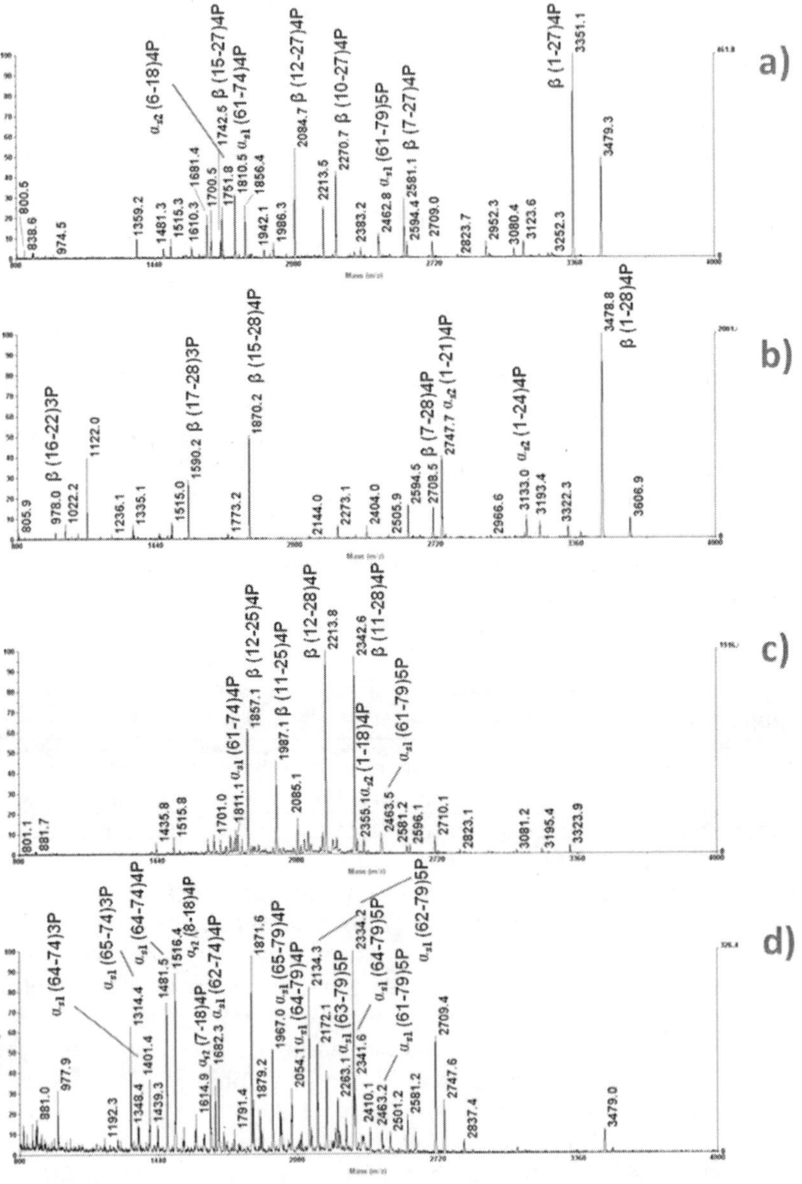

FIGURE 5. The MALDI spectra of CPP isolated by the addition of HA to pH 4.6 soluble fractions of Gorgonzola (a), Asiago (b), Provolone del Monaco (c), and Parmigiano Reggiano (d) cheeses. The inset magnifies the m/z values in the lower molecular mass peptide range 0.8-4 kDa.

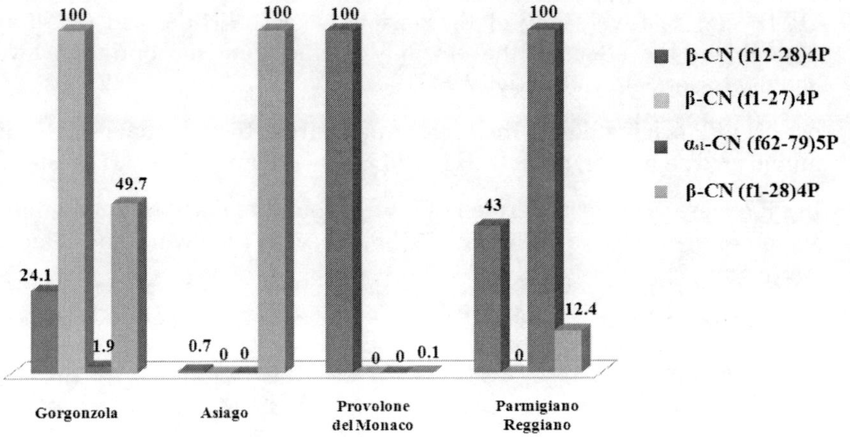

FIGURE 6. Histogram representation of CPP at 100% relative intensity and their performance in four cheeses.

β-CN (f1-28)4P (3478.4 Da) in AA cheese was the most abundant signal of the MALDI spectra and it was partially hydrolyzed into the shorter peptides β-CN (f7-28)4P (2708.5 Da), β-CN (f15-28)4P (1869.7 Da) and β-CN (f17-28)3P (1589.6 Da) (Figure 5b and 6). Considering exclusively the CPP molecular mass in the 3-3.5 kDa range of AA and GR cheese, the intensity of a high number of peptides transformed into a number of progressively lower molecular weight CPP, with accompanying liberation of peptides (Figure 5). The most common group of CPP occurred in the mass range of 1.7-2.9, reaching the maximum intensity for β-CN (f12-28)4P (2212.0 Da) in PM and α_{s1}-CN (f62-79)5P (2332.9 Da) in PR cheese (Figure 5c-d and 6). The presence of β-CN (f16-22)3P (977.7 Da) was discovered in both AA and PR cheeses and was not detected in the PM and GR cheeses (Figure 5). Our results show that longer plasmin-mediated peptides degraded into shorter CPP. These peptides became more evident when the chymosin retained in the cheese was largely inactivated by cooking the curd at high temperatures (~55 °C). The hydrolysis of CN by chymosin was covered by that of plasmin, which became the principal proteolytic enzyme in the cheese. This phenomenon is particularly evident in PM raw milk cheese for which the plasmin-mediated β-CN (f1-28)4P peptide, representing ~0.1% of the CPP, was almost completely hydrolyzed into the shorter peptides β-CN (f11-28)4P (2341.1 Da), β-CN (f12-28)4P (2212.0 Da), β-CN (f11-25)4P (1985.7 Da) and β-CN (f12-25)4P (1856.6 Da) (Figure 5c). When comparing the PR and PM cheese, the former had a high extent of β-CN (f1-28)4P as judged by the higher levels of the peptide. This demonstrates that CPP of PR cheese are progressively transformed into a number of lower-molecular-weight peptides. In contrast, the quasi-total absence in the PM cheese of β-CN (f1-28)4P and relatively few of the various sizes β-CN-derived CPP (Figure 6) could be the effect of the enzyme decline from the optimum level of activity to zero enzyme activity.

α_{s1}-CN CPP originated for the greater parts from the internal regions of the amino acid sequence, namely αs1-CN (f61-79)5P and αs1-CN (f33-60)3P.

In PR cheese, 23 casein-derived CPP were found to derive from the internal region of α_{s1}-CN, *i.e.*, α_{s1}-CN (f59-79)5P, whereas they were not detected in AA (Figure 7).

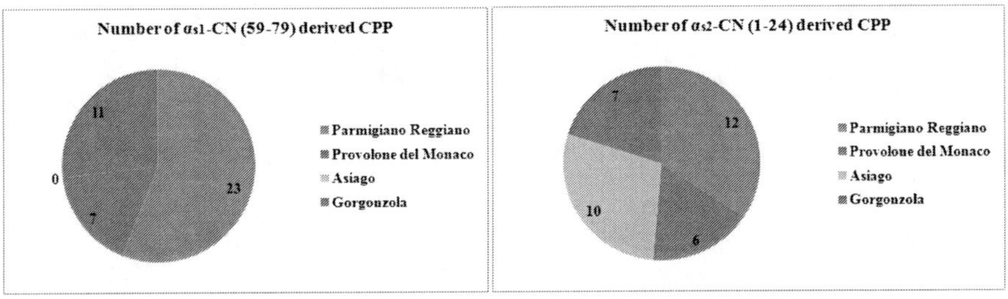

FIGURE 7. The number of CPP derived from α_{s1}-CN (f59-79)5P and α_{s2}-CN (f1-24)4P in Parmigiano Reggiano, Provolone del Monaco, Asiago and Gorgonzola cheeses.

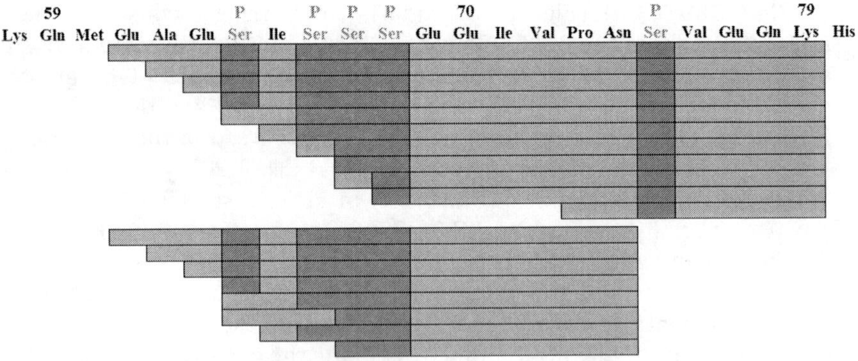

FIGURE 8. Amino acid sequence of the α_{s1}-CN (f59-79) 4P peptide and the CPP identified in PR cheese. Phosphoserine residues are indicated by red boxes.

The profile of the CPP depicts the mechanisms of both the proteolysis and dephosphorylation in a long ripened cheese. Peptide α_{s1}-CN (f61-79)5P, most likely arising from the parent peptide α_{s1}-CN (f1-79)7P through cleavage at Met^{60}-Glu^{61}, was dephosphorylated and concurrently hydrolyzed into shorter peptides. Alkaline and/or acid phosphatases acting on SerP residue dephosphorylated CPP. N-terminal Ser was then exposed to aminopeptidase and released as a free amino acid. Bacterial CPase or exopeptidase such as cathepsin D/chymosin release α_{s1}-CN (f61-74)4P and other derived CPP through cleavage at Asn^{74}-$SerP^{75}$ (Figure 8). In PR aged 30 months, α_{s1}-CN (f61-79)5P (2462.1 Da), α_{s1}-CN (f62-79)5P (2332.9 Da), α_{s1}-CN (f63-79)5P (2261.9 Da), α_{s1}-CN (f64-79)5P (2132.8 Da), α_{s1}-CN (f64-79)4P (2052.8 Da), α_{s1}-CN (f65-79)4P (1965.7 Da), α_{s1}-CN (f62-74)4P (1681.3 Da), α_{s1}-CN (f64-74)4P (1481.2 Da), α_{s1}-CN (f64-74)3P (1401.2 Da) and α_{s1}-CN (f65-74)3P (1314.1 Da) were the dominant CPP (Figure 5d and 8). Indeed, only 7 CPP for PM cheese and 11 CPP for GR were derived from α_{s1}-CN (f59-79)5P, namely α_{s1}-CN (f61-79)5P (2462.1 Da) and α_{s1}-CN (f61-74)4P (1810.5 Da) (Figure 5a and c and 7). Considering the α_{s2}-CN peptide, the α_{s2}-CN (f1-24)4P CPP were similar in number but significantly different in the case of the four cheese varieties (Figure 6). The dominant α_{s2}-CN-derived CPP were α_{s2}-CN (f1-24)4P (3132.9 Da) and α_{s2}-CN (f1-21)4P (2747.6 Da) in AA cheese (Figure 5b). The most abundant α_{s2}-CN-derived CPP were α_{s2}-CN (f1-18)4P (2355.1 Da) for PM and α_{s2}-CN (f6-18)4P (1751.4 Da), a shortened form of the primary CPP α_{s2}-CN (f1-18)4P, for GR cheese (Figure 5a and c). α_{s2}-CN (f7-18)4P (1614.3 Da) and α_{s2}-CN (f8-18)4P (1515.1 Da) characteristically accumulated in PR cheese (Figure 5d). Similar and discrete phosphorylated CPP derived species for α_{s2}-CN (f1-24) (3P and 4P) and β-CN (f1-28) (3P and 4P) occurred in all cheeses. For the other casein fractions, primary CPP are fully phosphorylated, such as α_{s1}-CN (f61-79)5P, whereas the derived peptides show a level of phosphorylation less than native form as observed in PM and GR. A higher dephosphorylation level characterized the CPP profile of PR cheese (Figure 8). The different profile of CPP could derive from the different length of ripening and from the cheese variety.

CPP. in ovine cheeses

Cheeses that contain CPP are also manufactured from ovine milk. Proteolytic enzymes in Pecorino cheese originate from chymosin, pepsin and other clotting preparations such as paste rennet. These enzymatic activities are complemented by those secreted by the vegetative spores of *Penicillium roqueforti* during the maturation of blue-veined cheeses. The CPP patterns of Pecorino and Roquefort cheese have been characterized and main components identified. The sequence alignment of the CPP released throughout the hydrolysis of the β-CN (f1-28)4P in Pecorino and Roquefort cheeses are compared in Figure 9.

In PR cheese of different ages, the released CPP were progressively degraded at C- and N-terminal ends. CPases work from the C-terminal end and aminopeptidases from the N-terminal end, both removing the terminal amino acid residues incrementally. LAB does not produce CPases; thus, the ability to liberate the carboxyterminal amino acid and peptides is typical of the mold. The N-terminal amino acid seemed to be released faster than the C-terminal residues because of the lower activity of CPases. There are negligible differences in the CPP level of different cheese lots primarily because of the action of the enzymes from *P. roqueforti* after its sporulation in blue-veined cheese. More long-chain CPP β-CN, such as β-CN (f1-28)5P-4P, β-CN (f1-27)5P-4P and β-CN (f1-24)5P-4P, were detected in Pecorino cheese, whereas β-CN (f7-28)5P-4P and β-CN (f7-27)5P-4P resulted from the longer CPP in Roquefort cheese (Figure 9). In these cheese varieties, both α_{s1}- and β-CN have been described as completely hydrolyzed at the end of ripening. This contradicts other findings indicating ~50% CN hydrolysis. Plasmin, NSLAB, and Lactobacilli contaminating flora proteinases are mainly responsible for extensive proteolysis in Parmigiano-Reggiano cheese, which is ripened for ~24 months at ~18-20 °C [116]. Here, chymosin is denatured by the high cooking temperature used during the manufacture of cheese. Molds develop at approximately 2 to 5 weeks of ripening, concurrently degrading CN into peptides of various sizes [117]. A similar mechanism for β-CN-derived CPP was found in Grana Padano cheese. Ser was proteolytically cleaved by aminopeptidases, and SerP hindered cleavage by the latter and continued its action after dephosphorylatation of SerP.

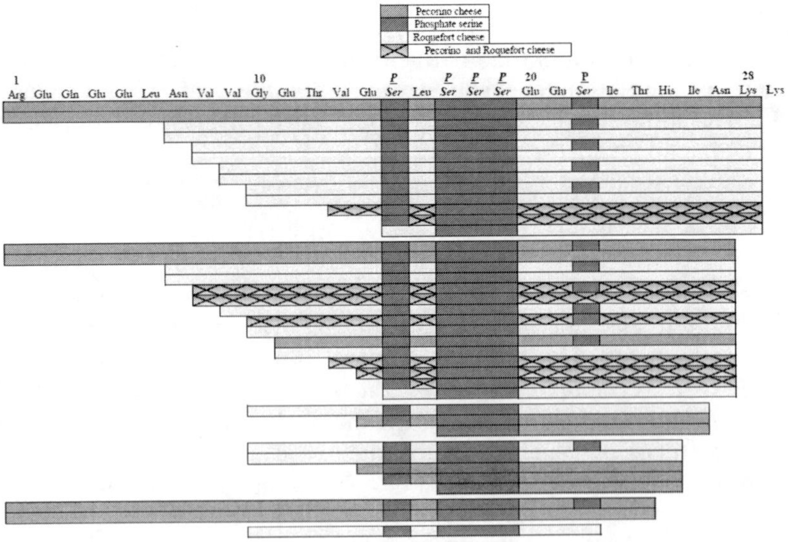

FIGURE 9. Amino acid sequence of the β-CN (f1-28)4P peptide and the CPP identified in Pecorino (blue line) and Roquefort (yellow line) cheeses. CPP common to the two cheeses are indicated by crosses, and phosphoserine residues are indicated by red boxes.

5. ANTICARIOGENIC EFFECT OF CPP IN YOGURT AND CHEESE

Due to proteolytic activity, a great number of CPP are formed in raw milk cheese. In contrast, the enzymatic digestion of proteins to peptides can be reduced by milk pasteurization. Notwithstanding this, yogurt remains a consistent source of bioavailable CPP even if milk is heated to high temperatures (90 °C, 30 min) to create an inoculation medium in which the bacteria can grow and produce lactic acid. LAB provide plenty of CPP in the form of soluble complexes with Ca^{2+} that are effective in avoiding the life-threatening calcium phosphate precipitation, enhancing the intestinal absorption of minerals and retention in the human body [34,118]. The mineral-binding power of CPP also depends on the number of binding sites and their relative accessibility [119]. Dairy products such as cheese and yogurt are both rich in multi-phosphorylated peptides capable of interacting with colloidal calcium phosphate to manifest anticariogenic properties in human and animal [120-121,13]. The mechanism of anticariogenicity might be due to a direct chemical effect of casein and calcium phosphate components [122]. Tooth enamel is a polymeric substance consisting of crystalline calcium phosphate, embedded in a protein matrix. Thus, CPP can significantly enhance localization of ACP at the tooth surface, inhibiting enamel demineralization and promoting remineralization. In the development of teeth and bone, CPP act as hydroxyapatite nucleator and control the growth of the crystals, resulting in unique crystal morphology. A new calcium phosphate remineralization technology has been recently developed based on the complex CPP-ACP [Recaldent™ CASRN691364-49-5] [66]. This preparation is claimed to stabilize calcium and phosphate ions in high concentrations by binding ACP to pellicle and plaque of the tooth surface. Moreover, CPP-ACP inhibited the adhesion of *Streptococcus mutans* to the tooth surface producing a copious reservoir of bioavailable calcium ions [67]. Cheese and yogurt CPP have the ability to stabilize calcium phosphate in solution, forming small CPP-ACP nanocomplexes. The calcium-binding ability of CPP has been applied by clinical dentists to show that CPP stabilize high concentrations of calcium, phosphate, and fluoride ions on the tooth surface by binding them to pellicle and plaque [66,123]. In dental plaque, CPP-ACP binds onto the surface components of the intercellular plaque matrix. Incorporation of CPP-ACP into the plaque will increase the calcium and phosphate content by forming a stable supersaturated solution of calcium phosphate. Thus, the availability of calcium in plaque provides a natural anticaries protective effect, either by suppression of demineralization promoted by fermentative acids in mouth, through an increased remineralization by binding calcium ions to teeth enamel, or possibly a combination of both. An inverse relation between plaque calcium and caries incidence has been evidenced [124-125]. Reynolds (1997) [126] has demonstrated that CPP-ACP can actually remineralize subsurface lesions in human enamel, and this is indeed the basis of one claim of his patent [127]. The diffusion of available CPP-ACP in the mouth is controlled by two main factors: i) the molecular weight of the diffusing species, the square of which is inversely proportional to the diffusion coefficient; and ii) the binding characteristics of the diffusing species, which dictate how much CPP-ACP is free to diffuse at a given time [128]. At neutral pH, calcium diffusion is limited by the quantity of bound calcium, reducing the effective diffusion coefficient (De) and creating a measurable restricted effective diffusion coefficient (rDe), where rDe =De/(R + 1) and R is the ratio of bound to free calcium. A large number of potential binding sites for calcium can have significant effects on the calcium diffusion coefficient; this effect is maintained at low pH, although overall diffusion is slightly faster [129-130]. Conversely, one may infer that ACP may also bind to dental plaque and tooth enamel, thus having beneficial effects on teeth remineralization [67]. In addition, SerP residues of the CCP-HA complex are exposed to intestinal alkaline phosphatase, which favors metal ion bioavailability by releasing inorganic phosphate. Milk, ice cream, and cheese have been observed to lower the incidence of dental caries in rats [131]. Elderly people that eat cheese several times per week had a lower incidence of root surface caries development [132]. CPP of yogurt have been observed to have an inhibitory effect on demineralization and are able to promote remineralization of dental enamel [133]. Moreover, anticariogenic activity has also been reported for egg phosphopeptides (viz. phosvitin and phosphophorin) [134-135]. The heating process affects the bioavailability of CPP; for example, milk sterilization can induce dephosphorylation of phosphoseryl residues and dehydroalanine residue formation [136].

6. CPP AS NUTRACEUTICAL INGREDIENTS IN FUNCTIONAL FOODS

6.1. Examples of Commercial CPP Preparation

Some commercial products have been developed containing moderately hydrolyzed milk proteins as the sole protein source. Hypoallergenic formulas are based on partially or extensively hydrolyzed proteins. Both formulae are better tolerated by small premature infants than native cow milk protein. Ultrafiltered micellar casein and microfiltered whey protein concentrate are known to slow down the digestive process. Typical composition data of commercial phosphopeptide preparations (m/m) include 91.3% (TN x 6.47) or 94.8 (dry basis) protein, 16.0% CPP, and 6.0% free amino acids. Compared with the expected CPP composition, the commercial preparation contained ~5% undigested protein, ~30% peptides in the intermediate molecular mass 5000-20000 Da range, and ~48% of molecular mass in the 500-5000 Da range (Tatua, New Zealand). The preparation is generally very complex and dependent on the procedure used to perform casein hydrolysis. Therefore, commercial products can be considered to be an enriched-CCP preparation containing 16% CPP without specification of the peptide size and phosphorylation degree. Preliminary analysis performed by MALDI-TOF analysis indicated that the signal in the mass spectra originated exclusively from the β-CN digestion. The intensity of non-CPP such as β-CN (f191-209), β-CN (f184-209), β-CN (f177-209) and β-CN (f170-209) was sufficiently high to obscure other CN peptides (MALDI spectrum not shown). As reported above, a number of available techniques allow the separation of CPP and non-CPP. To reduce the large dynamic range of non-CPP, HA was used for CPP enrichment [90,94,102]. CPP included multiply phosphorylated peptides (up to 4 phosphorylation residues). CPP β-CN (f1-28)4P and 3P, β-CN (f1-25)4P, 3P and 2P and β-CN (f2-25)4P, 3P and 2P were found and may be indicative of the progressive CPP dephosphorylation (MALDI spectrum not shown). Interestingly, in addition to β-CN, the commercial CPP (Tatua, NZ) also displayed α_{s1}- or α_{s2}-CN-derived CPP, invisible or weakly visible before the sample was treated with HA (Table 6).

Finally, a method for reducing the complexity of peptide mixtures was the separation of non-CPP and CPP by trapping CPP on HA under neutral conditions. This could be the principle for an industrially based production of CPP.

6.2. Examples of Industrial Methods for CPP Preparation

Starting from CN, the overall preparation process gave 16% CPP (Figure 10), less than the theoretical yield of 23% (Y_{theor} =22.8%), which means approximately 20% yield on the basis of weight, a yield higher than that obtained by other researchers [138-140]. In their production experiments, the authors obtained a CPP preparation as high as 18.8% degree of hydrolysis (DH).

TABLE 6. CPP identified in a commercial CPP preparation (Tatua Co-operative Dairy Company Ltd) by tandem MS sequencing.

Parent Protein	Molecular Mass (Da)		Start	End	Peptide Sequence	Peptide Modifications
	Measured	Theoretical				
α_{s1}-CN	1138.4	1138.4	115	123	(N)SAEERLHSM(K)	1P
	1526.7	1526.7	35	47	(K)EKVNELSKDIGSE(S)	1P
	1831.9	1831.8	75	89	(N)SVEQKHIQKEDVPSE(R)	1P
	1859.9	1859.9	75	89	(N)SVEKHIQKEDVPSER(Y)*	1P
	1926.5	1926.7	43	58	(K)DIGSESTEDQAMEDIK(Q)	2P
	1950.9	1950.9	104	119	(K)YKVPQLEIVPNSAEER(L)	1P
	1988.0	1987.9	75	90	(N)SVEQKHIQKEDVPSER(Y)	1P
	2054.7	2054.7	43	59	(K)DIGSESTEDQAMEDIKQ(M)	2P
	2185.8	2185.8	43	60	(K)DIGSESTEDQAMEDIKQM(E)	2P
	2400.9	2400.9	41	60	(L)SKDIGSESTEDQAMEDIKQM(E)	2P
	2597.0	2597.0	37	58	(K)VNELSKDIGSESTEDQAMEDIK(Q)	2P
	2856.1	2856.1	37	60	(K)VNELSKDIGSESTEDQAMEDIKQM(E)	2P
	3113.2	3113.3	35	60	(K)EKVNELSKDIGSESTEDQAMEDIKQM(E)	2P
	3193.2	3193.3	35	60	(K)EKVNELSKDIGSESTEDQAMEDIKQM(E)	3P
	3699.6	3699.5	61	90	(M)EAESISSSEEIVPNSVEQKHIQKEDVPSER(Y)	4P
	3779.6	3779.5	61	90	(M)EAESISSSEEIVPNSVEQKHIQKEDVPSER(Y)	5P
	3974.5	3974.6	59	90	(K)QMEAESISSSEEIVPNSVEQKHIQKEDVPSER(Y)	4P; Oxidation (Met)
	4479.0	4479.0	23	60	(R)FFVAPFPEVFGKEKVNELSKDIGSESTEDQAMEDIKQM(E)	2P
α_{s2}-CN	1432.6	1432.6	135	146	(N)SKKTVDMESTEV(F)	1P
	1538.6	1538.6	126	137	(R)EQLSTSEENSKK(T)	2P
	1578.6	1578.6	123	134	(T)LNREQLSTSEEN(S)	2P
	1694.7	1694.7	125	137	(N)REQLSTSEENSKK(T)	2P
	1793.7	1793.7	123	136	(T)LNREQLSTSEENSK(K)	2P
	1921.8	1921.8	123	137	(T)LNREQLSTSEENSKK(T)	2P
	2715.3	2715.3	115	137	(R)NAVPITPTLNREQLSTSEENSKK(T)	2P
	3051.3	3051.2	1	24	KNTMEHVSSSEESIISQETYKQEK(N)	3P
	3131.2	3131.2	1	24	KNTMEHVSSSEESIISQETYKQEK(N)	4P
	3786.6	3786.6	115	146	(R)NAVPITPTLNREQLSTSEENSKKTVDMESTEV(F)	3P
	4034.6	4034.8	115	148	(R)NAVPITPTLNREQLSTSEENSKKTVDMESTEVFT(K)	3P
β-CN	1600.7	1600.7	29	40	(K)KIEKFQSEEQQQ(T)	1P
	1785.7	1785.7	35	48	(Q)SEEQQQTEDELQDK(I)	1P
	1945.8	1945.8	29	43	(K)KIEKFQSEEQQQTED(E)	1P
	2559.2	2559.1	29	48	(K)KIEKFQSEEQQQTEDELQDK(I)	1P
	2805.2	2805.2	1	24	RELEELNVPGEIVESLSSSEESIT(R)	2P
	2885.2	2885.2	1	24	RELEELNVPGEIVESLSSSEESIT(R)	3P
	2906.3	2906.3	29	51	(K)KIEKFQSEEQQQTEDELQDKIHP(F)	1P
	2961.3	2961.3	1	25	RELEELNVPGEIVESLSSSEESITR(I)	2P
	2965.3	2965.2	1	24	RELEELNVPGEIVESLSSSEESIT(R)	4P
	3041.2	3041.3	1	25	RELEELNVPGEIVESLSSSEESITR(I)	3P
	3053.4	3053.4	29	52	(K)KIEKFQSEEQQQTEDELQDKIHPF(A)	1P
	3121.1	3121.3	1	25	RELEELNVPGEIVESLSSSEESITR(I)	4P
	3396.5	3396.5	1	28	RELEELNVPGEIVESLSSSEESITRINK(K)	3P
	3476.5	3476.5	1	28	RELEELNVPGEIVESLSSSEESITRINK(K)	4P

The example of 2000 L Na-caseinate solution containing 180 kg protein and 1 kg trypsin yielded 29 kg of calcium-enriched CPP, corresponding to a yield of ~16% (w/w) [141]. A variety of raw materials such as acid casein, sodium caseinate, and calcium caseinate, as well as skimmed raw milk, milk concentrate by ultrafiltration, pasteurized and UHT milk near the limit date for consumption, may be used as the substrate for CPP production. The optimal parameters for the hydrolysis with trypsin were 37 °C and pH between 7.5 and 8.5. The casein hydrolysate solution was roughly fractioned by ultrafiltration with appropriate membranes to obtain soluble CPP both in the "permeate" and "retentate". Lower-molecular peptides/phosphopeptides occurred in the "permeate" and larger in the "retentate". The diafiltrate containing the tryptic casein digest was loaded

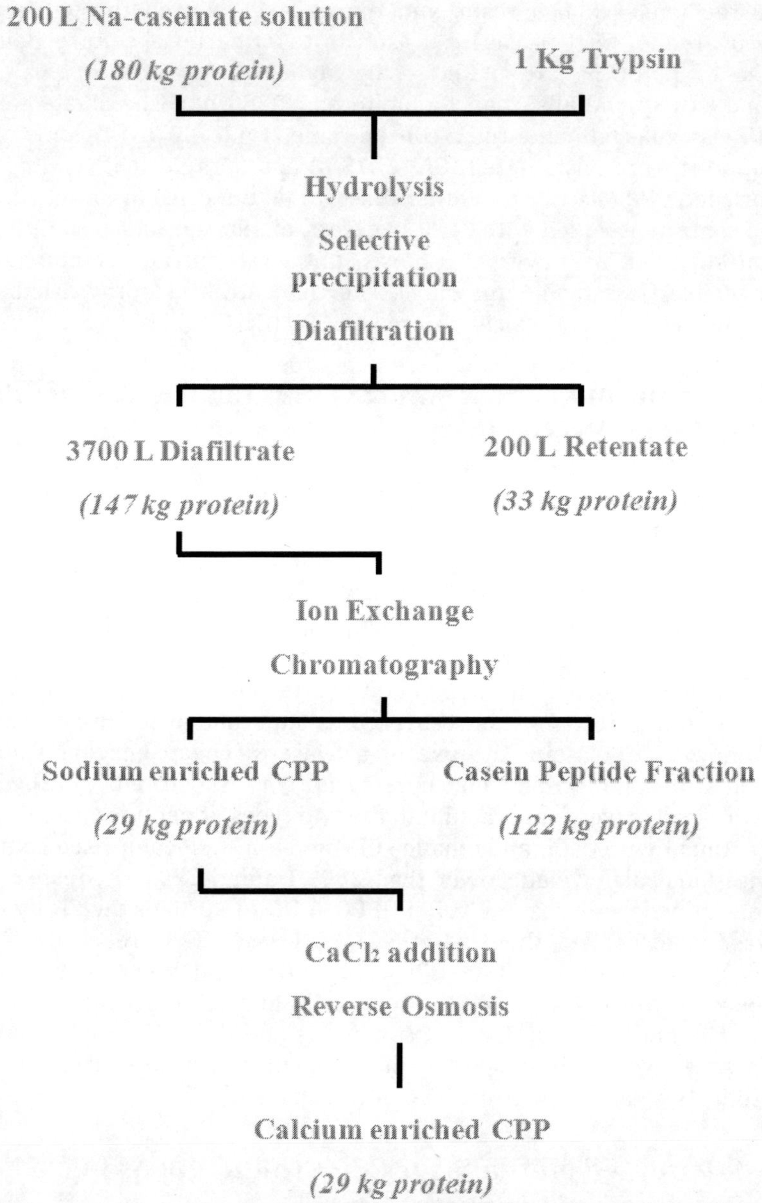

FIGURE 10. Schematic representation of the process-scale isolation of tryptic CPP from caseinate, showing the protein flow through the process (source: [141]).

on the ion exchange resin, and the non-CPP flowed through and were recovered for further use, *e.g.*, as substrate for bacterial culture. Bound CPP, free of non-CPP, are eluted using sodium hydroxide, *e.g.*, 0.2 M, and conductivity and absorbance monitored at 280 nm. The eluate containing CPP is collected and then concentrated, typically by reverse osmosis. Peptides were pasteurized (85 °C, 15 s) and spray dried, yielding sodium enriched CPP (Na-CPP). In other cases, to obtain calcium-enriched CPP, the concentrate is added with $CaCl_2$ in excess, diafiltered, and then CPP solution concentrated by reverse osmosis until the filtrate conductivity was negligible (less than 3 mS cm^{-1}). The concentrate is spray dried, and the product is labeled as calcium-enriched CPP (Ca-CPP).

6.3. Traditional and New Processes for The Use of CPP In Alimentary Products

A Ca^{2+}/ethanol selective precipitation procedure was used to produce a CPP and non-CPP concurrently from an alcalase digest of whole casein in which the traditional and new processes for CPP production were reported [142]. CN is trypsinized, and the pH of the solution is adjusted to 4.6 to separate the non-peptide material. The CPP-Ca^{2+} aggregation was induced by ethanol addition in the supernatant and recovered as precipitate for freeze drying. In the novel process, the step of non-peptide material removal was omitted, and non-CPP (CNPP) was recovered as supernatant for use in alimentary products. For casein, the use of alcalase, a cheap enzyme suitable for industrial application, for hydrolysis was suggested [142]. The CN hydrolysates were separated into the two types of peptides using combined treatment with $CaCl_2$ and ethanol. CPP and non-CPP comprised components with molecular weight lower than 2509 Da and 2254 Da, respectively, as determined using size exclusion HPLC. A DH of 20% for the CN hydrolysate was achieved. At the end, the recovery of CPP reached 24%. The phosphorus component of CPP was 3.08%, and nitrogen recovery was approximately 76% [142]. CPP generally had an improved solubility and transparency even under acid conditions and could be used as ingredient for beverages such as sport drinks, soft drinks, health drinks, fermented products, vitamin concentrates, fruit or fruit fractions.

6.4. Patented Methods for CPP Production as Ingredients for Alimentary Products

A method for the preparation of selected anticariogenic CPP comprised the steps of complete digestion of CN as soluble monovalent cation salt with a proteolytic enzyme: the addition of a mineral acid to the solution to adjust the pH to approximately 4.7; the removal of any produced precipitate; the addition of $CaCl_2$ to a concentration of approximately 1.0% (w/v) to cause the aggregation of CPP; the separation of the aggregated CPP from the solution through a filter with a molecular weight exclusion limit within the range 10000 to 20000 while passing the bulk of the remaining CPP in solution; the diafiltration of the separated CPP with water through a filter; the concentration of the solution; and the drying the retentate [143]. Peptides not included in the aggregation were removed by ultrafiltration/diafiltration. By this means, anticariogenic CPP at purity greater than 90% were obtained [143]. CPP including calcium, magnesium or both salts (or zinc, ferric or other salts) are produced by submitting CN to proteolytic enzyme hydrolysis, ultrafiltering the resulting hydrolyzate to produce a permeate containing CPP, adding a bivalent cation salt to the peptides to form CPP aggregates, and separating by ultrafiltration the CPP aggregates and non-CPP [144]. When CPP salts need to be converted to free phosphopeptides, they can be restored by acidification with HCl; the solution is then diafiltered extensively through a 1000 molecular weight cut-off membrane to remove excess calcium chloride.

6.5. Example of CPP Applications as Nutraceuticals in Functional Foods

The CPP-salts complex can be added to different foods. A stable acidic beverage or other alimentary products can be obtained by digesting casein with trypsin, precipitating the insoluble components at acid pH, adjusting the pH of the obtained supernatant to approximately 6.0, then adding calcium chloride and ethanol to recover an acid-soluble calcium complex of CPP and the reaction product and adding them to a soft drink [145]. The acid-soluble calcium complex of CPP enhances calcium absorption from food because calcium may be absorbed by the body in the form of soluble calcium. Acid-soluble CPP is a mixture of α_{s1}-, α_{s2}- and β-CPP, forming essentially no turbidity in solution at pH of 3.0 or less, and having purity greater than 90%, molecular weights between approximately 2500 and 4600 Da, and the ability to solubilize at least 100 ppm calcium at a concentration of 0.5 mg/ml CPP. The acid-soluble CPP produced *in vitro* also have a solubilizing capability on iron. It is widely believed that iron must be solubilized for absorption through the small intestine. Accordingly, health may also be enhanced by the absorption of the soluble iron in the drink by the human body. Similarly, magnesium may be solubilized in a drink or edible product of this type [145]. Therefore, in addition to drinks, preparation of CPP with high negative charge could be used as additive for healthy foods or for dietetic or pharmaceutical compositions, as they are capable of increasing the *in vivo* absorption of calcium or other ions [146]. Another interesting invention relates to processes and the compositions that are useful to remineralize the teeth of mammals, particularly humans, and impart acid resistance thereto. These compositions included a gum base or carrier, sweetening agents, CPP-ACP preparation and food-grade acids [147]. Because many chewing gum and confectionery products usually contain acids, many consumers enjoying chewing gum and confectionery products ingest acids causing demineralization of the tooth surface. CPP-calcium phosphate complexes are known to have anticariogenic teeth strengthening effects that could be used address the problem of dissolution or demineralization of tooth enamel and the resultant formation of dental caries. Exogenous CPP-ACP preparations have also been added to milk in 2.0-5.0 g/L amounts to remineralize enamel subsurface lesions, which actually increased with respect to the control [148].

These inventions create functional foods with undoubted beneficial effects on human health, possibly promoting recalcification of bones, protecting the tooth enamel from decay and other possible health benefits.

7. CONCLUSION

Proteins are no longer considered merely nutritional components because they possess encrypted peptides with possible biological properties [149-153]. The cited literature has highlighted that bioactive peptides may be released *in vitro* or *in vivo* by digestive and bacterial enzymes starting from casein or generally inactive precursors [151,154-155]. We have examined the case of casein-derived phosphopeptides that can be applied as dietary supplements in "functional foods" and produced on an industrial scale. With this, it has been demonstrated that, each time a health-enhancing nutraceutical is required for a functional food, an appropriate enzymatic hydrolysis of casein needs to be designed [156-157]. In addition to the anticariogenic activity, the most important function displayed by CPP is that soluble and encrypted casein phosphopeptides can arrive without modification of the phosphorylated sequence to the brush border membrane. It is postulated that short-sequence peptides reach their putative receptors in many tissues without modification. Whether they enter blood circulation or whether their action is restricted to a peripheral circle are current questions that await response. In conclusion, this review has considered milk and cheese CPP for specific food ingredients. The contribution of *in vitro* casein digests by gastric proteases to potential biologically active substances in the intestine must be simultaneously considered. Scientists are currently involved in investigations to define the in vivo fate of all of the bioactive peptides. The in vitro studies have allowed the scientists to compare the predicted and the experimental sequence of CPP. This step is preliminary to the clinical investigations designed to determine the bioactivity of the milk hydrolysates. These findings open an industrial perspective that will permit the unrestricted use of CPP in healthy promoting food application.

ACKNOWLEDGMENTS

Publication in partial fulfilment of the requirements for PhD in Sciences and Technologies of Food Productions - XXV Cycle, University of Naples 'Federico II'. The Authors gratefully acknowledge the American Journal Experts Association for the text revision (http://www.journalexperts.com/). This work was partly supported by the financial aid to C.L. from MIUR, Program PRIN-2008 HNHAT7-004.

REFERENCE

1. Hong. J. Lambert, G. Y. Yang, J. Liao, C. S. Yang, Inhibition of Carcinogenesis by Polyphenols: Evidence from Laboratory Investigations.American Society for Clinical Nutrition 2005284S EOF291S EOF
2. Shishodia. S. Aggarwal, Molecular Targets of Dietary Agents for Prevention and Therapy of CancerBiochemical Pharmacology2006711013971421
3. Chang DS.Preparations for Sustained Release of Nutraceuticals and Methods of Controllably Releasing Nutraceuticals. US Patent 07115283.
4. R. Cartwright, L. E. Hendricks, Composition. U. S. Nutraceutical, Patent 20030091552.
5. S. Torres-Giner, A. Martinez-Abad, Lagaron. J. M. Ocio, Stabilization of a Nutraceutical Omega-3 Fatty Acid by Encapsulation in Ultrathin Electrosprayed Zein ProlamineJournal of Food Science2010N6979
6. A. Mora-Gutierrez, M. H. Gurin, Bioactive Complexes Compositions and Methods of Use Thereof. US Patent 7780873.
7. Y. Policker, Sweetener. Nutraceutical, U. S. Composition, Patent 2010
8. Schanbacher FL, Talhouk RS, Murray FA, Gherman LI, Willett LB. Milk-Borne Bioactive Peptides.International Dairy Journal 199885393403
9. S. J. Husson, B. Landuyt, T. Nys, G. Baggerman, K. Boonen, E. Clynen, M. Lindemans, T. Janssen, L. Schoofs, Comparative Peptidomics of Caenorhabditis Elegans versus C. Briggsae by LC-MALDI-TOF MSPeptides2009303449457
10. S. Fosset, D. Tomé, from. Nutraceuticals, In. Milk, H. Roginski, J. W. Fuquay, P. F. Fox, (ed, Dairy. Enciclopedia, London. Science, Press. Accasemy, 200321082112
11. Silva SV, Malcata FX.Caseins as Source of Bioactive PeptidesInternational Dairy Journal2005151115
12. Sanchez. I. Recio, Bioactive Peptides Identified in Enzymatic Hydrolyzates of Milk Caseins and Method of Obtaining Same.US 2010
13. FitzGerald RJ.Potential Uses of Caseinophosphoeptides. International Dairy Journal 1998
14. Luo SJ, Wong LL.Oral Care Confections and Method of Using. US Patent 6733818.

15. E. C. Reynolds, F. Cai, P. Shen, G. D. Walker, Retention in Plaque and Remineralization of Enamel Lesions by Various Forms of Calcium in a Mouthrinse or Sugar-free Chewing Gum.Journal of Dental Research2003823206211
16. H. Korhonen, A. Pihlanto, Peptides. Bioactive, Production, Functionality, International Dairy Journal 2006169945960
17. J. Law, A. Haandrikman, Proteolytic Enzymes of Lactic Acid BacteriaInternational Dairy Journal199771111
18. M. Gatti, Mucchetti. G. Fornasari, F. Addeo, E. Neviani, Presence of Peptidase Activities in Different Varieties of CheeseLetters in Applied Microbiology1999285368372
19. J. Léonila, V. Gagnaire, D. Mollé, S. Pezennec, S. Bouhallab, Application of Chromatography and Mass Spectrometry to the Characterization of Food Proteins and Derived Peptides. Journal of Chromatography A 2000
20. Christensen JE, Dudley EG, Pederson JA, Steele JL.Peptidases and Amino Acid Catabolism in Lactic Acid Bacteria.Antonie Van Leeuwenhoek1999217 EOF46 EOF
21. E. R. S. Kunji, I. Mierau, A. Hagting, B. Poolman, W. N. Konings, The Proteolytic Systems of Lactic Acid Bacteria.Antonie Van Leeuwenhoek1996187 EOF221 EOF
22. M. K. Doeven, J. Kok, B. Poolman, Specificity and Selectivity Determinants of Peptide Transport in Lactococcus Lactis and Other MicroorganismsMolecular Microbiology2005573640649
23. Rul. F. Fernandez-Espla, S. Pep, Streptococcus. from, A. thermophilus, Member. New, the. of, T. Aminopeptidase, of. Family, Bacteria. Thermophilic, European Journal of Biochemistry19992632502510
24. V. Gagnaire, M. Piot, B. Camier, J. P. Vissers, G. Jan, J. Léonil, Survey of Bacterial Proteins Released in Cheese: a Proteomic Approach.International Journal of Food Microbiology2004942185201
25. Saz JM, Marina ML. Application of Micro- and Nano-HPLC to the Determination and Characterization of Bioactive and Biomarker Peptides.Journal of Separation Science2008313446458
26. L. Miclo, É. Roux, M. Genay, É. Brusseaux, C. Poirson, N. Jameh, C. Perrin, A. Dary, Variability of Hydrolysis of β-, αs1-, and αs2-Caseins by 10 Strains of Streptococcus thermophilus and Resulting Bioactive Peptides. Journal of Agricultural and Food Chemistry 2012602554565
27. A. Pihlanto-Leppälä, T. Rokka, H. Korhonen, I. Angiotensin, Enzyme. Converting, Peptides. Inhibitory, from. Derived, Milk. Bovine, Proteins, International Dairy Journal 199884325331
28. Pélissier JP. Protéolyse des Caséines.Sciences des Aliments 198441135
29. H. Teschemacher, G. Koch, V. Brantl, Milk Protein-Derived Opioid Receptor Ligands.Biopolymers199743299117
30. H. Meisel, Chemical Characterization and Opioid Activity of an Exorphin Isolated from in Vivo Digests of Casein.FEBS Letters19861962223227
31. Ondetti MA, Cushman DW.Enzymes of the Renin-Angiotensin System and their Inhibitors.Annual Review of Biochemistry198251283
32. P. Bruneval, N. Hinglais, F. Alhenc-Gelas, I. Angiotensin, Enzyme. Converting, Human. in, Intestine, Kidney, Ultrastructural Immunohistochemical Localization. Histochemistry19868517380
33. H. Meisel, Casokinins as Bioactive Peptides in the Primary Structure of Casein. In Schwenke KD, Mothes R. (ed.) Food Proteins: Functionality. New York: VCH-Weinheim; 19936775
34. R. Sato, T. Noguchi, H. Naito, phosphopeptide. . C. P. P. Casein, Calcium. Enhances, from. Absorption, Ligated. the, of. Segment, Small. Rat, Intestine, Journal of Nutritional Science and Vitaminology19863216776
35. Reynolds EC.Remineralization of Enamel Subsurface Lesions by Casein Phosphopeptide-Stabilized Calcium Phosphate Solutions.Journal of Dental Research199776915871595

36. H. Meisel, C. Olieman, Estimation of Calcium binding Constants of Casein Phosphopeptides by Capillary Zone ElectrophoresisAnalytica Chimica Acta1998291 EOF
37. I. Hala, S. Higashiyama, H. Otani, Identification of a Phosphopeptide in Bovine αs1-Casein Digest as a Factor Influencing Proliferation and Immunogloblin Production in Lymphocyte Cultures. Journal of Dairy Research 1998654569578
38. I. Hala, J. Ueda, H. Otani, Immunostimulatory Action of a Commercially Available Casein Phosphopeptide Preparation, CPP-III. in Cell Cultures. Milchwissenschaft 199954137
39. S. Loukas, D. Varoucha, C. Zioudrou, R. Streaty, W. A. Klee, Opioid Activities and Structures of alpha-Casein-Derived Exorphins.Biochemistry198322194567
40. S. Loukas, F. Panetsos, E. Donga, C. Zioudrou, δ. Selective-Antagonist, Analogs. Peptides, α. of-Casein, Probes. Exorphin, the. for, Receptor. Opioid, Nyberg. F. In, Brantl V. (ed.) β-Casomorphins and Related Peptides. Uppsala: Fyris-Tryck AB; 1990143149
41. A. Pihlanto-Leppälä, P. Antila, P. Mäntsälä, J. Hellman, Opioid Peptides Produced by in Vitro Proteolysis of Bovine Caseins. International Dairy Journal 199444291
42. E. Miquel, J. Á. Gómez, A. Alegría, R. Barberá, R. Farré, I. Recio, Identification of Casein Phosphopeptides after Simulated Gastrointestinal Digestion by Tandem Mass Spectrometry. European Food Research and Technology 2006
43. G. Picariello, P. Ferranti, O. Fierro, G. Mamone, S. Caira, A. Di Luccia, S. Monica, F. Addeo, Peptides Surviving the Simulated Gastrointestinal Digestion of Milk Proteins: Biological and Toxicological ImplicationsJournal of Chromatography BAnalytical Technologies in the Biomedical and Life Sciences 2010295 EOF308 EOF
44. H. M. Farrell, R. Jr Jimenez-Flores, G. T. Bleck, E. M. Brown, J. E. Butler, L. K. Creamer, C. L. Hicks, C. M. Hollar, K. F. Ng-Kwai-Hang, H. E. Swaisgood, Nomenclature of the Proteins of Cows' Milk-Sixth Revision.Journal of Dairy Science200487616411674
45. A. J. Kenny, S. Maroux, Topology of Microvillar Membrance Hydrolases of Kidney and Intestine.Physiological Reviews198262191128
46. M. Yoshioka, R. H. Erickson, J. F. Woodley, R. Gulli, D. Guan, Y. S. Kim, Role of Rat Intestinal Brush-Border Membrane Angiotensin-Converting Enzyme in Dietary Protein Digestion.American Journal of Physiology-Gastrointestinal and Liver Physiology 1987Pt 1) G781G786.
47. R. H. Erickson, I. S. Song, M. Yoshioka, R. Gulli, S. Miura, Y. S. Kim, Identification of Proline-Specific Carboxypeptidase Localized to Brush Border Membrane of Rat Small Intestine and its Possible Role in Protein Digestion.Digestive Diseases and Sciences1989343400406
48. O. Mellander, G. Folsch, Enzyme Resistance and Metal Binding of Phosphorylated Peptides. In: Bigwood EJ. (ed.) Protein and Amino Acid Function. New York: Pergamoon Press Inc; 1972569579
49. R. Boutrou, E. Coirre, J. Jardin, J. Léonil, Phosphorylation and Coordination Bond of Mineral Inhibit the Hydrolysis of the β-Casein (1–25) Peptide by Intestinal Brush-Border Membrane Enzymes. Journal of Agricultural and Food Chemistry 2010581379557961
50. I. B. Kibangou, S. Bouhallab, G. Henry, F. Bureau, S. Allouche, A. Blais, P. Guérin, P. Arhan, D. L. Bouglé, Milk Proteins and Iron Absorption: Contrasting Effects of Different Caseinophosphopeptides.Pediatric Research2005584731734
51. J. M. Pérès, S. Bouhallab, C. Petit, F. Bureau, J. L. Maubois, P. Arhan, D. Bougle, Improvement of Zinc Intestinal Absorption and Reduction of Zinc/Iron Interaction Using Metal Bound to the Caseinophosphopeptide 1-25 of β-casein. Reproduction, Nutrition & Development 1998384465472
52. T. Kasai, R. Iwasaki, M. Tanaka, S. Kiriyama, . C. P. P. Caseinphosphopeptides, Feces. in, in. Contents, Tract. Digestive, Rats. of, Casein. Fed, C. P. P. Preparations, BioscienceBiotechnology and Biochemistry 19955912630

53. A. Ferraretto, A. Signorile, C. Gravaghi, A. Fiorilli, G. Tettamanti, Casein Phosphopeptides Influence Calcium Uptake by Cultured Human Intestinal HT-29 Tumor Cells. Journal of Nutrition 2001131616551661
54. S. Richter, R. Bergmann, J. Pietzsch, T. Ramenda, J. Steinbach, F. Wuest, Fluorine-, of. Labeling, A. Phosphopeptides, Approach. Potential, the. for, of. Evaluation, Metabolism. Phosphopeptide, Vivo. In, Biopolymers2009926479488
55. D. Erba, S. Ciappellano, G. Testolin, Effect of Caseinphosphopeptides on Inhibition of Calcium Intestinal Absorption due to PhosphateNutrition Research2001214649
56. Marcus CS, Lengemann FW. Absorption of Ca45 and Sr85 from Solid and Liquid Food at Various Levels of the Alimentary Tract of the Rat.Journal of Nutrition 1962772155160
57. Marcus CS, Lengemann FW.Use of Radioyttrium to Study Food Movement in the Small Intestine of the Rat.Journal of Nutrition 1962762179182
58. H. Meisel, H. Bernard, S. Fairweather-Tait, Gerald. R. J. Fitz, R. Hartmann, . C. N. Lane, D. Mc Donagh, B. Teucher, J. M. Wal, Detection of Caseinophosphopeptides in the Distal Ileostomy Fluid of Human Subjects.British Journal of Nutrition 2003893351359
59. H. Meisel, H. Frister, Chemical Characterization of a Caseinophosphopeptide Isolated from in Vivo Digests of a Casein Diet.Biological Chemistry Hoppe-Seyler19883691212751279
60. B. Chabance, P. Marteau, J. C. Rambaud, D. Migliore-Samour, M. Boynard, P. Perrotin, R. Guillet, P. Jollès, A. M. Fiat, Casein Peptide Release and Passage to the Blood in Humans during Digestion of Milk or Yogurt.Biochimie1998802155165
61. H. Meisel, Biochemical Properties of Regulatory Peptides Derived from Milk Proteins.Biopolymers1997432119128
62. R. Sato, T. Noguchi, H. Naito, phosphopeptide. . C. P. P. Casein, Calcium. Enhances, from. Absorption, Ligated. the, of. Segment, Small. Rat, Intestine, Journal of Nutritional Science and Vitaminology19863216776
63. A. Ferraretto, C. Gravaghi, A. Fiorilli, G. Tettamanti, Bioactive. Casein-Derived, Role. Phosphopeptides, Phosphorylation. of, Structure. Primary, Promoting. in, Uptake. Calcium, H. by, T-, Cells. F. E. B. Tumor, FEBS Letters200392 EOF8 EOF
64. S. Cosentino, C. Gravaghi, E. Donetti, Lombardi. G. Donida, M. Bedoni, A. Fiorilli, G. Tettamanti, A. Ferraretto, Caseinphosphopeptide, Caseinphosphopeptide-Induced Calcium Uptake in Human Intestinal Cell Lines HT-29 and Caco2 is Correlated to Cellular DifferentiationJournal of Nutritional Biochemistry 2010213247254
65. Mrak. E. Donida, C. Gravaghi, I. Villa, S. Cosentino, E. Zacchi, S. Perego, A. Rubinacci, A. Fiorilli, G. Tettamanti, A. Ferraretto, Casein Phosphopeptides Promote Calcium Uptake and Modulate the Differentiation Pathway in Human Primary Osteoblast-like CellsPeptides2009301222332241
66. Reynolds EC.Casein Phosphopeptide-Amorphous Calcium Phosphate: The Scientific Evidence.Advances in Dental Research20092112529
67. Rose RK.Binding Characteristics of Streptococcus mutans for Calcium and Casein Phosphopeptide. Caries Res 2000345427431
68. A. L. Mc Cormack, D. M. Schieltz, B. Goode, S. Yang, G. Barnes, D. Drubin, J. R. I. I. I. . Yates, Direct Analysis and Identification of Proteins in Mixtures by LC/MS/MS and Database Searching at the Low-femtomole Level Analytical Chemistry 1997694767776
69. Eng JK, McCormack AL, Yates JR III.An Approach to Correlate Tandem Mass Spectral Data of Peptides with Amino Acid Sequences in a Protein Database. Journal of The American Society for Mass Spectrometry 1994511976989
70. A. Stensballe, S. Andersen, O. N. Jensen, Characterization of Phosphoproteins from Electrophoretic Gels by Nanoscale Fe(III) Affinity Chromatography with Off-line Mass Spectrometry Analysis. Proteomics 200112207222
71. A. G. Craig, Miller. C. L. Hoeger, T. Goedken, J. E. Rivier, W. H. Fischer, Monitoring Protein Kinase and Phosphatase Reactions with Matrix-Assisted Laser Desorption/Ionization Mass Spectrometry and Capillary Zone Electrophoresis:

Comparison of the Detection Efficiency of Peptide-Phosphopeptide Mixtures. Biological Mass Spectrometry 1994238519528
72. R. S. Annan, Verma. R. Huddleston, R. J. Deshaies, S. A. A. Carr, Electrospray. M. S. Multidimensional-Based, to. Approach, Mapping. Phosphopeptide, Analytical Chemistry 2001733393404
73. W. Manson, W. D. Annan, The Structure of a Phosphopeptide Derived from β-casein. Archives of Biochemistry and Biophysics 197114511626
74. S. P. Gygi, G. L. Corthals, Y. Zhang, Y. Rochon, R. Aebersold, Evaluation of two-Dimensional Gel Electrophoresis-Based Proteome Analysis Technology. PNAS 200097179390
75. Y. Oda, T. Nagasu, B. T. Chait, Enrichment Analysis of Phosphorylated Proteins as a Tool for Probing the Phosphoproteome. Nature Biotechnology 2001194379382
76. H. Zhou, Aebersold. R. A. Watts, Approach. Systematic, the. to, of. Analysis, Phosphorylation. Protein, Nature Biotechnology 2001194375378
77. M. C. Posewitz, P. Tempst, Gallium. I. I. I. Immobilized, Chromatography. Affinity, Phosphopeptides. of, Analytical Chemistry 1999711428832892
78. M. Kokubu, Y. Ishihama, T. Sato, T. Nagasu, Y. Oda, Specificity of Immobilized Metal Affinity-Based IMAC/C18 Tip Enrichment of Phosphopeptides for Protein Phosphorylation Analysis. Analytical Chemistry 2005771651445154
79. M. Machida, H. Kosako, K. Shirakabe, M. Kobayashi, M. Ushiyama, J. Inagawa, J. Hiran, T. Nakano, Y. Bando, E. Nishida, S. Hattori, Purification of Phosphoproteins by Immobilized Metal Affinity Chromatography and its Application to Phosphoproteome Analysis. FEBS Journal 2007274615761587
80. Thingholm TE, Jensen ON.Enrichment and Characterization of Phosphopeptides by Immobilized Metal Affinity Chromatography (IMAC) and Mass Spectrometry. Methods in Molecular Biology 200952747
81. M. W. H. Pinkse, P. M. Uitto, Ooms. B. Hilhorst, A. J. Heck, Isolation. Selective, the. at, Level. Femtomole, Phosphopeptides. of, Proteolytic. from, Using. . D. Digests-Nano, L. C. -E, S. I. -M, S. M. S. , Oxide. Titanium, Precolumns, Analytical Chemistry 200476143935
82. S. Rinalducci, M. R. Larsen, S. Mohammed, L. Zolla, Novel Protein Phosphorylation Site Identification in Spinach Stroma Membranes by Titanium Dioxide Microcolumns and Tandem Mass Spectrometry. Journal of Proteome Research 200654973982
83. Thingholm TE, Jørgensen TJ, Jensen ON, Larsen MR.Highly Selective Enrichment of Phosphorylated Peptides using Titanium Dioxide. Nature Protocols 2006141929
84. Dunn JD, Reid GE, Bruening ML.Techniques for Phosphopeptide Enrichment prior to Analysis by Mass Spectrometry. Mass Spectrometry Reviews 20102912954
85. K. Moser, F. M. White, Phosphoproteomic Analysis of Rat Liver by High Capacity IMAC and LC-MS/MS. Journal of Proteome Research 20065198104
86. M. R. Larsen, T. E. Thingholm, O. N. Jensen, P. Roepstorff, T. J. D. Jorgensen, Highly Selective Enrichment of Phosphorylated Peptides from Peptide Mixtures Using Titanium Dioxide Microcolumns. Molecular & Cellular Proteomics 200547873886
87. B. Bodenmiller, L. N. Mueller, M. Mueller, B. Domon, R. Aebersold, Reproducible Isolation of Distinct, Overlapping Segments of the Phosphoproteome. Nature Methods 200743231237
88. N. Sugiyama, T. Masuda, K. Shinoda, A. Nakamura, M. Tomita, Y. Ishihama, Phosphopeptide Enrichment by Aliphatic Hydroxyl Acid-Modified Metal Oxide Chromatography for Nano-LC-MS/MS in Proteomics Applications. Molecular & Cellular Proteomics 20076611031109
89. McLachlin DT, Chait BT.Analysis of Phosphorylated Proteins and Peptides by Mass Spectrometry. Current Opinion in Chemical Biology 200155591602
90. G. Pinto, S. Caira, M. Cuollo, S. Lilla, O. Fierro, F. Addeo, Hydroxyapatite as a Concentrating Probe for Phosphoproteomic Analyses. Journal of Chromatography B 20108782826692678

91. H. Y. Zhou, E. Salih, Glimcher, Isolation of a Novel Bone Glycosylated Phosphoprotein with Disulphide Cross-Links to Osteonectin. Biochemical Journal 199833031423
92. S. R. Schmidt, F. Schweikart, Andersson, Current Methods for Phosphoprotein Isolation and Enrichment. Journal of Chromatography B 2007
93. T. Kawasaki, Hydroxyapatite as a Liquid Chromatographic Packing. Journal of Chromatography A 54417147184
94. G. Mamone, G. Picariello, P. Ferranti, F. Addeo, Hydroxyapatite Affinity Chromatography for the Highly Selective Enrichment of Mono- and Multi-Phosphorylated Peptides in Phosphoproteome Analysis. Proteomics 2010103380
95. C. Cirulli, G. Chiappetta, G. Marino, P. Mauri, A. Amoresano, Identification of Free Phosphopeptides in Different Biological Fluids by a Mass Spectrometry Approach. Analytical and Bioanalytical Chemistry 2008
96. L. Hu, H. Zhou, Y. Li, S. Sun, L. Guo, M. Ye, X. Tian, J. Gu, S. Yang, H. Zou, Profiling of Endogenous Serum Phosphorylated Peptides by Titanium (IV) Immobilized Mesoporous Silica Particles Enrichment and MALDI-TOFMS Detection. Analytical Chemistry 200981194104
97. Y. Li, D. Qi, C. Deng, P. Yang, X. Zhang, Cerium Ion-Chelated Magnetic Silica Microspheres for Enrichment and Direct Determination of Phosphopeptides by Matrix-Assisted Laser Desorption Ionization Mass Spectrometry. Journal of Proteome Research 20087417671777
98. M. C. Maurer, J. L. Peng, S. S. An, J. Y. Trosset, A. Henschen-Edman, H. A. Scheraga, Structural Examination of the Influence of Phosphorylation on the Binding of Fibrinopeptide A to Bovine Thrombin. Biochemistry 1998
99. L. M. Matrisian, G. W. Sledge, S. Jr Mohla, Extracellular Proteolysis and Cancer: Meeting Summary and Future Directions Cancer Research 2003631961056109
100. P. Resmini, L. Pellegrino, G. Battelli, Accurate Quantification of Furosine in Milk and Dairy Products by a Direct HPLC Method. Italian Journal of Food Science 199023173183
101. N. Corzo, R. López-Fandiño, T. Delgado, M. Ramos, A. Olano, Changes in Furosine and Proteins of UHT-Treated Milks Stored at High Ambient Temperatures. Zeitschrift fur Lebensmittel-Untersuchung und-Forschung 19941984302306
102. G. Pinto, S. Caira, M. Cuollo, O. Fierro, Chianese. L. Nicolai, F. Addeo, Lactosylated Casein Phosphopeptides as Specific Indicators of Heated Milks. Analytical and Bioanalytical Chemistry 2012402519611972
103. M. R. Mendoza, A. Olano, M. Villamiel, Chemical Indicators of Heat Treatment in Fortified and Special Milks. Journal of Agriculture and Food Chemistry 200553829952999
104. J. Mottar, A. Bassier, M. Joniau, J. Baert, Effect of Heat-Induced Association of Whey Proteins and Casein Micelles on Yogurt Texture. Journal of Dairy Science 198972922472256
105. Ö. Tokusoglu, Unal. K. Akalin, Rapid High Performance Liquid Chromatographic Detection of Furosine (epsilon- N-2-furoylmethyl-L-lysine) in Yogurt and Cheese Marketed in Turkey. Journal of Food Quality 20062913846
106. S. Bouhallab, N. A. Oukhatar, D. Molle, G. Henry, J. L. Maubois, P. Arhan, D. Bougle, Sensitivity of Beta-Casein Phosphopeptide-Iron Complex to Digestive Enzymes in Ligated Segment of Rat Duodenum. The Journal of Nutritional Biochemistry 19991012723727
107. J. M. Pérès, S. Bouhallab, P. Bureau, D. Neuville, J. L. Maubois, G. Devroede, P. Arhan, D. Bouglé, Mechanisms of Absorption of Caseinophosphopeptide Bound Iron. The Journal of Nutritional Biochemistry 1999104215222
108. Adamson NJ, Reynolds EC.Characterization of Multiply Phosphorylated Peptides Selectively Precipitated from a Pancreatic Casein Digest. Journal of Dairy Science 1995781226532659
109. F. A. M. Klaver, F. Kingma, A. H. Weerkamp, Growth and Survival of Bifidobacteria in Milk. Netherlands Milk Dairy Journal 1993

110. F. Roudot-Algaron, D. Le Bars, L. Kerhoas, J. Einhorn, J. C. Gripon, Phosphopeptides from Comté Cheese. Nature and Origin. Journal of Food Science 1994593544
111. P. Ferranti, F. Barone, L. Chianese, F. Addeo, A. Scaloni, L. Pellegrino, P. Resmini, Phosphopeptides from Grana Padano: Nature, Origin and Changes during Ripening. Journal of Dairy Research 1997644601615
112. M. Lund, Y. Ardö, Purification and Identification of Water Soluble Phosphopeptides from Cheese using Fe(III) Affinity Chromatography and Mass Spectrometry. Journal of Agricultural and Food Chemistry 2004522166166622
113. L. Pellegrino, G. Battelli, P. Resmini, P. Ferranti, F. Barane, F. Addeo, Effects of Heat Load Gradient Occurring in Moulding on Characterization and Ripening of Grana Padano. Lait 1997772217228
114. P. Walstra, R. Jenness, Dairy Chemistry and Physics. New York: John Wiley and Sons; 1984
115. A. Pirisi, G. Pinna, M. Addis, G. Piredda, R. Mauriello, S. De Pascale, F. Addeo, L. Chianese, Relationship Between the Enzymatic Composition of Lamb Paste Rennet and Proteolytic and Lypolitic Pattern and Texture of PDO Fiore Sardo Cheese. International Dairy Journal 2007172143156
116. B. Battistotti, C. Corradini, Cheese. Italian, Fox. P. F. In, (ed, Chemistry. Cheese, Physics, Volume. I. I. Microbiology, Cheese. Major, London. Groups, Applied. Elsevier, Science, 1993221243
117. D. D. Voigt, F. Chevalier, M. C. Qian, A. L. Kelly, Effect of High-Pressure Treatment on Microbiology, Proteolysis, Lipolysis and Levels of Flavour Compounds in Mature Blue-Veined Cheese. Innovative Food Science & Emerging Technologies 201011168
118. Mykkanen HM, Wasserman RH.Enhanced Absorption of Calcium by Casein Phosphopeptides in Rachitic and Normal Chicks. Journal of Nutrition 19801101121412148
119. H. Meisel, Overview on Milk Protein-Derived Peptides. International Dairy Journal 1998
120. Reynolds EC, Johnson IH.Effect of Milk on Caries Incidence and Bacterial Composition of Dental Plaque in the Rat. Archives of Oral Biology 1981265445451
121. S. Rosen, D. B. Min, Harper. W. J. Harper, E. X. Beck, F. M. Beck, Effect of Cheese with and without Sucrose, on Dental Caries and Recovery of Streptococcus mutans in Rats. Journal of Dental Research 1984636894896
122. Osborne. J. C. Harper, J. J. Hefferren, R. Clayton, Cariostatic Evaluation of Cheeses with Diverse Physical and Compositional Characteristics. Caries Research 1986202123130
123. G. D. Walker, F. Cai, P. Shen, G. G. Adams, C. Reynolds, E. C. Reynolds, Casein Phosphopeptide Amorphous Calcium Phosphate Incorporated into Sugar Confections Inhibits the Progression of Enamel Subsurface Lesions in Situ. Caries Research 2010441340
124. L. Shaw, J. J. Murray, C. K. Burchell, J. S. Best, Calcium and Phosphorus Contents of Plaque and Saliva in Relation to Dental Caries. Caries Research 1983176543548
125. Margolis HC, Moreno EC.Composition of Pooled Plaque Fluid from Caries-Free and Caries-Positve Individuals Following Sucrose Exposure. Journal of Dental Research 1992711117761784
126. Reynolds EC.Remineralization of Enamel Subsurface Lesions by Casein Phosphopeptide-Stabilized Calcium Phosphate Solutions.Journal of Dental Research199776915871595
127. Reynolds EC. Anticariogenic Phosphopeptides. US Patent 5015628.
128. Rose RK, Turner SJ.Fluoride-Induced Enhancement of Diffusion in Streptococcal Model Plaque Biofilms. Caries Research 1998323227232
129. Rose RK, Dibdin GH.Calcium and Water Diffusion in Streptococcal Model Plaques. Archives of Oral Biology 1995405385391
130. Rose RK, Turner SJ, Dibdin GH. Effect of pH and Calcium Concentration on Calcium Diffusion in Streptococcal Model Plaque Biofilms.Archives of Oral Biology19974212795800

131. Shaw JH, Ensfield BJ, Wollman DH.Studies on the Relation of Dairy Products to Dental Caries in Caries-Susceptible Rats.Journal of Nutrition 1959672253273
132. Joshi. A. Papas, A. J. Belanger, R. L. Kent, Paola. P. F. Palmer De, Dietary Models for Root CariesAmerican Journal of Clinical Nutrition1995417S EOFS.
133. G. F. Ferrazzano, T. Cantile, M. Quarto, A. Ingenito, L. Chianese, F. Addeo, Protective Effect of Yogurt Extract on Dental Enamel Demineralization in Vitro. Australian Dental Journal2008534314319
134. E. C. Reynolds, P. F. Riley, N. A. Adamson, Precipitation. Selective, Procedure. Purification, Multiple. for, Peptides. Phosphoseryl-Containing, for. Methods, Indentification. their, Analytical Biochemistry 19942172277284
135. A. Tirelli, I. De Noni, P. Resmini, Bioactive Peptides in Milk Products.Italian Journal of Food Science 1997929198
136. H. Meisel, H. B. Andersson, K. Buhl, H. F. Erbersdobler, E. Schlimme, Heat, Heat-Induced Changes in Casein-Derived Phosphopeptides.Zeitschrift Fur Ernahrungswissenschaft 1991303227232
137. P. Ferranti, S. Lilla, L. Chianese, F. Addeo, Alternative Nonallelic Deletion is Constitutive of Ruminant αs1-Casein. Journal of Protein Chemistry 1999185595602
138. Baechler. R. Juillerat, R. Berrocal, S. Chanton, J. C. Scherz, R. Jost, Tryptic Phosphopeptides from Whole Casein. I. Preparation and Analysis by Fast Protein Liquid ChromatographyJournal of Dairy Research1989564603611
139. Adamson NJ, Reynolds EC.Characterization of Tryptic Casein Phosphopeptides Prepared under Industrially Relevant Conditions.Biotechnology and Bioengineering1995453196204
140. A. Goepfert, H. Meisel, Semi, Semi-Preparative Isolation of Phosphopeptides Derived from Bovine Casein and Dephosphorylation of Casein Phosphopeptides.Nahrung 1996405245248
141. K. H. Ellegard, C. Gammelgard-Larsen, E. S. Sørensen, S. Fedosovc, Process Scale Chromatographic Isolation, Characterization and Identification of Tryptic Bioactive Casein PhosphopeptidesInternational Dairy Journal 199999639652
142. L. Zhao, Z. Wang, S. Xu, Preparation of Casein Phosphorylated Peptides and Casein non-Phosphorylated Peptides using AlcalaseEuropean Food Research and Technology2007579 EOF584 EOF
143. Reynolds EC.Production of Phosphopeptides from Casein. US Patent 6448374.
144. Brule G, Roger L, Fauquant J, Piot M. Casein Phosphopeptide Salts. US Patent 5028589.
145. H. Naito, T. Noguchi, R. Sato, K. Tsuji, H. Hidaka, Transparent Acid Drink Containing Acid-Soluble Casein Phosphopeptides. US Patent 5405756.
146. L. Ramalingam, L. B. Messer, E. C. Reynolds, Adding Casein Phosphopeptide-Amorphous Calcium Phosphate to Sports Drinks to Eliminate in Vitro Erosion.Pediatric Dentistry20052716167
147. D. Tancredi, D. Ming, S. Holme, Calcium Phosphate Complex in Acid Containing Chewing Gum. US Patent 8133475.
148. G. Walker, F. Cai, P. Shen, C. Reynolds, B. Ward, C. Fone, S. Honda, M. Koganei, M. Oda, E. Reynolds, Increased Remineralization of Tooth Enamel by Milk Containing Added Casein Phosphopeptide-Amorphous Calcium PhosphateJournal of Dairy Research20067317478
149. H. Kostyra, E. Kostyra, Biologically Active Peptides Derived from Food Proteins. Polish Journal of Food and Nutrition Sciences 1992424517
150. H. Meisel, E. Schlimme, Proteins. Milk, of. Precursors, Peptides. Bioactive, Trends in Food Science and Technology 1990141
151. T. Rokka, E. L. Syväoja, J. Tuominen, H. Korhonen, Release of Bioactive Peptides by Enzymatic Proteolysis of Lactobacillus GG Fermented UHT MilkMilchwissensch 199752675
152. B. Jarmolowska, E. Kostyra, S. Krawczuk, H. β. Kostyra-Casomorphin, from. Isolated, Cheese. Brie, Journal of the Science of Food and Agriculture 1999791317881792

153. Muehlenkamp MR, Warthesen JJ. β-Casomorphins: Analysis in Cheese and Susceptibility to Proteolytic Enzymes from Lactococcus lactic ssp.cremoris. Journal of Dairy Science 19967912026
154. E. Smacchi, M. Gobbetti, Bioactive Peptides in Dairy Products: Synthesis and Interaction with Proteolytic EnzymesFood Microbiology2000172129141
155. P. Ferranti, M. V. Traisci, G. Picariello, A. Nasi, Falconi. C. Boschi, L. Chianese, F. Addeo, Casein Proteolysis in Human Milk Tracing the Pattern of Casein Breakdown and the Formation of Potential Bioactive Peptides.Journal of Dairy Research20047117487
156. A. Pihlanto-Leppälä, Bioactive Peptides Derived from Bovine Whey Proteins: Opioid and Ace-Inhibitory Peptides. Trends in Food Science and Technology 2000
157. L. Bitri, Optimization Study for the Production of an Opioid-like Preparation from Bovine Casein by Mild Acidic HydrolysisInternational Dairy Journal2004146535539

CHAPTER 3

The Feasibility of Harmonizing Gluten ELISA Measurements

Malgorzata Rzychon, Marcel Brohée, Fernando Cordeiro, Reka Haraszi[1] , Franz Ulberth, Gavin O'Connor

European Commission, Joint Research Centre, Institute for Reference Materials and Measurements, Retieseweg 111, Geel 2440, Belgium

ABSTRACT

Many publications have highlighted that routine ELISA methods do not give rise to equivalent gluten content measurement results. In this study, we assess this variation between results and its likely impact on the enforcement of the EU gluten-free legislation. This study systematically examines the feasibility of harmonizing gluten ELISA assays by the introduction of: a common extraction procedure; a common calibrator, such as a pure gluten extract and an incurred matrix material. The comparability of measurements is limited by a weak correlation between kit results caused by differences in the selectivity of the methods. This lack of correlation produces bias that cannot be corrected by using reference materials alone. The use of a common calibrator reduced the between-assay variability to some extent, but variation due to differences in selectivity of the assays was unaffected. Consensus on robust markers and their conversion to "gluten content" are required.

Keywords: Gluten; Wheat; Food allergen; Celiac disease; ELISA; Gluten-free; Reference material; Commutability

INTRODUCTION

Gluten is defined for legislative purposes as "a protein fraction from wheat, rye, barley, oats or their crossbred varieties and derivatives thereof, to which some persons are intolerant and that is insoluble in water and 0.5 M NaCl" CODEX STAN 118-1979 (Codex Alimentarius, 1979 rev. 2008). The proteins that form gluten are major storage proteins and represent between 70% and 80% of the total protein content of grain (Scherf, Koehler, & Wieser, 2015). Consequently, gluten is found in many staple foods in the Western diet. Due to its unique functional properties, wheat gluten is also widely used in the food industry as an ingredient essential for formulating high-quality baked goods. It is also used in a range of foodstuffs other than bakery products such as meats, noodles, sausages, breakfast cereals, meat and cheese replacements and other products (Day, Augustin, Batey, & Wrigley, 2006).

Gluten is responsible for a number of immune-mediated disorders, which include celiac disease (CD), dietary wheat allergy (WA) and gluten sensitivity (GS) (Sapone et al., 2012). The spectrum of clinical manifestations includes gastrointestinal symptoms (such as abdominal pain, nausea and diarrhoea), inflammatory disease of the small intestine, which can result in nutrient malabsorption and/or allergic reactions including anaphylaxis. To date the clinically relevant thresholds for WA and GS have not been established. However, in the case of CD patients, damage to the small intestinal mucosa can be induced even when trace concentrations of gluten are present (Sapone et al., 2012). For patients suffering from gluten-related disorders, the only available treatment is the life-long elimination of gluten from their diet.

The prevalence of CD is approximately 1% in regions populated by individuals of European origin. However, due to the increase in popularity of a western style diet, which is gluten-rich, the diagnosis of CD is increasing globally (Lionetti, Gatti, Pulvirenti, & Catassi, 2015). WA affects roughly 1% of the world's population (Scherf et al., 2015). The prevalence of GS is unknown, although estimates range from 0.6% to 6% of the population (Czaja-Bulsa, 2015).

To assist patients with gluten-related disorders in making sound dietary choices, food labelling legislation exists in many countries. These classify foods according to the gluten level they contain. The Codex Alimentarius standard defines foods as 'gluten-free' if the gluten level does not exceed 20 mg kg^{-1} in total and recommends the threshold of 100 mg kg^{-1} for the labelling of low gluten foodstuffs that have been specially processed to reduce their gluten content (Codex Alimentarius, 1979 rev. 2008).

The 20 mg kg^{-1} gluten free threshold has been adopted for foods bearing a 'gluten-free' label by the regulatory bodies in the European Union, the United States of America and Canada. However, in Australia and New Zealand a 'gluten-free' claim can only be made when there is no detectable gluten present. Less restrictive supplementary standards have been established in the EU at the threshold of 100 mg kg^{-1} ('very low gluten') and in Australia and New Zealand at 200 mg kg^{-1} ('low-gluten') (ANZFA, 2012; EC, 2009; FDA, 2013 ; Health Canada, 2012).

The global market for gluten-free products was valued at USD 3.4 billion in sales in 2014 and is expected to grow (Scherf et al., 2015).The conformity of food products with "gluten-free legislation" is most commonly assessed with immunoassays. In recent years, several enzyme-linked immunosorbent assays (ELISA) have been developed and commercialized. However, there is a wide variation in the reported measurement results between different commercially available test kits (Bruins Slot *et al.*, 2015; Diaz-Amigo and Popping, 2012 ; Sharma, 2012). This variation is well illustrated by the necessity to separate the results of proficiency tests into sub-sets according to the ELISA kits used. The separate treatment of results is necessary because ELISA kits from different manufacturers form different populations and no single statistical assessment of all the results can be carried out (FAPAS, 2012; Scharf *et al.*, 2013 ; Sykes *et al.*, 2012).

The consequences of these discrepancies for the risk management of gluten ingestion are worrying. People with health conditions which require the elimination of gluten from the diet, who confidently buy food products labelled as 'gluten-free', are put at risk of suffering from adverse effects of gluten exposure. Market surveys aimed at assessing the rate of gluten contamination in products sold as gluten-free suggest between 1 and 20% of the products do not comply with current legislation (Lee *et al.*, 2014; NSWFA, 2011 ; Sharma *et al.*, 2015). However, these claims are derived from measurements using the very kits that are known to provide highly variable results. This may result in unnecessary actions such as market recalls or legal claims. In 2015, the EU-wide Rapid Alert System for Food and Feed received 126 allergen notifications. Twenty-three of these concerned gluten or wheat and resulted in product recalls, withdrawals or border rejections (RASFF Portal, accessed 10/12/2015). The lack of reliable monitoring of cross-contamination during food production adds to the financial burden of preventive measures which go beyond the legislative requirements. Moreover, this contributes to the excessive use of precautionary ('may contain') labelling which results in confusion among the consumers. A further concern is the verification of the absence of gluten cross-contamination in naturally gluten-free foodstuffs. These are commonly used in a gluten-free diet but are not necessarily labelled as 'gluten free'.

The supposed reasons for the observed discrepancies between results of different gluten assays have been extensively discussed (Bruins Slot *et al.*, 2015 ; Diaz-Amigo and Popping, 2012) and cover differences in sample preparation protocols, choice of kit buffer, antibody specificities and calibration procedures. Much focus is put on the absence of reference materials, whose availability is often perceived as the remedy to the current problems with gluten quantification (Diaz-Amigo and Popping, 2012 ; Sykes *et al.*, 2012).

In this paper, we evaluate the feasibility of harmonizing gluten ELISA assays. We assess the current variation between results, both in quantitative and qualitative terms and perform correlation studies for gluten assays. The study also systematically examines the effect of: the introduction of a common extraction procedure; a common pure protein-based calibrator; and an incurred matrix reference material, on the comparability of gluten results.

2. MATERIALS AND METHODS

2.1. Materials

The gluten material used in this study was a commercially available, industrially extracted wheat gluten (GluVital™) batch from the year 2010 (Cargill, Bergen op Zoom, NL).

BME-IGRM-0, BME-IGRM-10 and BME-IGRM-50 are candidate reference materials produced at the Budapest University of Technology and Economics, Hungary (Bugyi et al., 2012). The materials consist of a blank baked cookie and cookies incurred with 10 mg kg^{-1} and 50 mg kg^{-1} of the WGPAT gliadin. WGPAT gliadin is an isolate from a mixture of 28 European wheat varieties prepared by the Working Group on Prolamin Analysis and Toxicity (van Eckert et al., 2006).

In the first study, we used a set of surrogate samples spiked and incurred with the industrial wheat gluten extract at 20 mg kg^{-1} and 100 mg kg^{-1}. All ingredients used in their preparation, were purchased from local grocery stores.

In the second study, we used real food samples and test items from proficiency testing studies. They represented the food categories, as reported in market surveys, most commonly used by people following a gluten-free diet (Lee *et al.*, 2014; NSWFA, 2011 ; Sharma *et al.*, 2015) and contained cereals, bakery products, snacks, diet foods and food supplements, meat, confectionery and condiments.

Most of the food samples chosen fell in the 5th, lower-left sector of the food matrix organisational scheme, of the AOAC Food Triangle (Wolf & Andrews, 1995), which represents food matrices high in carbohydrates and low in fat and proteins. Three samples represented three other sectors with higher relative levels of fat and proteins (sector 4, 6 and 9).

The food samples studied were routine test samples submitted to control laboratories for gluten testing and were kindly provided by the Western Regional Public Analyst's Laboratory (Galway, Ireland), Nuscana Biotechnika Laboratoryjna (Poznan, Poland) or were purchased from local grocery stores. The test items included T27142AQC (Cake mix); T27142BQC (Cake mix); T27109BQC (Cake Mix) and T27106B (Infant Soya Formula) from FAPAS (Food Analysis Performance Assessment Scheme, Fera Science Ltd., Sand Hutton, York, UK). Between 30 g and 500 g of each sample were homogenised in a kitchen blender, vacuum packed and stored at −20 °C until it was required for analysis.

2.2. Characterisation of gluten material

The total protein content and homogeneity of GluVital™ vital wheat gluten was assessed by the Dumas method. A conversion factor of 6.25 was used to convert from the determined nitrogen to total protein content (International Dairy Federation, 2006). Ten aliquots of 0.2 g were analysed with a Vario Max CN elemental analyser (Elementar Analysensysteme GmbH) using L-Glutamic acid (>99.5%, Sigma) as a reference material. The total gluten content was 79.4% with a relative standard deviation (RSD) of 0.24% indicating an adequate homogeneity for the purpose of this study.

2.3. Spiked flours preparation

Three blank flours consisting of soy, corn and rice were used in the study. The gluten content of each blank flour was determined to be below the limit of detection (LOD) of 3 mg kg^{-1} when measured with a Ridascreen Gliadin ELISA test kit (R-biopharm, Darmstadt, Germany).

The blank flours were spiked with 100,000 mg kg^{-1} of the vital wheat gluten. This stock was further mixed with the blank flour to yield the 100 mg kg^{-1} and 20 mg kg^{-1} gluten mixtures. The homogeneity of the spiked flours was assessed with a single ELISA test kit. Ten 1 g aliquots of the spiked flours were extracted and each extract was analysed in duplicate. For each flour the RSDs of all results were less than 10% indicating that, for the purpose of this study, the spiked flours were sufficiently homogenous with respect to gluten content.

The spiked flours were vacuum packed and stored at −20 °C until required for analysis.

2.4. Incurred cookie preparation

Soy and corn cookies were prepared with an incurred gluten content of 100 mg kg^{-1} and 20 mg kg^{-1} by mixing egg (75 g), sugar (200 g), butter (200 g), blank flour (400 g minus the appropriate amount of gluten flour if added) and the appropriate amount of flour that had been spiked with 1000 mg kg^{-1} of gluten.

Cookies were baked at 180 °C for 16 min, ground in a kitchen blender, vacuum packed and stored at -20 °C until required for analysis.

The gluten content of all the food ingredients used in the preparation of the cookies was below the LOD (0.3 mg kg^{-1}) when measured with a Wheat Protein (Morinaga) ELISA test kit.

The homogeneity of the incurred cookies was tested with a single ELISA test kit. Ten separate 1 g aliquots were extracted and analysed in triplicate. The RSDs were less than 10% indicating that the homogeneity, with respect to gluten content of all the cookie materials, was fit for the purpose of this study.

2.5. Extraction with generic buffer and preparation of a universal calibrator

Vital wheat gluten, 0.25 g, was extracted with 24.75 ml of a generic extraction buffer (Akiyama, Imai, & Ebisawa, 2011) (PBS pH 7.4 containing 0.5% SDS, 2% β-mercaptoethanol) by shaking overnight at room temperature. Additionally, 0.25 g of gluten was extracted with 24.75 ml of each ELISA kit's buffer according to the kit's protocol. The extracts were centrifuged (4500*g*, 20 min), filtered (0.45 μm), diluted 1:10 in PBS pH 7.4 and further diluted 1:2 in PBS containing 0.2% BSA to obtain a standard solution of 500 μg mL^{-1}. The standard solution was further diluted with each ELISA kit's original buffer according to the kit's dilution protocol. A standard calibration curve of 7 points spanning the kit's quantification range was constructed in each kit's respective buffer, using consecutive 1:2 dilutions.

2.6. ELISA measurements

Seven commercial sandwich ELISA test kits were used in the assessment of the spiked flours and real food samples: AllerTek Gluten ELISA Assay

(ELISA Technologies, Inc., Gainesville, FL, USA), BioKits Gluten Assay Kit (Neogen Corporation, Lansing, MI, USA), Ridascreen Gliadin (R-biopharm, Darmstadt, Germany), Wheat Protein ELISA kit (Morinaga Inst. of Biological Science, Yokohama, Japan), AgraQuant Gluten G12 Assay (Romer Labs UK Ltd., Runcorn, Cheshire, UK), Transia Plate Prolamins (BioControl Systems, Inc. Seattle, WA, USA) and GlutenTox ELISA Sandwich (Biomedal Diagnostics, Sevilla, Spain). The homogeneity of the spiked and incurred materials was tested with a single test kit (sandwich Gluten-E Nutrilinia, Transia GmbH, Butzbach, Germany).

All measurements were performed by a single analyst in a single laboratory. The analytical protocols provided by each test kit manufacturer were strictly followed unless stated otherwise. Each of the flour samples incurred at the 20 mg kg^{-1} and 100 mg kg^{-1}were extracted twice and each extract was measured in triplicate on a single ELISA plate from each manufacturer. For the food samples, we used 3 independent extractions for each sample with measurements again being performed in triplicate. Each of the 3 extracts were analysed on a separate plate from each manufacturer. Results from the first run were used to optimize the dilution range that was then applied to samples run on the remaining two plates. This approach resulted in 3 valid datasets from 3 runs for most kits with the exception of Morinaga and AllerTek (2 datasets each) and Neogen (1 dataset).

2.7. Data analysis

For both studies only values falling between the lowest non-zero calibrator and the highest calibration standard for each test kit were used. However, for the BioKits (Neogen) in the second study all readings above the kit LOD were included in our calculations. This was beneficial from the point of view of completeness of the study and had no influence on the conclusions drawn when compared to calculations performed when the points between the LOD and LOQ were excluded.

The data analyses were performed using Microsoft Excel and Analyse-it Method Validation edition (Analyse-it Software, Leeds, UK).

The results of absorbance readings were analysed according to the kit manufacturer's instructions. A polynomial (cubic or quadratic) regression was used in all cases. For each calibration curve, the combined uncertainty in the polynomial regression model was calculated (Jeter, 2003 ; Kragten, 1994).

3. RESULTS AND DISCUSSION

3.1. Current status of gluten ELISA measurements

The equivalence of gluten measurements from seven different test kits was assessed. These kits were representative of the variety present on the market in terms of antibody specificities, sample extraction protocols and calibration procedures. They include official test methods recognized by Codex Alimentarius, AOAC International, the American Association of Cereal Chemists International AACCI and authorities in Japan (Akiyama *et al.*, 2011 ; Scherf and Poms, 2015). The characteristics of the test protocols are presented in Table 1.

Table 1. Characteristics of the ELISA test kits used. Precision data are from kit brochures (k), collaborative studies reports (l) (Halbmayr-Jech *et al.*, 2015 ; Immer and Haas-Lauterbach, 2012) or are calculated based on our experiments (e). Different extraction procedures include 1 or 2 steps. RT indicates room temperature, ® – proprietary extraction buffer components undisclosed by the manufacturer, 2-ME – 2-mercaptoethanol, SDS – sodium dodecyl sulphate, PBS – phosphate buffer saline, EtOH – ethanol, mAb – monoclonal antibody, pAb – polyclonal antibody, RSD (relative standard deviation).

		Neogen	AllerTek	Agraquant	Biomedal	Rbiopharm	Transia	Morinaga
Antibody	coating detecting	401.21 (Skeritt) mAb	401.21 (Skeritt) mAb	G12 mAb	A1 mAb G12 mAb	R5 mAb	R5 mAb	anti-wheat protein pAb
	target	HMW-glutenins, ω-gliadins	HMW-glutenins, ω-gliadins	gliadins	gliadins	gliadins	gliadins	wheat protein
Quantification	range unit	3 – 50 ppm gluten	5 – 80 ppm gluten	4 - 200 ppm gluten	1.56 - 25 ng/ml gliadin	5 – 80 ppb gliadin	1.56 – 25 ng/ml gliadin	0.78 – 50 ng/ml wheat protein
Calibrant		Gluten(vital wheat gluten extract)	Gluten(in cooked breads)	Gluten(vital wheat gluten extract)	Gliadin	Gliadin	Gliadin	Wheat protein standard
Traceability		Kjeldhal method (N x 6.25)	–	WGPAT gliadin	–	WGPAT gliadin	–	–
Extraction.	Sample to buffer ratio	1:10	1:10	1:40	1:10	1:40	1:40	1:19
	Buffer 1st	*, 40% EtOH	*, 40% EtOH	*	*	*	*	*, 2-ME
	2nd	–	–	80% EtOH	–	40% EtOH	80% EtOH	–
	Time 1st	90 s	15 min	40 min	40 min	40 min	40 min	12 h
	2nd	–	–	1 h	–	1 h	1 h	–
	Temp 1st	RT	45 °C	50 °C	50 °C	50 °C	50 °C	RT
	2nd	–	–	RT	–	RT	RT	–
Repeatability (intra-assay) RSD$_{wa}$%	12.5-15% (k)	–	16% (l)	–	27% (l)	15% (k)	<10% (k)	
Reproducibility (inter-assay) RSD$_{ba}$%	–	–	32% (l)	–	37% (l)	23% (k)	<10% (k)	
Repeatability (within-run) RSD$_{wrx}$ (e)	10%	12%	12%	7%	5%	14%	9%	
Between-run precision RSD$_{br}$% (e)	12%	17%	7%	11%	14%	20%	11%	

In order to assess the performance of each kit the data were treated in a way that is typically used in proficiency testing (PT) schemes, handling each test kit as a single laboratory.

The data were first assessed for trueness and precision according to ISO 5725–2 (ISO., 1994). In our first study, we analysed samples that consisted of spiked blank flours and incurred cookies. For samples spiked/incurred with 20 mg kg^{-1} of gluten extract, the observed results ranged from 4 to 101 mg kg^{-1} and from 24 to 574 mg kg^{-1} at the 100 mg kg^{-1} level. The computation according to ISO 5725-2 was done separately for the 20 mg kg^{-1} and 100 mg kg^{-1} levels. The analysis indicated that one test kit (AllerTek) produced results which were markedly different from the others, at both concentrations. However, as the goal was to evaluate the current comparability of measurements by ELISA, data from all seven test kits were retained. The calculated repeatability (s_r, within-kit) and reproducibility (s_R, between-kit) standard deviations were $s_r = 21\%$, $s_R = 101\%$ at the 20 mg kg^{-1} level and $s_r = 11\%$, $s_R = 102\%$ at 100 mg kg^{-1}. When the results from the outlying kit were removed from the dataset, $s_r = 16\%$ and $s_R = 47\%$ at 20 mg kg^{-1} and $s_r = 19\%$, $s_R = 41\%$ at 100 mg kg^{-1}.

In the second study, we performed a comparison of measured gluten content in food samples. Again, in many cases the test kits generated values that differed by 10–20 times but the analysis did not identify any outlying test kit (at the 95% confidence level). The average repeatability (s_r, within-kit) standard deviation was 17% (ranging from 3% to 32% depending on the sample) and average reproducibility (s_R, between kit) standard deviation was 76% (varying from 46% to 124%).

The data generated by both studies were treated following the procedure recommended in ISO 13528 (ISO, 2015). The performance of each ELISA test kit was evaluated based on z scores where $z = (\chi \ X)/\sigma$: if the calculated $|z| \leq 2$ the result differed from the reference value by less than 2 times the standard deviation used for proficiency assessment and was therefore considered satisfactory. The z scores were derived using the gravimetric values (first study) or robust averages (second study) for X and standard deviations (σ) set at a value of 25% of the reference value. A value of 25% in this context defines the maximum acceptable standard uncertainty and was chosen based on values reported earlier by providers of allergen PT schemes (DLA FAPAS, 2012; GbR, 2014 ; National Measurements Institute, & Australian Government, 2012).

In the first study, 70% of results achieved the satisfactory level of performance. Two test kits exhibited non-satisfactory performance for most of the samples. In the second study, only 43% of results achieved satisfactory performance. Youden plots presented in Fig. 1 indicate the general lack of equivalence of results among the kits. The distribution of points relative to the 45-degree reference line suggests the presence of a systematic bias: one test kit is significantly biased low in both studies (Neogen) and one test kit is significantly biased high (AllerTek). However, in the second study, this bias is less pronounced because of a presence of a higher number of extreme points in both the top right-hand and lower left-hand quadrant of the graph which represent biased results generated by other test kits (mainly the positively biased Morinaga and negatively biased Agraquant and Biomedal). The differences in the pattern of the distribution of the results between the first and second study may be attributed to a batch-to-batch variability of kits from the same manufacturer. Alternatively, they may also result from differences in responses obtained for different types of samples (depending on the matrix, level of processing etc.). In the second study, where a greater variety of sample types were used, the individual points are more scattered and ellipses enclosing the majority of results generated by individual test kits tend to overlap to a larger extent.

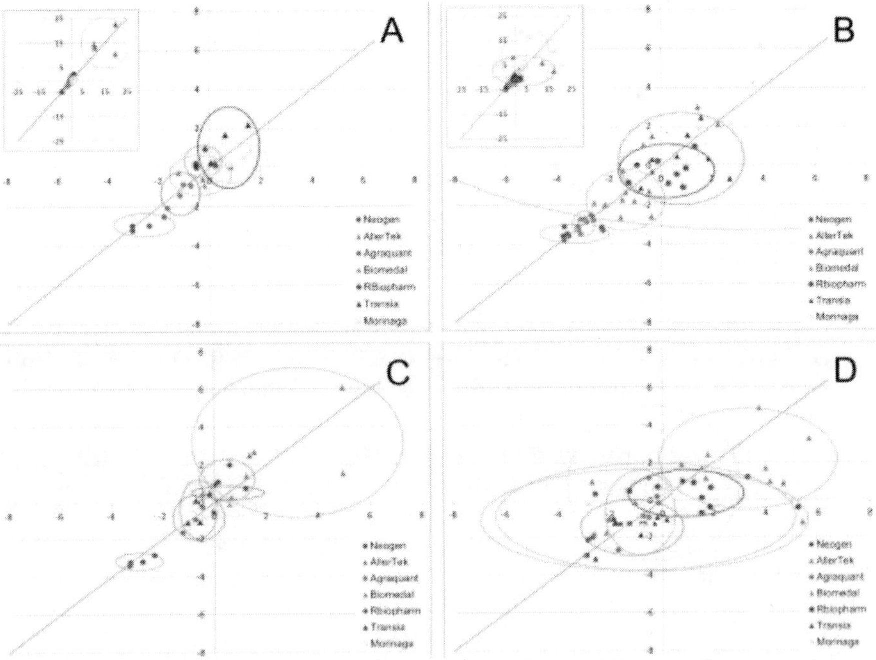

Fig. 1. Youden plots showing the z scores derived from robust averages and standard deviation set at 25% of the reference value. The pairs of samples are matched for similar matrices. The z scores obtained for one of the samples are plotted against the z scores obtained for the other sample. Ellipses represent the mean ± 2 standard deviations for the results of each kit. A: 1st study data; B: 2nd study data; C: 1st study data corrected with universal calibrator, D: 2nd study data corrected with BME-IGRM. Two samples with largest variability in results resulting in extreme z scores (meat sample and cereals sample) were excluded from the graphs presenting 2nd study data.

3.1.1. Assessment of qualitative agreement between results

The lack of equivalence in gluten measurements between different ELISA kits is most problematic when placed in the context of interpreting results in respect to the legislative thresholds.

The results of the second study were classified as "gluten free", "low gluten" or "contains gluten" from the values determined by each kit, based on the 20 mg kg^{-1} and 100 mg kg^{-1} legislative thresholds (Fig. 2). The measurement uncertainty was incorporated applying a decision rule chosen in accordance with Eurachem/CITAC (Eurachem/CITAC., 2007) recommendations to provide at least 97.5% probability that the gluten concentration in a sample is less than the threshold (i.e. a low risk of false acceptance). Therefore, to be classified as "gluten free" or "low gluten" the reported value would have to be less than the threshold value minus two times the result's standard measurement uncertainty. Out of 24 analysed samples only 4, which were found negative for gluten by all kits, had the same classification in all assays. For the remaining 20 samples, the highest degree of agreement was 5 identical classifications out of a possible 7 and this was observed for only 4 samples. A pairwise comparison of the qualitative agreement between kits (Table 2) revealed that only 6 out of 21 possible combinations produced identical statements, for more than 70% of the samples. Only one pair of kits (R-Biopharm and Transia) reached 100% agreement.

To examine the impact of these discrepancies on the level of consumer protection afforded by the results, we plotted the percentage of all kit results above the legislative thresholds versus the median concentration observed for each sample (Fig. 2C). As the true value of gluten concentration in the samples from the second study was not known, the median of all results from each sample was considered the best estimate available. A logistic regression was used to provide a fit to the data. Based on the obtained response curve, an estimate of the reliability of the ELISA test kits' results at any gluten concentration of interest can be determined (Magnusson & Ornemark, 2014).

The data in Fig. 2C suggests that with the current gluten measurement system, as modelled by this study, the threshold of 20 mg kg^{-1} ensures that the gluten concentration in food products labelled as gluten-free would not exceed a level between 80 and 90 mg kg^{-1}, independent of the test kit used. This is the concentration below which the current gluten measurement system becomes unreliable or agreement between different test kits becomes less likely. At 30 mg kg^{-1} the probability of being classified as gluten-free, within the current measurement system is nearly 40%. However, from the perspective of a food producer, to obtain a 95% probability that the product will conform to the current requirements for gluten-free labelling by any of the test kits, the actual gluten content in the sample has to be well below 5 mg kg^{-1}.

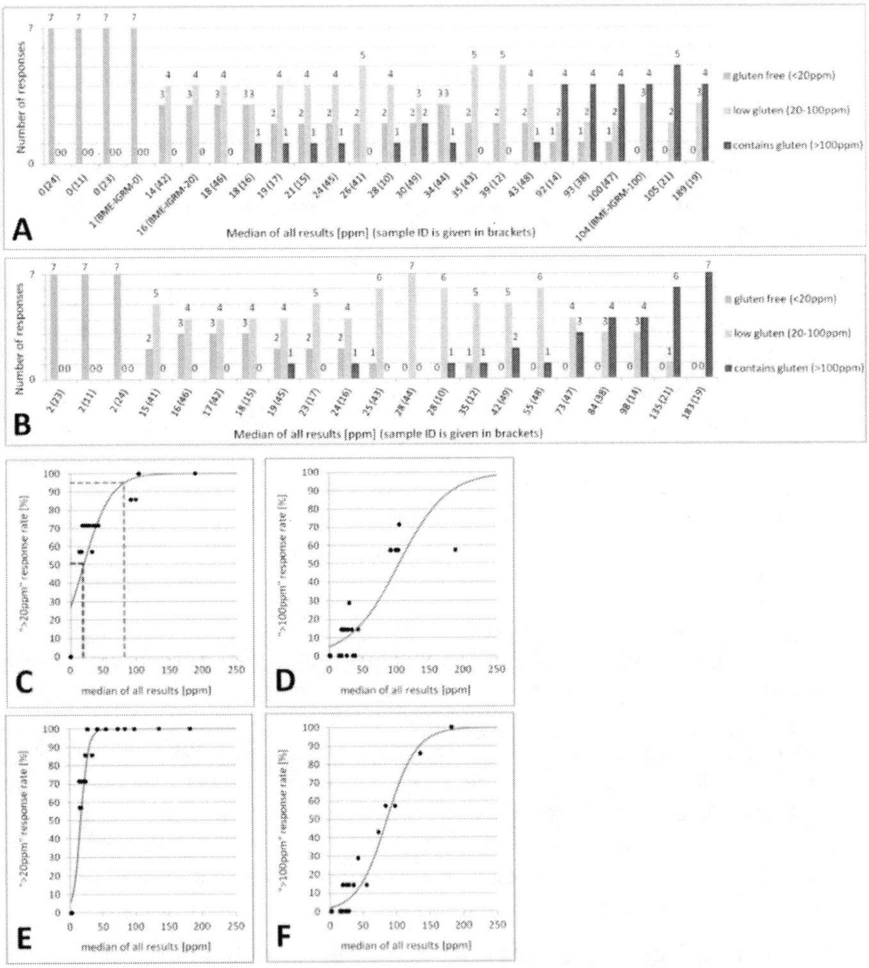

Fig. 2. Assessment of the qualitative agreement between results before and after correction with BME-IGRM. Column plots show number of statements with regards to the gluten content (gluten-free, low gluten, contains gluten) per sample before (A) and after (B) correction with a BME-IGRM matrix material. Scatter plots show the response curves representing rate of "> 20 mg kg^{-1}" or "> 100 mg kg^{-1}" responses before (C, D) and after (E, F) correction with BME-IGRM.

Table 2. Comparison of agreement between pairs of assays for results of the food samples (2nd study). Correlation: Pearson's r values. Values of r that are above 0.8 are in bold. Quantitative and qualitative agreement: without any correction (No corr), after mathematical recalibration (Mat rec) and after correction with BME-IGRM material (BME-IGRM). Quantitative agreement: average between-assay RSD_{ba} in%. Qualitative agreement: number of identical statements produced by the two methods with regards to the gluten content (gluten-free, low gluten, contains gluten) per number of measured samples (in %).

Correlation[Pearson's r]	AllerTek			Agraquant			Biomedal			Rbiopharm			Transia			Morinaga		
Neogen	**0.96**			0.72			0.73			0.61			**0.87**			0.30		
AllerTek				0.62			0.59			0.55			0.78			0.25		
Agraquant							0.77			**0.88**			**0.91**			0.66		
Biomedal										**0.81**			**0.95**			0.25		
Rbiopharm													**0.85**			**0.87**		
Transia																0.66		
Quantitative agreement RSD_{ba} [%]	*No corr*	*Mat rec*	*BME-IGRM*	*No corr*	*Mat rec*	*BME-IGRM*	*No corr*	*Mat rec*	*BME-IGRM*	*No corr*	*Mat rec*	*BME-IGRM*	*No corr*	*Mat rec*	*BME-IGRM*	*No corr*	*Mat rec*	*BME-IGRM*
Neogen	120	11	13	57	31	43	79	38	62	98	39	61	108	35	49	118	40	61
AllerTek				96	34	44	69	33	60	57	37	58	49	32	50	54	38	57
Agraquant							48	28	48	80	29	34	91	22	31	110	32	36
Biomedal										46	25	31	57	22	64	87	33	41
Rbiopharm													22	20	39	55	27	29
Transia																48	26	44
Qualitative agreement [%]	*No corr*	*Mat rec*	*BME-IGRM*	*No corr*	*Mat rec*	*BME-IGRM*	*No corr*	*Mat rec*	*BME-IGRM*	*No corr*	*Mat rec*	*BME-IGRM*	*No corr*	*Mat rec*	*BME-IGRM*	*No corr*	*Mat rec*	*BME-IGRM*
Neogen	0	94	78	78	67	44	28	83	33	6	78	39	6	83	56	0	61	39
AllerTek				0	61	44	39	78	28	78	72	28	78	78	44	61	56	28
Agraquant							50	72	67	0	89	83	0	83	67	0	61	61
Biomedal										50	83	83	50	89	67	33	78	61
Rbiopharm													100	94	83	72	72	78
Transia																72	78	72

The second legislative threshold of 100 mg kg^{-1} (Fig. 2C) ensures consumer protection at the practical level of 210–220 mg kg^{-1}. For samples below 210 mg kg^{-1}, the responses provided by gluten measuring assays are unreliable and at the 100 mg kg^{-1} level the probability of an incorrect classification is 50% and still exceeds 20% at 150 mg kg^{-1}. If a food producer wishes to have a 95% assurance that his product will not be classified in the EU as "contains gluten", the actual concentration of gluten in a product needs to be less than 20 mg kg^{-1}.

This directly translates into a risk model of consuming gluten-containing food when a dietary choice is made based on labelling information assured by the current measurement system. The above illustrates the current difficulties in providing protection to consumers based on the legislative levels using the current ELISA based measurement infrastructure.

3.1.2. Assay performance – Boundaries for harmonisation

The achievable degree of equivalence between results from different test kits is very much dependent on the performance characteristics of each individual assay. The validation data provided by manufacturers for the repeatability and reproducibility of the kits used in this study are presented in Table 1, which also includes our assessment of the repeatability and intermediate precision as observed in this study. The relative standard deviation (RSD) of averages within (first study, spiked samples and cookies) and between (second study, food samples) ELISA plates for spiked and incurred samples was calculated. In addition, the within-plate variation of multiple absorbance readings for the same extracts was evaluated. This within-plate variation was low, 5% on average, for all the ELISA assays studied. The within-run RSD_{wr} (independent extracts of the same sample analysed on the same ELISA plate) was higher with an average value of 10% (7% to 14%). The average between-run RSD_{br} (independent extracts of the same sample analysed on two different ELISA plates from the same manufacturer's batch) was 13% (7–20%). The experimental setup did not provide data on the reproducibility of individual kits. According to the reports from collaborative studies, it is expected to be in the range of 10% to 37%, however, data are not available for all the kits studied.

The precision attainable by individual kit assays sets the limit for possible harmonisation of gluten measurements. This can only be improved by kit producers by further optimization of their analytical protocols and manufacturing processes. In particular, the repeatability of extraction should be scrutinized, as the variation between extracts seems to constitute one of the most prominent components of within-kit variability. In our study, however, we focused on the aspects of the measurement procedure that can be challenged on a more global scale such as the use of universal calibrators, extraction procedure and the provision of matrix reference materials.

3.2. Evaluation of impact of uniform extraction procedure

All ELISA measurement protocols aim to quantify the total gluten content in a food sample, but full extraction of all gluten proteins is problematic due to differences in the solubility of the two main classes of proteins in gluten (Diaz-Amigo & Popping, 2013). The extraction protocols included in commercial ELISA test kits are not harmonized and comprise proprietary extraction buffers, of largely undisclosed composition and use diverse incubation times and temperatures (Table 1).

In this study, as inspired by the Japanese official method for allergen detection, a generic extraction procedure was applied (Akiyama et al., 2011). The vital wheat gluten was extracted with each kit's extraction procedure as well as with the generic extraction procedure. The obtained extracts were consecutively diluted to cover each kit's quantification range and measured by ELISA according to each kit's protocol.

The results presented in Fig. 3 show that for most of the kits the efficiency of extraction with the kits' buffers is roughly equal to the efficiency of extraction with the generic buffer, within the limits of the observed measurement uncertainty. For two kits (AllerTek and Neogen) the generic buffer extracted approximately 10 times more gluten than the respective proprietary buffers. These are the two kits which produced the most extreme gluten concentration values in our first and second studies. However, these two kits produced values which were at the opposite ends of the observed range of results. Furthermore, the group of kits that showed similar extraction efficiency did not produce comparable results for gluten content. The comparison reveals that differences in efficiencies of extraction by themselves do not explain the observed differences in the reported results for gluten content. Therefore, the introduction of a harmonized extraction protocol, in isolation, would not immediately improve the degree of equivalence between results of gluten measurements. During the development of an ELISA assay there might be a number of correction factors introduced by a manufacturer to compensate for issues such as non-quantitative extraction or antibody specificity.

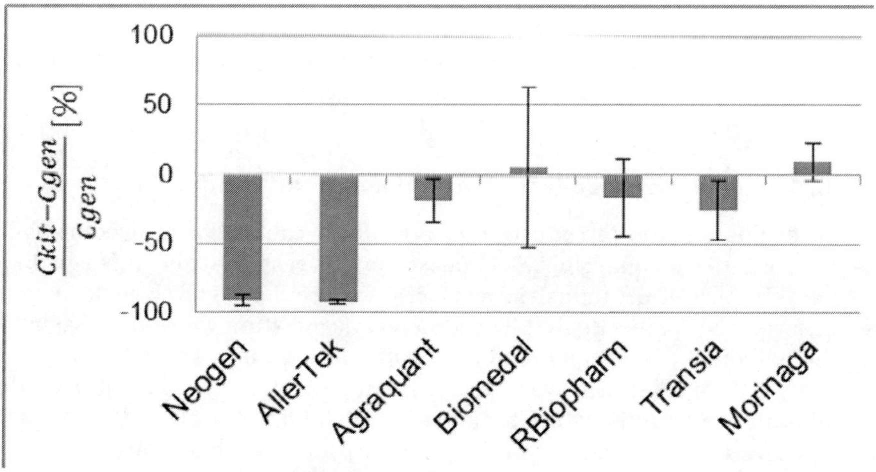

Fig. 3. Comparison of efficiency of extraction with the kit-supplied and generic procedure. Relative differences between gluten concentration measured following the kit buffer extraction (Ckit) and gluten concentration measured following the generic buffer extraction (Cgen). Error bars represent the expanded uncertainty. The results were expressed as the averages of the relative differences between gluten content measured following the kit buffer extraction and gluten content measured following the generic buffer extraction.

3.2.1. Evaluation of the impact of calibration procedures

The gluten ELISA test kits currently on the market use different calibrators ranging from extracts of vital wheat gluten or gliadin to a processed matrix material (Table 1). There is often no detailed information provided with regards to the origin of the calibrators, their varietal composition or value assignment procedure. A number of test kits claim traceability of measurement results to a material called WGPAT gliadin (van Eckert et al., 2006). In many cases the calibration protocol includes a correction factor which defines how a calibrator relates to total gluten in a sample.

The calibration curves, as generated using kit supplied calibrators and measurement procedures, were examined. All resulting plots showed increasing curvature at higher concentrations. For sample extracts with high gluten concentrations but still within the range of quantification stated by the manufacturer, a small variation in measured absorbance has a large impact on the determined concentration. This effect, combined with the estimation of the uncertainty for the polynomial regression, resulted in a large measurement uncertainty estimate for gluten concentrations that were determined from the upper portion of the polynomial curve. For example, the uncertainty contribution from the fitting of the polynomial regression model calculated for a concentration at ¼ of the calibration range was between 2% and 8%, this increased to between 4% and 12% when calculated at ¾ of the calibration range.

3.2.2. Assessment of effect of the universal calibrator

To evaluate if the introduction of a uniform calibration procedure would improve the comparability of measurement results from different ELISA kits, the use of an industrial extract of vital wheat gluten as a universal calibrator was investigated. The gluten concentration value was assigned to the universal calibrator based on results of measurements of total protein content. The calibrator was extracted with each kit's extraction procedure and then consecutively diluted to cover each kit's quantification range and measured in accordance with each kit's protocol. The determined gluten mass fractions (mg kg^{-1}) were plotted against the assigned values and a linear regression was used to provide a fit to the data. The resulting regression coefficients were used to correct the data set obtained in the first study.

The corrected results were treated according to ISO 5725-2. The complete range of corrected measurement results was from 3 to 46 mg kg^{-1} for samples spiked/incurred with 20 mg kg^{-1} of gluten and from 16 to 215 mg kg^{-1} for samples spiked/incurred at the 100 mg kg^{-1} level. Contrary to the uncorrected data, the statistical analysis did not identify an individual test kit that consistently produced outlying results. The average repeatability and reproducibility standard deviations were 17% and 53% at 20 mg $kg^{-1}$and 8% and 47% at the 100 mg kg^{-1} level, respectively. The comparison of precision statistics obtained for corrected and uncorrected data sets showed that the effect of the introduction of the universal calibrator was similar to the effect of removal of results from the kit identified as an outlier in the assessment of the uncorrected data (Fig. 1C). This suggests that the most prominent component of the variation of results observed in the first study was largely the biased calibration procedure of one test kit (AllerTek). However, there is an unknown source of bias present that was not reduced by the use of harmonized calibration procedures and accounts for the limited level of precision observed in the corrected data set.

Analysis of corrected data according to the International Standard 13528 showed little improvement in the overall performance statistics compared to the uncorrected results. The proportion of results with a satisfactory performance score was 77% compared to 70% prior to correction. Correction induced a shift in distribution of satisfactory performance between the kits (Fig. 1). The kit which scored unsatisfactory performance scores in all cases before correction (AllerTek), scored satisfactory performance for 50% of samples after correction. However, the negative bias observed before correction for another kit (Neogen) increased as a result of correction. This confirms that harmonisation of the calibration procedure reduced the bias that could be attributed only to the calibration procedure itself and that other significant, so far unidentified, sources of bias still persists.

3.2.3. Assessment of the effect of the candidate matrix reference material BME-IGRM

The question remains whether a pure gluten extract was sufficiently fit-for-purpose to be used for the correction of results of measurements in matrix samples. To address this issue the suitability of a matrix material BME-IGRM, (Bugyi et al., 2012) consisting of baked cookies incurred with the WGPAT gliadin at the levels corresponding to 0, 20 and 100 mg kg^{-1} gluten, was evaluated. BME-IGRM was measured along with food samples during the second study. The measurement results obtained from this reference material were plotted against the incurred concentrations. The coefficients of the fitted regression lines were used to correct the values obtained for the food samples.

The corrected results were treated according to ISO 5725-2. For the food samples, no individual test kit was identified as providing discrepant results. The average repeatability standard deviation was 19% (ranging from 5% to 55%) and average reproducibility standard deviation was 57% (25–105%). Analysis according to the ISO 13528 yielded 70% satisfactory performance expressed as z scores.

In comparison to the results obtained from the uncorrected data set, there was a reduction in the observed reproducibility standard deviation (from 76% to 57%) and a reduction in the number of non-satisfactory performance z scores (from 57% to 30%). Although correction helped to improve the performance statistics for some test kits (Agraquant, Neogen, Morinaga), it didn't bring any substantial change for the others. For kits where improvements were not observed, the correction modified the pattern of distribution of unsatisfactory results between the kits and between the samples (Fig. 1D). Furthermore, the negative or positive biases became less evident and more scattered and the results that were not recognised as biased before correction received non-satisfactory performance scores afterwards.

The impact of normalising the results of the second study to the values obtained for the BME-IGRM was evaluated by assigning each sample as "gluten-free", "low gluten", or "contains gluten" (Fig. 2B). Only 5 out of 21 samples resulted in the same classification by all test kits after correction. Three of these were found to contain no detectable gluten. The pairwise comparison of this qualitative agreement between kits, in relation to the uncorrected data, revealed improvements for some pairs of kits. However, for others the agreement became even poorer. The most remarkable increase in agreement was observed for a combination of AllerTek and Neogen (from 0% to 78%) and the combination of Agraquant with R-Biopharm, Transia and Morinaga (from 0% to 83%, 67% and 61% respectively). However, overall there was no change in the number of kits resulting in identical statements for greater than 70% of the samples. Not a single pair of kits reached 100% agreement after correction compared with 1 pair before correction.

The inspection of the curves plotted for qualitative responses observed after correction for BME-IGRM at each legislative threshold (Fig. 2D) revealed some improvement of the reliability of the gluten measurement system as approximated by this experimental design. As a consequence of correction, the gluten concentration at which there is <5% probability of false classification as "gluten free" decreased from 80–90 mg kg^{-1} to 30–40 mg kg^{-1}, and from 210–220 mg kg^{-1} to 150–160 mg kg^{-1} for "low gluten".

3.3. Did the reference materials improve the comparability of results?

3.3.1. Correlation analysis

To gain more insight into how the results from different ELISA kits are inter-related, a correlation analysis of the results obtained in the second study was performed. The scatter plots were used as graphical indicators of the association between the values generated by each pair of kits (data not shown). The Pearson's product-moment correlation coefficient (r) was used to quantify the strength of the hypothesized relationship. A cut-off value of 0.8 was chosen to discriminate between meaningful and meaningless correlation. This cut-off was chosen because at correlation coefficients below 0.8 the scatter of points around the correlation line and confidence intervals estimated for Pearson's r values at 95% probability, were so broad that the association between results was no longer significant.

The correlation coefficients for all 21 assay comparisons ranged from 0.25 to 0.96, indicating no correlation to a high degree of correlation between measurement results (Table 2). For 8 assay pairs, the value of r was above 0.8. These results highlight a group of four kits that were correlated (Agraquant, Biomedal, RBiopharm and Transia) and three pairs (Neogen-AllerTek, R-Biopharm-Morinaga, Transia-Neogen) that also showed a high degree of correlation.

A visual inspection of plots revealed that there were individual samples that behaved differently from most other samples. These outlying samples were excluded and the Pearson's correlation coefficients were recalculated. The calculations were repeated to include only samples of similar matrix composition. In both cases, the resulting r and confidence interval values led to the same conclusions with respect to the significance of the calculated correlations.

The poor correlation between results generated by different assays makes it impossible for gluten measurements to achieve full comparability by using a reference material for calibration alone. To assess the degree of agreement potentially achievable, by using a reference material, a mathematical recalibration of the assays was performed. A Passing-Bablok regression line was fitted to the data for each individual assay from the second study versus all assay medians and the regression parameters were used to correct the results (Table 2).

Before mathematical recalibration, the average between-assay RSD_{ba} was 98% for all assays, this was reduced to 39% after applying the correction factors. The same operation was applied separately to groups of assays for which the correlation factors were higher than 0.8. This resulted in a reduction of average RSD_{ba} from 120% to 11% for the comparison of Allertek and Neogen, from 54% to 28% for R-Biopharm and Morinaga, and from 63% to 28% for Agraquant, Biomedal, RBiopharm and Transia.

Although this approach is very basic, these numbers provide an idea of the best case scenario that can be achieved with the current measurement system and what could be achieved if all the assays were correlated.

The pattern of distribution of meaningful correlations is largely reflected in how the use of the BME-IGRM reference material influenced the degree of agreement between pairs of assays. After recalibration with BME-IGRM, the most remarkable increase in qualitative agreement was observed for the combinations of assays with the highest Pearson's r values (Table 2): from 0% to 78% for the pair Neogen-Allertek (r = 0.96), from 0% to 83% for RBiopharm-Agraquant (r = 0.88) and from 0% to 67% for Agraquant-Transia (r = 0.91). In the absence of meaningful correlation, the recalibration brought little or no improvement of agreement e.g. from 28% to 33% for Neogen-Biomedal (r = 0.73) or decrease from 78% to 28% for AllerTek-RBiopharm (r = 0.55).

3.3.2. Commutability

There were some pairs of assays that had high Pearson's r values but correction with BME-IGRM material did not improve the comparability of their results (e.g. for a combination of Biomedal and Transia, the r value was 0.95 but the degree of qualitative agreement changed from 50% to only 67% after correction). This is attributed to the relationship between the values measured for BME-IGRM and the test samples and the concept of commutability. The reference material is commutable when its behaviour towards a given measurement procedure is equivalent to the behaviour observed in routine test samples (ISO, 2009). In this study, routine test samples were represented by the real food samples included in the second study.

As the full commutability study of incurred BME-IGRM material is not the primary subject of this paper, a basic approach for its assessment was adopted. A simple regression analysis was used to establish the relationship between the results obtained with each pair of assays and a 95% prediction interval was calculated, as implemented in Analyse-It, to describe the distribution of the ratio of results expected for the routine samples. Results, where the standardised residual was greater than three, were excluded as outliers and a new linear regression performed. The material was deemed commutable if the result ratio obtained for a reference material using the two methods was consistent with the prediction interval (ISO, 2009) (Fig. 4). The commutability of the BME-IGRM material was verified for the 7 pairs of assays for which the Pearson's r value was higher than 0.8 and where the scatter of points on the correlation plots was not too large to calculate the meaningful prediction intervals. For 5 assay combinations, the commutability of the BME-IGRM material was acceptable. Whereas for 2 pairs of assays, the BME-IGRM material showed remarkable non-commutability (Transia-Biomedal, Transia-RBiopharm). The lack of commutability of BME-IGRM for these two combinations of methods correlates with a lack of improvement of agreement between the results of these methods, after the correction with this material. A similar picture is reproduced in the configuration of points in the Youden plots before and after correction with BME-IGRM (Fig. 1D). The ellipses representing results generated by individual assays overlap more closely for the groups of correlated assays after the recalibration with a commutable material whereas the ellipse representing results of Transia is even more separated from the ellipses of R-Biopharm and Biomedal after correction. The lack of commutability of BME-IGRM between some assay pairs also explains the differences in agreement achievable globally for all assays, as estimated by average between-assay RSD_{ba} for the second study after mathematical recalibration (39%) and the degree of agreement achieved after recalibration with BME-IGRM (RSD_{ba} 57%).

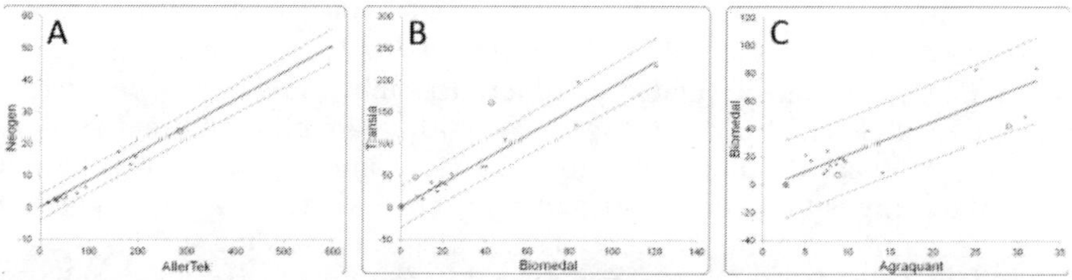

Fig. 4. Examples of results for assessing the commutability of the BME-IGRM material for: A) where the material is commutable; B) where the material is non-commutable; and C) where the assessment is ambiguous due to a large spread of results around the regression line. Results for food samples are shown as black squares, results for BME-IGRM are shown as empty circles. Solid lines represent the regression line plotted for the food samples, dashed lines represent the 95% prediction interval.

The concept of commutability was developed in the field of clinical testing where many routine methods are based on measurement of various sub-components of complex biomolecular species that are assumed to relate directly to the level of the clinically relevant molecule in a sample (ISO,

2009). Gluten measurements pose similar challenges because the measurand is a combination of complex molecules which can be present in different forms and in different proportions in a sample, depending on its varietal composition and degree of alterations arising from food processing or sample preparation. Such structural differences between samples with the same total gluten content may introduce a bias between results obtained with one measurement procedure relative to the results obtained with other procedures. Moreover, the sensitivity of a measurement procedure to interferences resulting from differences between the reference material matrix and sample matrices needs consideration. Samples routinely analysed for gluten content represent different food products and thus different matrices. The identification of a complete set of routine samples, in relation to which the commutability of a reference material is assessed, is problematic. Therefore, in the food testing field, it may be necessary to develop a series of reference materials representing different matrix types.

In this study, we observed that some samples were outliers in many pairwise assay comparisons and showed exceptionally high between-kit variability in their results. One of these materials was a meat sample which falls in the 4th sector of the AOAC food matrix triangle which was different from most of the samples represented in this study. For the two other cases, the outlying samples were breakfast cereals which could be considered as representative of most of the samples studied. In the case of the meat sample, one could speculate that differences in responses may be attributed to matrix effects. However, for the two other samples, different responses were produced despite similar food matrices and may be due to differences in the composition and/or structure of the analyte. This may suggest that the basic assumption in food analysis that foods falling within the same sector of the AOAC food matrix triangle should be chemically similar and thus behave in a similar analytical manner (Wolf & Andrews, 1995), is questionable in the case of gluten analysis.

4. CONCLUSIONS

There were three main objectives of this study: to confirm previous observations that the measured gluten content of the same sample can differ greatly on different ELISA platforms; to understand the effect this spread has on the enforcement of current EU legislation; to observe the effects that standardisation would have on making gluten results comparable.

The observed discrepancies between kits in determining the gluten content were large, with the reproducibility (between-kit RSDs) ranging from 46%

to 124%, depending on the sample. This was despite the precision of individual kits being below 20% and the fact that all results in this study were generated in one laboratory. The only samples that gave rise to the same outcome, from a food labelling perspective, were those that had no detectable gluten present. None of the other twenty samples would have resulted in the same legislative outcome being suggested by all kits.

This study suggests that the current measurement system for gluten quantification cannot fully support the gluten-free claims as defined by legislative requirements. A qualitative response model revealed that there is a 50% probability that a food product deemed compliant at the 20 mg kg^{-1} threshold on the basis of measurements performed by commercial ELISA test kits may, in reality, contain up to 80–90 mg kg^{-1} of gluten. While measurements of gluten at the supplementary 100 mg kg^{-1} threshold for "low gluten" resulted in a 50% probability that the real gluten content was higher than 100 mg kg^{-1} and could even be as high as 210–220 mg kg^{-1}.

The lack of correlation between pairs of test kits suggests that, not only do the kits target different gluten markers but, these markers may not be present in gluten at a constant ratio and/or the effect of food processing may alter these markers. The introduction of reference materials can do little to improve measurements where this lack of correlation is observed. Where correlation between pairs of test kits was proven to be significant, the inclusion of a reference material in the measurement system was seen to improve the comparability of results. However, these beneficial effects were not observed equally by all kits or even for all matrices on the same kit.

The measurement of gluten is proving to be a complex issue. The lack of comparable results could result in a potential detrimental exposure to those suffering from a gluten intolerance/allergy or an unnecessary product recall by food industries. Without agreement on a single common gluten marker, it is clear that the availability of a matrix reference material will not improve the situation for all kits and may even make it worse if used with some kits. Therefore, a greater need for consensus on acceptable robust markers and the conversion of the quantities of these markers to "gluten content" is required before conventional standardisation practices will have the desired effects.

ACKNOWLEDGEMENTS

The authors would like to thank Ingrid Zegers (IRMM) for valuable comments, Alain Maquet (IRMM) for participation in the early stage of the study, Andrew Flanagan from the Western Regional Public Analyst's Laboratory, Galway, Ireland and Adam Dabrowski from Nuscana Biotechnika Laboratoryjna, Poznan, Poland for providing the test samples, Olivier De Rudder (IRMM) for Dumas analysis, Sandor Tomoskozi from Budapest University of Technology and Economics for provision of the matrix reference material.

REFERENCES

1. Akiyama, H., Imai, T., & Ebisawa, M. (2011). Japan Food allergen labeling regulationhistory and evaluation. Advances in Food and Nutrition Research, 62, 32.
2. ANZFA. (2012). Australian New Zealand Food Standards Code: Standard 1.2.1 – Labelling and other information requirements; Standard 1.2.8 - Nutrition information requirements. Australia New Zealand Food Agency, Government of Australia.
3. Bruins Slot, I. D., Bremer, M. G. E. G., van der Fels-Klerx, I., & Hamer, R. J. (2015). Evaluating the performance of gluten ELISA test kits: the numbers do not tell the tale. Cereal Chemistry, 92(5), 9.
4. Bruins Slot, I. D., van der Fels-Klerx, H. J., Bremer, M. G., & Hamer, R. J. (2015). Immunochemical detection methods for gluten in food products: where do we go from here? Critical Reviews in Food Science and Nutrition, 16.
5. Bugyi, Z., Török, K., Hajas, L., Adonyi, Z., Poms, R. E., Popping, B., ... Tömösközi, S. (2012). Development of incurred reference material for improving conditions of gluten quantification. Journal of AOAC International, 95(2).
6. Codex Alimentarius. (1979 rev. 2008). Codex standard for foods for special dietary use for persons intolerant to gluten. CODEX STAN 118-1979.
7. Czaja-Bulsa, G. (2015). Non coeliac gluten sensitivity. A new disease with gluten intolerance. Clinical Nutrition, 34(2), 6.
8. Day, L., Augustin, M. A., Batey, I. L., & Wrigley, C. W. (2006). Wheat-gluten uses and industry needs. Trends in Food Science and Technology, 17, 9.
9. Diaz-Amigo, C., & Popping, B. (2012). Gluten and gluten-free: Issues and considerations of labeling regulations, detection methods, and assay validation. Journal of AOAC International, 95(2), 12.
10. Diaz-Amigo, C., & Popping, B. (2013). Accuracy of ELISA detection methods for gluten and reference materials: A realistic assessment. Journal of Agricultural and Food Chemistry, 61, 8.
11. DLA GbR. (2014). Allergens III: Beta-lactoglobulin and gluten in infant food. Evaluation report proficiency test. Dienstleistung Lebensmittel Analytik GbR.
12. EC. (2009). Commission regulation (EC) No 41/2009 of January 2009 concerning the composition and labelling of foodstuffs suitable for people intolerant to gluten. European Commission. Eurachem/CITAC. (2007). Eurachem / CITAC Guide: Use of uncertainty information in compliance assessment. In).
13. FAPAS (2012). FAPAS report 27109. Gluten in cake mix. Food Analysis Performance Assessment Scheme.
14. FDA. (2013). Food labelling; Gluten-free labelling of foods. Food and Drug Administration Agency, Federal Register/Rules and Regulations, 78(150).
15. Halbmayr-Jech, E., Rogers, A., Don, C., & Prinster, M. (2015). Gluten in rice flour and baked rice products by G12 sandwich ELISA: First action 2014.03. Journal of AOAC International, 98(1).
16. Health Canada. (2012). Health Canada's Position on Gluten-Free Claims.

17. Immer, U., & Haas-Lauterbach, S. (2012). Gliadin as a measure of gluten in foods containing wheat, rye, and barley - enzyme immunoassay method based on a specific monoclonal antibody to the potentially celiac toxic amino acid prolamin sequences: Collaborative study. Journal of AOA C International, 95(4).
18. International Dairy Federation (2006). Comprehensive review of scientific literature pertaining to nitrogen protein conversion factors. Bulletin of the International Dairy Federation, 405, 11.
19. ISO. (1994). International Standard ISO 5725–2:1994. Accuracy (trueness and precision) of measurement methods and results. Part 2: Basic method for the determination of repeatability and reproducibility of a standard reference method. International Organization for Standardization.
20. ISO (2015). International standard ISO 13528:2015. Statistical methods for use in proficiency testing by interlaboratory comparisons. International Organization for Standardization.
21. ISO (2009). International standard ISO guide 34:2009. General requirements for the competence of reference material producers. International Organization for Standardization.
22. Jeter, S. (2003). Evaluating The Uncertainty Of Polynomial Regression Models Using Excel. In ASEE PEER Annual Conference). Nashville, Tennessee: https://peer.asee.org/12110.
23. Kragten, J. (1994). Calculating standard deviations and confidence intervals with a universally applicable spreadsheet technique. Analyst, 119, 5.
24. Lee, H. J., Anderson, Z., & Ryu, D. (2014). Gluten contamination in foods labeled as "gluten free" in the United States. Journal of Food Protection, 77(10), 4.
25. Lionetti, E., Gatti, S., Pulvirenti, A., & Catassi, C. (2015). Celiac disease from a global perspective. Best Practice and Research Clinical Gastroenterology, 29, 23.
26. Magnusson, B., & Ornemark, U. (2014). Eurachem Guide: The fitness for purpose of analytical methods – A laboratory guide to method validation and related topics. In 2nd ed.). www.eurachem.org. National Measurements Institute, & Australian Government. (2012). Proficiency study AQA 12–10. Allergens in food.
27. NSWFA. (2011). Presence of gluten in foods labelled 'gluten-free'. New South Wales Food Authority, retrieved on November 2015 from www.foodauthority. nsw.gov.au, 15.
28. RASFF Portal. (accessed 10/12/2015). Rapid Alert System for Food and Feed, http://ec. europa.eu/food/safety/rasff/portal.
29. Sapone, A., Bai, J. C., Ciacci, C., Dolinsek, J., Green, P. H., Hadjivassiliou, M., ... Fasano, A. (2012).
30. Spectrum of gluten-related disorders: Consensus on new nomenclature and classification. BMC Medicine, 10(13).
31. Scharf, A., Kasel, U., Wichmann, G., & Besler, M. (2013). Performance of ELISA and PCR methods for the determination of allergens in food: An evaluation of six years of proficiency testing for soy (Glycine max L.) and wheat gluten (Triticum aestivum L.). Journal of Agricultural and Food Chemistry, 61, 12.
32. Scherf, K. A., Koehler, P., & Wieser, H. (2015). Gluten and wheat sensitivities – An overview. Journal of Cereal Science. in press.
33. Scherf, K. A., & Poms, R. E. (2015). Recent developments in analytical methods for tracing gluten. Journal of Cereal Science, in press, 11.
34. Sharma, G. M. (2012). Immunoreactivity and detection of wheat proteins by commercial ELISA kits. Journal of AOAC International, 95(2), 8.
35. Sharma, G. M., Pereira, M., & Williams, K. M. (2015). Gluten detection in foods available in the United States – A market survey. Food Chemistry, 169, 7.
36. Sykes, M., Anderson, D., & Parmar, B. (2012). Normalisation of data from allergens proficiency tests. Analytical and Bioanalytical Chemistry, 403, 7.
37. van Eckert, R., Berghofer, E., Ciclitira, P., Chirdo, F., Denery-Papini, S., Ellis, H., & Mamone, G. (2006). Towards a new gliadin reference material-isolation and characterisation. Journal of Cereal Science, 43(3), 10.

38. Wolf, W. R., & Andrews, K. W. (1995). A system for defining reference materials applicable to all food matrices. Fresenius Journal of Analytical Chemistry, 352, 4.

CHAPTER 4

Effect of Soxhlet Method Extraction on Characterization of Pectin of Pumpkin Peels

Awadia Aldow Ramadan Hamed[1*], Abdeen Elsiddig Eltyeb Elkhedir[1] and Salma Elghali Mustafa[2]

[1]*Food Industry Department, Industrial Research and Consultancy Center (IRCC), Khartoum, Sudan*

[2]*Department of Food Science and Technology, Faculty of Agricultural Studies, Sudan University of Science and Sechnology, Khartoum, Sudan*

ABSTRACT

With the present study was focused on the potential of pumpkin peel as a source of pectin, it was aim to determine a pratical follow-up to the extraction of pectin from pumpkin peels Extraction (Cucurbita spp.) using soxhlet with two different acids. Moreover to investigate the influence of time on pectin yield and to characterize the output, determinations were performed of methoxyl content, acetyl content, equivalent weight and degree of esterification in a laboratory, on a small scale. The result showed higher average yield of pectin obtained by using soxhlet acid extraction (7.72% for nitric acid and 6.80% for citric acid), while the lower yield were obtained with acid extraction without using soxhlet (6.24% for nitric acid and 5.36% for citric acid). Equivalent weight and acetyl contain of extracted pectin with both nitric and citric acids were (1250 g/mol and 0.43) respectively, while methoxyl content was (6.20% and 7.23 %), the degree of esterification was (66.53% and 66.57%) for nitric and citric acid respectively. The results showed that pumpkin peels are a promising commercial source of pectin.

Keywords: Pumpkin peels; Extraction; Pectin

INTRODUCTION

Pumpkin (*Cucurbita maxima*) belongs to the family Cucurbitaceae, is an excellent resource of carotenoids, precursors of vitamin A, and has been regarded as a functional food [1,2]. It is also a low cost source of pectin [3]. Pumpkin peels, which are produced during processing and discarded as agricultural by-products, are rich in pectin. Accordingly, information about the yield, structure and properties of pectin from pumpkin peels could be of importance from both industrial and scientific points of view. Pectic substances in fruits were discovered by the French chemist Louis Nicolas Vauquelin in tamarind fruit. The term "pectin" was introduced by Henri Braconnot due to the gelling properties of these substances [4,5]. Pectins are structural polysaccharides present within all dicotyledonous plant cell walls. The primary structural feature of pectin is a linear 1,4-α linked DGalacturonic acid chain with varying degrees of methylation, which are responsible for different physiological processes [6]. In the cell walls they serve as one of the main agents cementing the cellulose fibrils and may be linked covalently to other polymers. Intracellular pectins provide the channels for passage of nutrients and water [7]. The main raw materials used to produce commercial pectin are apple pomace and citrus peel [6] sugar beet and sunflower heads [7]. Several studies on the extraction methods of pumpkin pectin have been performed using different extraction methods, such as an acidic extraction method and enzymatic extraction method [8,9]. Commercial pectins are extracted at low pH and high temperature. Pectins are widely used as food additives (E440) with gelling and stabilizing properties in jams, jellies, marmalades, milks and confectionery products [10]. The study is aimed to determine the proximate composition of pumpkin peels (moisture, crude protein, ash, fat, crude fiber and carbohydrate), to examine the influence of soxhlet method extraction with different acids (nitric and citric acids) on pectin obtained and to characterize the obtained pectin.

MATERIALS AND METHODOLOGY

Materials

Mature pumpkin was purchased from local market, Khartoum North; the peel was separated from the flesh, dried, and the rind was ground to uniform size using a food processor at high speed.

Methods

The proximate analysis of pumpkin peels (Moisture, Ash, Protein, Fats, Fiber content and Carbohydrate content) were performed according to the methods described by [11].

The soxhlet acid extraction of pectin was carried out using the method of [12], the acid extraction was carried out as suggested by [13]. Moisture and ash content were determined depending on the method of [12].

Equivalent weight and Methoxyl Content, Acetyl content and the Degree of esterification (DE) were determined according to the methods of [14].

Statistical analysis

All statistical analyses were performed using SAS version 9.1 (SAS Institute Inc., Cary, NC, USA). Analysis of variance (ANOVA) was performed using the general linear models (GLM) procedure to determine significant differences among the samples. Means were compared by using Fisher's least significant difference (LSD) procedure. Significance was defined at the 5% level.

RESULTS AND DISCUSSION

The proximate analysis of pumpkin peels:

The result of the proximate analysis of pumpkin peels were showed in **Table 1** expressed on a dry basis.

Table 1: Proximate analysis of pumpkin peels on dray basis.

Parameter	Content (%)
Moisture	20.1
Ash	7.1
Protein	3.2
Fiber	10. 2
Fats	2.3
Carbohydrate	57. 2

The moisture, ash, protein, fiber, fat and carbohydrates were 20.1, 7.1, and 3.2, 10.15, 2.3 and 57.15% respectively.

The results in **Figure 1** revealed that the yields of pectin using soxhlet acid extraction method were 7.72% and 6.80% for nitric and citric acid, respectively, while acid extraction without using soxhlet showed lower value (6.24%) for nitric and (5.36%) for citric acid respectively. The results were also supported by other researcher who concluded the amount of pectin extracted from pumpkin peel 253-0.233 % obtained using Alcohol insoluble by Hamed [15].

The effects of extraction time on pectin yield were shown in **Figure 2**. There was increasing on pectin yield from 30 to 60 min in pectin extraction times.

The moisture content of pectin from pumpkin peels was 5.54% using nitric and 5.42% using citric acids as shown in **Table 2**. Significantly different from each other at $p \leq 0.05$; It was 7.88-8.96% in grapefruit peels pectin [14]. 5.03-5.04 % as in mango pulp pectin as reported by [16].

3.17 for pectin obtained from pumpkin peel's using nitric acid and 2.96% using citric acids as **Table 2**, lemonorange and apple 1.65%-1.56 , 0.49-0.81% and 0.49-8.05%, respectively [15,16].

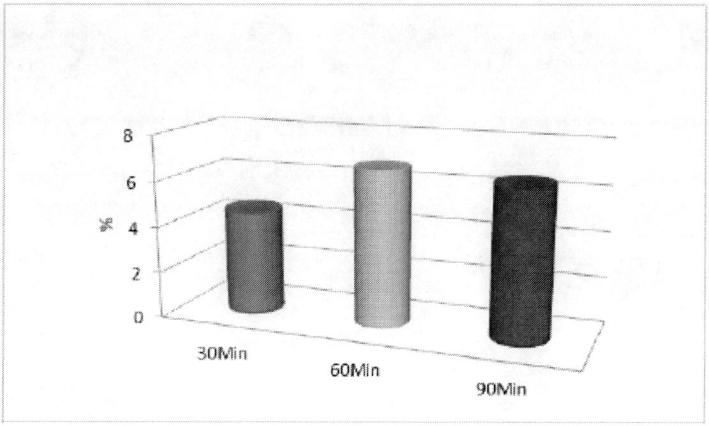

Figure 1: Effect of time of extraction on the yield of pumpkin peels pectin.

Figure 1: Effect of time of extraction on the yield of pumpkin peels pectin.

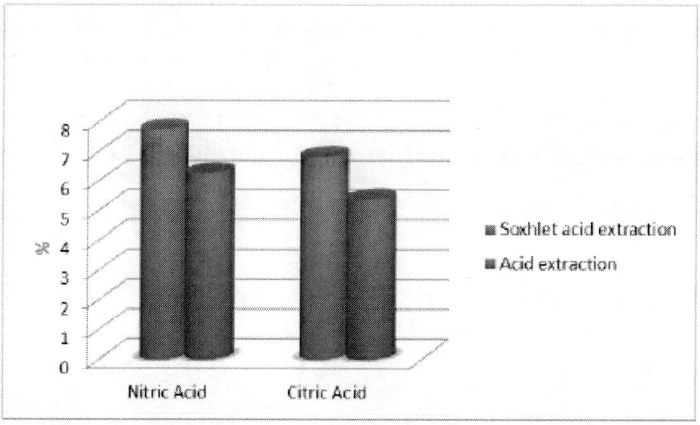

Figure 2: Effect of extraction methods on pumpkin peels pectin yield (%).

Figure 2: Effect of extraction methods on pumpkin peels pectin yield (%).

Table 2: Characterization of pumpkin peel pectin.

Parameter	Pectin extracted by nitric acid (%)	Pectin extracted by citric acid (%)	Level of significance (P-value)(%)
Moisture content	5.45 (± 0.01)	5.42 (± 0.01)	0.0**
Ash content	3.17 (± 0.006)	2.96 (± 0.006)	0.0011*
Acetyl	0.43 (± 0.01)	0.43 (± 0.03)	1.0 n.s
Methoxyle	6.2 (± 0.10)	7.23 (± 0.89)	0.2102 n.s
Equivalent weight	1250 (± 0.0)	1250 (± 0.0)	1.0 n.s
Degree of esterification	66.53 (± 0.058)	66.57 (± 0.058)	0.6667 n.s
Values are mean ± SD; n.s. Not Significant; *= significant ($P \leq 0.05$); **= highly significant ($P \leq 0.01$)			

1250 g/molas (**Table 2**), cocoa husk pectin it was 510.68-645.19 g/mol Ramli [17] and lower than the range of 1389-2003,833.33-1666.30 and 263,000-303,000 g/mol for mango pulp, apple pomace and ambarella peels [16,18].

The methoxyl contents of pumpkin peel pectin were 6.20% using nitric acid and 7.23% using citric acids, as shown in **Table 2**. No significant differences at $p \leq 0.05$ were observed between the two samples. This was lower than lemon pomace, which was 10.25%, pomelo peel 8.57%, Lime 9.92% and passion 8.81-9.61% [19,20], but it was similar in mango peels pectin 7.33%, banana peels pectin 7.03% [20]. Methoxyl content is an important factor in controlling the setting time of pectin and the ability of the pectin to form gels [16].

The acetyl content of pumpkin peel pectin was 0.43 for both nitric and citric acids. The results were confirmed by other researchers who found 0.46-1.63% in grapefruit peel pectin whereas it was higher than 0.117% to 0.314%, recorded in mango pulp pectin [15,21]. The acetyl contain of pectin has the important role on the jelly formatting ability [21].

The degree of esterification of pectin pumpkin peel's was showed in table 2, 66.53% for nitric acid and 66.57% with citric acids. These results are similar to the values 66.9% reported for mango peel pectin but higher than 51.01-51.24% obtained for grapefruit peel pectin [15,22,23]. They are lower than the values 73.9, 87.0 and 79.51% reported for pumpkin pectin, mango pulp pectin and lemon pomace pectin respectively [21,22,24].

REFERENCES

1. Arima HK, Rodriguez-Amaya DB (1990) Carotenoid composition and vitamin A value of a squash and a pumpkin from north eastern Brazil. Archivos Latino americanos de Nutrición, 40, 284.

2. Adams GG, Imran S, Wang S, Mohammad A, Kok MS, et al. (2012) Extraction, isolation and characterisation of oil bodies from pumpkin seeds for therapeutic use. Food Chemistry 134: 1919-1925.
3. Mülleder M, Neunteufl H (2002) Carotenoid content in different varieties of pumpkins. J Food Compost Anal 15: 633-638.
4. Georgiev Y, Ognyanov M, Yanakieva I, Kussovski V, Kratchanova M (2012) Isolation, characterization and modification of citrus pectins. J Bio Sci Biotech 1: 223-233.
5. Devi EW, Shukla RN, Abraham A, Jarpula S, Kaushik U (2014) Optimized extraction condition and characterization of pectin from orange peel. IJREAT 2: 1-10.
6. Wang YC, Chuang YC, Hsu HW (2008) The flavonoids carotenoid and pectin content in peels of citrus cultivated in Taiwan. Food Chem 106: 277-284.
7. Kertesz ZI (1951) The pectin substances. Interscience, New York, USA, p.438.
8. Fissore EN, Ponce NM, Stortz CA, Rojas AM, Gerschenson LN (2007) Characterisation of fiber obtained from pumpkin (cucumis moschata duch) mesocarp through enzymatic treatment. Food Sci Tech Int 13: 141-151.
9. Ptichkina NM, Markina OA, Rumyantseva GN (2008) Pectin extraction from pumpkin with the aid of microbial enzymes. Food Hydrocolloids 22: 192-195.
10. Sakai T, Sakamato T, Hallaert J, Vandamme EJ (1993) Pectin, pectinase, and protopectinase: production, properties, and applications. Adv Appl Microbiol 39: 213-294.
11. Horwitz W, Chichilo P, Reynolds H (1970) Official methods of analysis of the Association of Official Analytical Chemists. Washington, AOAC 11: 1015.
12. Malviya R, Srivastava P, Bansal M, Sharma PK (2010) Mango Peel Pectin as natural polymer Agents. J Sci Ind Res 69: 688-690.
13. Crandall P, Braddock R, Rouse A (1978) Determining the yield and quality of pectin from fresh peel and pectin pomace. Proc Fla State Hort Soc 91: 109-111.
14. Owens H (1952) Methods used at Western Regional Research Laboratory for extraction and analysis of pectic materials. Albany, CA 340: 33
15. Hamed AAR (2015) Extraction and Assessment of pectin form Pumpkin Peels, Sudan University of Sciences and Technology.
16. Constenla D, Lozano J (2003) Kinetic model of pectin demethylation. Latin American applied research 33: 91-96.
17. Ramli N (2011) Effect of ammonium oxalate and acetic acid at several extraction time and pH on some physicochemical properties of pectin from cocoa husks (Theobroma cacao). Afr J Food Sci 5: 790-798.
18. Kumar A, Chauhan GS (2010) Extraction and characterization of pectin from apple pomace and its evaluation as lipase (steapsin) inhibitor. Carbohydrate Polymers 82: 454-459.
19. Azad A, Ali M, Akter MS, Rahman MJ, Ahmed M (2014) Isolation and characterization of pectin extracted from lemon pomace during ripening. J Food Nutr Sci 2: 30-35.
20. Madhav A, Pushpalatha P (2002) Characterization of pectin extracted from different fruit wastes. J Tropical Agriculture 40: 53-55.
21. Abdel-Rahman N (2002) Characterization of pectic substances of abu-samaka cultivar in relation to their influence on mango pulp concentrate.
22. El Tinay A, El-shafie A, Nour A (1982) A chemical study of pumpkin pectic substances. Dep of Bio, Faculty of Agriculture, University of Khartoum, Sudan.
23. Koubala B, Mbome L, Kansci G, Mbiapo FT, Crepeau MJ, et al. (
24. Physicochemical properties of pectins from ambarella peels (Spondias cytherea) obtained using different extraction conditions. Food Chemistry 106: 1202-1207.

CHAPTER 5

Novel Insights into Ascorbate Retention and Degradation During The Washing and Post-Harvest Storage of Spinach and other Salad Leaves

Rebecca A. Dewhirst [a,c] , Graham J.J. Clarkson [b,d] , Steve D. Rothwell [b] , Stephen C. Fry [a]

[a] *The Edinburgh Cell Wall Group, Institute of Molecular Plant Sciences, The University of Edinburgh, Daniel Rutherford Building, The King's Buildings, Max Born Crescent, Edinburgh EH9 3BF, UK*

[b] *Vitacress, Lower Link Farm, St Mary Bourne, Andover, Hampshire SP11 6DB, UK*

[c] *wildFIRE Lab, Hatherly Laboratories, University of Exeter, Prince of Wales Road, Exeter EX4 4PS, UK1*

[d] *Edward Vinson Ltd, 4 Ewell Barn, Graveney Rd, Faversham ME13 8UP, UK*

ABSTRACT

Post-harvest treatments of pre-packaged salad leaves potentially cause L-ascorbate loss, but the mechanisms of ascorbate degradation remain incompletely understood, especially *in planta*. We explored the extent and pathways of ascorbate loss in variously washed and stored salad leaves.

Ascorbate was assayed by 2,6-dichlorophenolindophenol titration, and pathways were monitored by ^{14}C-radiolabelling followed by high-voltage electrophoresis. All leaves tested showed ascorbate loss during storage: lettuce showed the greatest percentage loss, wild rocket the least. Spinach leaves were particularly prone to losing ascorbate during washing, especially with simultaneous mechanical agitation; however, washing in the presence of hypochlorite did not significantly increase ascorbate loss. In spinach, [^{14}C]oxalate was the major product of [^{14}C]ascorbate degradation, suggesting that commercial washing causes oxidative stress. This study highlights that ascorbate/dehydroascorbic acid are lost via the oxidative pathway during washing and post-harvest storage of salad leaves. Thus changes to washing procedures could potentially increase the post-harvest retention of ascorbate.

Keywords: Vitamin C; Spinach; Salad leaves; Washing; Post-harvest storage; Chlorine; Oxalic acid

1. INTRODUCTION

Vitamin C, comprising L-ascorbate and dehydro-L-ascorbic acid (DHA), is chemically the simplest vitamin. Unlike humans, plants can synthesise ascorbate, accumulating it at up to millimolar concentrations such that it accounts for up to 10% of the total water-soluble 'carbohydrates' (Noctor & Foyer, 1998).

Vitamin C participates in collagen synthesis (Mandl, Szarka, & Bánhegyi, 2009). Other reported health benefits include the treatment or prevention of diabetes, cardiovascular disorders, age-related diseases and cancer (Ames *et al.*, 1993 ; Mandl *et al.*, 2009). Ascorbate is an antioxidant, but also has numerous other roles in plants including as an enzyme co-factor (Gallie, 2013), and in regulating the cell cycle (Smirnoff, Wheeler, & Loewus, 2000). In plants, apoplastic ascorbate may also play a beneficial pro-oxidant role, generating reactive oxygen species e.g. hydroxyl radicals (Fry, 1998), which may serve to loosen the cell wall during fruit ripening (Airianah, Vreeburg, & Fry, 2016).

Up to 90% of our dietary vitamin C is plant-derived (Lee & Kader, 2000) but cooking generally destroys much of the ascorbate in food (Lee & Kader, 2000). Therefore raw salads are an invaluable source of ascorbate. The ascorbate content of salad plants varies hugely, e.g. from 110 mg (curly kale) to as little as 3 mg (wholehead iceberg head lettuce) per 100 g fresh weight (McCance & Widdowson, 1991). Ascorbate content can also vary between cultivars of the same species (Hodges and Forney, 2003; Koh *et al.*,

2012; Ren *et al.*, 2013), and younger plant tissues often have higher ascorbate concentrations than older ones, e.g. in spinach (Bergquist, Gertsson, & Olsson, 2006) and celery (Huang et al., 2016), presumably related to the ascorbate's role in plant growth.

The washing process of pre-packaged salads is also a potential source of ascorbate loss. Most commercial washing processes use recirculated water, treated with a sanitiser (e.g. chlorine-based). Iceberg lettuce washed in chlorinated water showed a marked decrease in ascorbate content after just one day's storage compared with lettuce washed in non-chlorinated water (Kenny & O'Beirne, 2009). However, spinach leaves washed in chlorinated water did not show any rapid ascorbate loss (Karaca & Velioglu, 2014). Conversely, washing with chlorine-based sanitisers caused considerably more ascorbate loss during subsequent storage than washing in peroxyacetic acid-based sanitisers, although a water-only control was not included (Gómez-López, Marín, Medina-Martínez, Gil, & Allende, 2013). Equally, spinach washed in chlorinated water showed greater loss of antioxidant activity than when washed in oxalic acid (Cefola & Pace, 2015), probably owing to the oxidising nature of chlorine-based sanitisers. These somewhat contradictory results suggest a need for further investigation of chlorine effects in the spinach.

Other steps in the processing of pre-packaged salads could also lead to ascorbate loss. For example, the slicing sometimes used on iceberg lettuce influences ascorbate content throughout shelf life, with hand-torn leaves showing higher ascorbate retention than blade-cut leaves (Barry-Ryan & O'Beirne, 1999), presumably because blades cause more severe wounding, leading to ascorbate consumption during the wound response.

Although vitamin C has been widely studied for many decades, much remains unclear about its degradation pathways. The first relatively stable degradation product of ascorbate is DHA. The oxidation reactions involved are effectively reversible in plants owing to the presence of DHA reductase and monodehydroascorbate reductase (Foyer and Halliwell, 1977; Truffault *et al.*, 2017). DHA can then be further oxidised to a range of products (Fig. 1), or hydrolysed to form diketogulonate (DKG), both these reactions representing a permanent loss of vitamin C from the plant tissue. DKG can itself be reduced to a redox-reactive substance with the formula $C_6H_6O_5$ (Kärkönen, Dewhirst, Mackay, & Fry, 2017), and DHA can be further oxidised, e.g. to oxalyl threonate (OxT) and cyclic oxalyl threonate (Fig. 1) (Green and Fry, 2005; Parsons *et al.*, 2011). Some plants accumulate ascorbate oxidation products, e.g. L-threarate (L-tartrate) in grapes (DeBolt, Hardie, Tyerman, & Ford, 2004) and oxalate in spinach (Yang & Loewus, 1975).

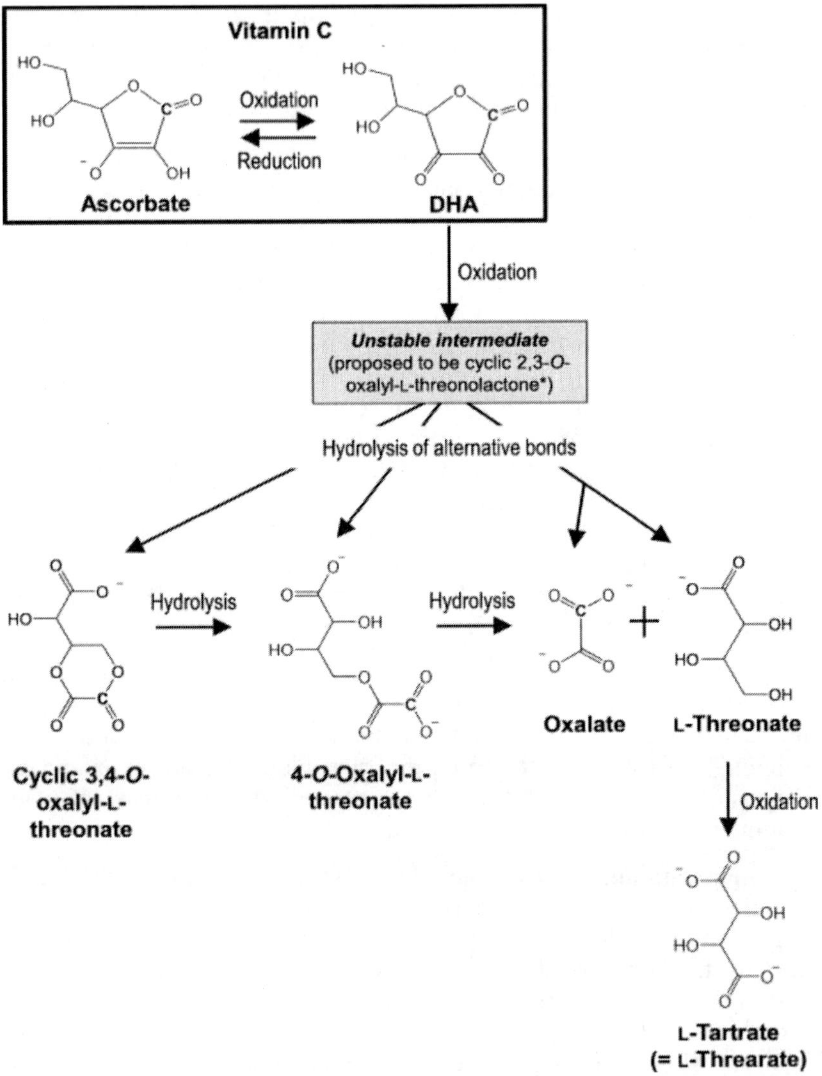

Fig. 1. The oxidation pathway of ascorbate. Vitamin C consists of ascorbate (AA) and dehydroascorbic acid (DHA). Further degradation of DHA, e.g. by the oxidation reactions shown here, results in a loss of vitamin C activity. The initial oxidation step between AA and DHA is effectively reversible in plants owing to the presence of DHA reductase. The C shown in the structural formulae indicates the radiolabelled carbon derived from the [1-^{14}C]AA used in this study. Pathway simplified from Parsons, Yasmin, and Fry (2011).

Increasing the ascorbate content of food (Hancock & Viola, 2005) would create more nutritious crops, as well as potentially making the crops themselves more tolerant of stress, such as oxidative stress. Although ascorbate can be easily synthesised chemically and then added to food, there is a general trend away from artificial food additives, creating a market for naturally ascorbate-enriched crops. An increase of ascorbate in crop plants could be achieved by either increasing the biosynthesis or decreasing the degradation. We are interested in the latter option. Therefore, this paper focuses on the post-harvest processing of young salad leaves as a potential area in which the loss of ascorbate could be minimised, as well as investigating the degradation pathways of ascorbate during post-harvest processing and storage.

2. MATERIALS AND METHODS

2.1. Plant material

Salad leaves used in experiments were grown commercially at Mullins Farm, Pewsey, Wiltshire and St Mary Bourne, Hampshire, UK, and processed on an industrial scale by Vitacress Salads Ltd, St Mary Bourne, from June to August 2013. The salad leaves studied were wild rocket (*Diplotaxis tenuifolia*), white wall 'wasabi' rocket (*Diplotaxis erucoides*), mizuna (*Brassica juncea var. japonica*), watercress (*Nasturtium officinale*), green Batavia (*Lactuca sativa*), iceberg lettuce (*Lactuca sativa*), spinach (*Spinacia oleracea*), red spinach (*Amaranthus dubius*), red chard (*Beta vulgaris,* subsp. *vulgaris*), pea shoots (*Pisum sativum*) and fennel (*Foeniculum vulgare*). In addition, a commercial variety (Toucan, from Rijk Zwaan) of spinach seeds was grown in soil at 21 °C (day) and 16 °C (night) with 16-h light levels of 150 μmol m^{-2} s^{-1}. Leaves were harvested for experiments 4 weeks after sowing.

2.2. Determination of ascorbate content by titration with DCPIP

Salad leaves (1 g) were ground in 5 ml of either 2% (w/v) 'metaphosphoric acid' (an incompletely defined mixture of polymeric acids with overall empirical formula HPO_3), 56 mM oxalic acid (Ponting, 1943) or 98 mM formic acid, with a pestle and mortar. The thoroughly ground sample was then vacuum-filtered on Whatman No. 1 filter paper and the filtrate collected, or the samples were centrifuged at 1000*g* for 10 min and the supernatant collected. Duplicate 1-ml aliquots of the filtrate or supernatant were titrated with 3.73 mM DCPIP (2,6-dichlorophenolindophenol), added in 10-µl shots until a pink colour remained for 10 s. The volume of DCPIP added was compared to a standard curve of ascorbic acid concentrations.

2.3. Postharvest washing procedures

Salad leaves harvested from Vitacress farms were stored in unsealed plastic packaging in the dark at 4 °C for up to 10 days. Watercress (*Nasturtium officinale*), spinach (*Spinacia oleracea*) and wild rocket (*Diplotaxis tenuifolia*) from the same harvest batch were sampled before and after the commercial washing process. The washing process at Vitacress consists of a counter-current flow system in which the leaves are washed in spring water. The leaves are spun dry and distributed into plastic packaging (microperforated 30-µm orientated polypropylene).

The industrial washing process was also simulated under laboratory conditions. Samples of spinach leaves (1 g, in triplicate) were incubated either in air (with no H_2O added), in still water or in shaken water (on a small orbital shaker) for 0.5, 1 or 2 h in the dark, then analysed for ascorbate. Samples (1 g in triplicate) were also analysed immediately after harvesting (time 0).

The effect of chlorine on the retention of ascorbate in spinach leaves during washing was investigated. Samples of spinach leaves purchased from a local supermarket (1 g in triplicate) were incubated in plastic vials either in air (with no H_2O added), in H_2O or in chlorinated H_2O (100 ppm active chlorine from sodium hypochlorite), all with gentle shaking at 7 °C for 1 h in the dark.

2.4. Infiltration of spinach leaves with [^{14}C]ascorbate

Each of three spinach leaves was placed with its petiole (cut at 90° with a razor blade) in a round-bottomed tube containing L-[1-^{14}C]ascorbate (8 kBq; Amersham Pharmacia Biotech UK Ltd) diluted to 50 µl with H_2O. The [^{14}C]ascorbate was taken up into the leaf by transpiration. After this initial solution has been taken up, three further aliquots (50 µl) of H_2O were added into the tube, to ensure all the [^{14}C]ascorbate was taken up. The presence of radioactivity in the lamina was confirmed with a Geiger counter. Leaf discs (1 cm diameter) were cut out from the lamina, avoiding the main veins (as shown in Fig. 6a). Sets of four equivalent discs were prepared and each disc within a set was subjected to a different treatment: analysed immediately (time 0) or incubated in either air, 5 ml still water or 5 ml shaken water (on a mini orbital shaker at 150 rpm with an orbit of 2 mm) at 7 °C. The two halves of the lamina are assumed to be identical, thus each disc within a set was presumed to contain equivalent levels of radiolabelled compounds. This allowed direct comparisons to be made between these four leaf discs.

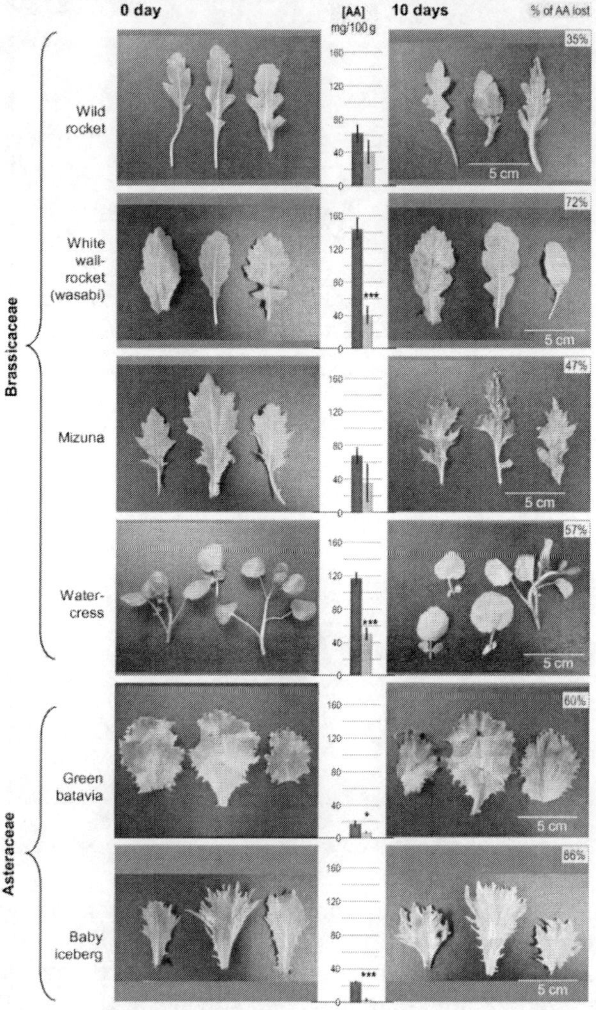

Fig. 2. Changes in visual characteristics and ascorbate loss in different salad leaves during storage. Various types of salad leaf were stored in the dark at 4 °C for 10 days, then photographed and assayed for ascorbate content. The histograms show mg ascorbic acid per 100 g fresh weight after 0 and 10 d (left and right bar, respectively). Each bar shows the mean of titrations of 6 separate extracts ± SE. Significant differences between 0 and 10 d are indicated (Student's *t*-test; $^{*}p < 0.05$, $^{**}p < 0.01$, $^{***}p < 0.001$, n = 6). Values in yellow boxes indicate the % loss of ascorbate between 0 and 10 d.

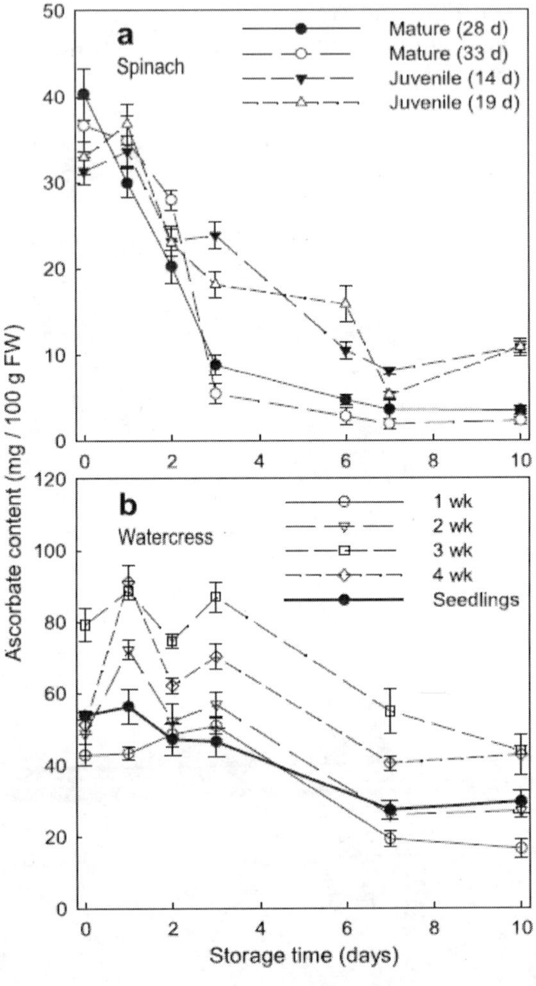

Fig. 3. Ascorbate loss in different growth stages of spinach and watercress. Plant material tested was: (a) Four batches of spinach (cultivar Toucan) were simultaneously harvested, at different growth stages: two relatively late (harvested 28 and 33 d after field sowing, at the 'vegetable-leaf' stage; described here as 'mature' spinach) and two relatively early (harvested 14 and 19 d after sowing, at the 'baby' salad-leaf stage; described here as 'juvenile' spinach). (b) Watercress seedlings (7–10 days after sowing), and plants harvested 1–4 weeks after these seedlings had been planted out. In each case, detached leaved were stored in the dark at 4 °C and assayed for ascorbate content after 0–10 d. Each data-point represents the mean of three leaf samples ± SE.

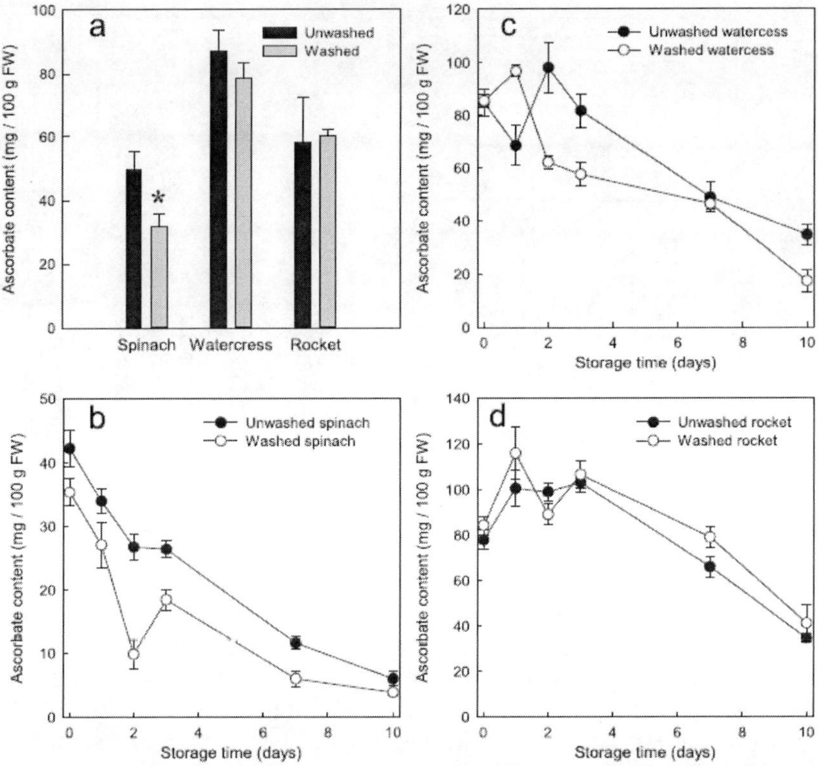

Fig. 4. Ascorbate content of salad leaves before and after washing and/or subsequent storage. (a) Salad leaves (spinach, watercress and wild rocket) from the same harvest batch were assayed for ascorbate before and immediately after the washing process. Data show the mean ± SE (n = 6); * = significant effect of washing (Student's *t*-test; $p < 0.05$, n = 6). (b)–(d) Samples of washed and unwashed spinach (b), watercress (c) and rocket (d) leaves were then stored in the dark at 4 °C in unsealed plastic packaging (standard commercial barrier salad film consisting of microperforated 30-µm orientated polypropylene, 80 g fresh weight per 25 × 20-cm bag) for up to 10 days. The ascorbate content was assayed at intervals. Data are mean ± SE (n = 3).

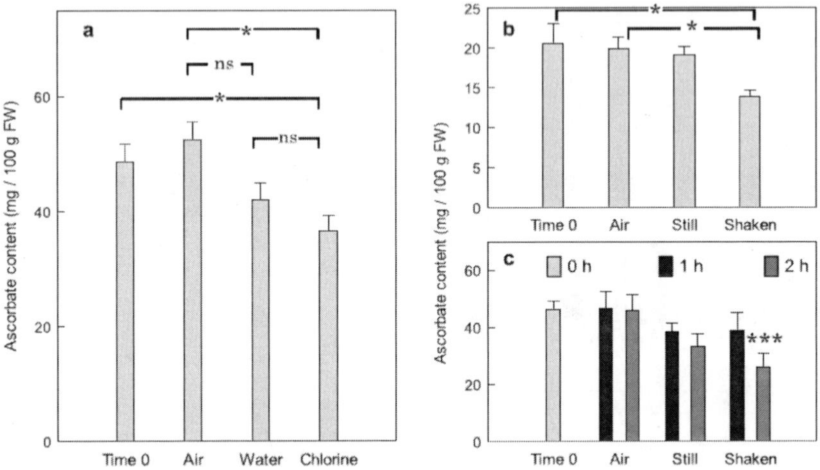

Fig. 5. Effect of chlorine and mechanical agitation on ascorbate loss during washing of spinach. (a) Effect of chlorine. Spinach leaves (1 g per sample, from a local supermarket) were incubated for 1 h either in air, or submerged in 20 ml H_2O (water), or submerged in 20 ml chlorinated H_2O (100 ppm active chlorine). The samples incubated in pure and chlorinated water were incubated shaken on a mini orbital shaker (150 rpm, with a 2-mm orbit) for 1 h. Ascorbate was determined initially (time 0) and after the 1-h incubation. Data show the mean (of 12 replicate leaf samples) ± SE. Statistically significant differences between conditions are indicated (one-way ANOVA, $n = 12$): $^{*}p < 0.05$; ns, difference not significant ($p > 0.05$). (b, c) Effect of mechanical agitation on leaves and leaf-discs. Whole spinach leaves (1*g* fresh weight) (b) or 1-cm-diameter leaf discs (250 mg fresh weight) (c) were incubated in 60-ml plastic beakers at 7 °C in the dark, either in air or submerged in 5 ml H_2O (still water) or submerged in 5 ml H_2O and shaken on a mini orbital shaker (150 rpm with an orbit of 2 mm) (shaken water) for 1 or 2 h, and then assayed for ascorbate. Data show the mean from three separate beakers ± SE. Significant differences from the corresponding time-0 sample are indicated (one-way ANOVA, $^{*}p < 0.05$, $^{***}p < 0.001$, $n = 3$).

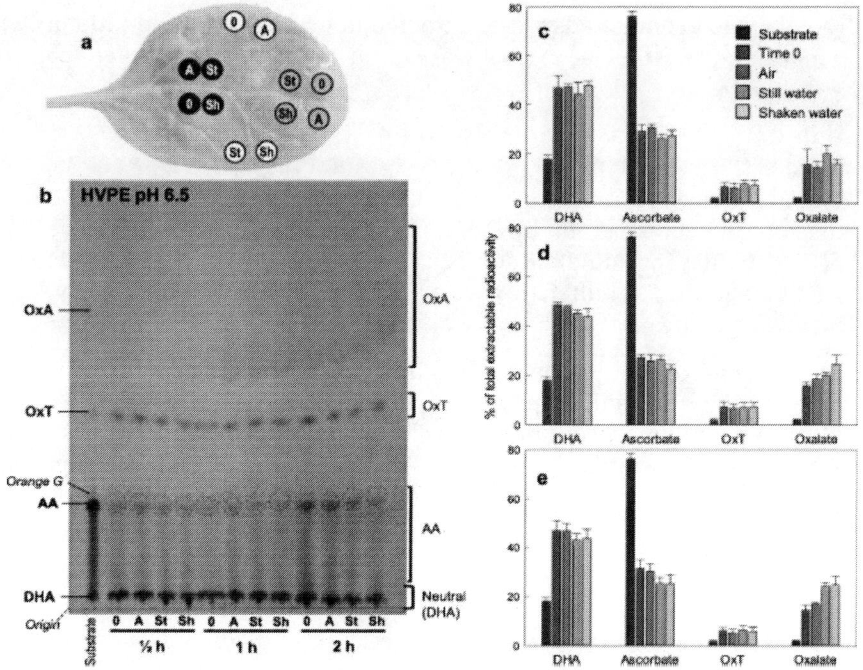

Fig. 6. The degradation of [^{14}C]AA in spinach leaf discs during washing. Three individual excised spinach leaves were fed [^{14}C]AA (50 µl; approximately 0.5 mM) through the cut petiole. Twelve discs were then punched from each lamina as shown in (a) (black, proximal; grey, distal; white, lateral), and treated for 0.5, 1 or 2 h at 7 °C in the dark in 20-ml plastic vials containing only air (A), 5 ml of still water (St), or 5 ml of shaken water (Sh). Time 0 is represented with '0'. Extracts from each of the 36 leaf discs were then run by HVPE at pH 6.5 (12 samples and a substrate-only control per sheet), and the papers was autoradiographed; a representative set of 12 is shown in (b). The radioactivity present in each spot was also quantified by scintillation counting (c–e, for 0.5, 1 and 2 h incubation respectively). Quantification of ascorbate includes the streak (visible in b), which comprises (electrophoretically mobile) AA that oxidised to (immobile) DHA during the electrophoresis. The bars in c–e each represent the mean of three individual leaf disc extracts ± SE; to avoid bias, the three replicate discs used for each average came from all three leaves and represented all three zones — for example, leaf-1 distal, leaf-2 proximal and leaf-3 lateral.

Five replicate sets (of the four treatments) of leaf discs were prepared for each of the three time-points (0.5, 1 and 2 h). After the appropriate incubation time the leaf disc was removed from the vial and the radiolabelled compounds were extracted in 0.5% formic acid (200 µl) with a mortar and pestle. The extract was stored at −80 °C until further analysis.

2.5. Analysis of radiolabelled extracts

The extracts were centrifuged at 2000*g* for 5 min, and samples (50 µl) of the supernatant were run by HVPE (high-voltage paper electrophoresis) in pH 6.5 buffer (pyridine/ acetic acid/ H_2O, 33:1:300 v/v/v) for 30 min at 2.5 kV or in pH 2.0 buffer (formic acid/acetic acid/ H_2O, 1:4:45, v/v/v) for 50 min at 2.5 kV.

The presence of [^{14}C]ascorbate and any degradation products formed was detected by autoradiography (3 week exposure to Kodak BioMax MR-1 film) or scintillation counting. Samples dried onto paper were assayed in 2 ml ScintiSafe scintillation fluid in a Beckman LS 6500 CE multi-purpose scintillation counter.

2.6. Statistical analyses

Statistical significance ($p > 0.05$) of the ascorbate content of salad leaves was assessed using Student's *t*-test or one-way analysis of variance (ANOVA) with a post-hoc Tukey test to determine the differences between treatments.

3. RESULTS

3.1. Ascorbate retention during salad shelf-life

The aim of this work was to determine changes in the concentration of ascorbate during the washing and postharvest storage of a range of packaged salad leaves. The mean initial ascorbate concentration among the eleven salad types surveyed was 80 mg/100 g fresh weight (~4.7 mM in total sap), but there was an 11-fold difference in initial ascorbate concentration between the highest (fennel) and lowest (Green Batavia lettuce) (Fig. 2, histograms). Regardless of its initial concentration, endogenous ascorbate proved unstable during 10 days' storage at 4 °C in the dark. All types of salad leaf tested showed a significant decrease in ascorbate over the 10 days, apart from wild rocket and mizuna; however, even these showed a trend towards ascorbate loss, albeit not statistically significant (Fig. 2). The mean ascorbate loss during the 10-d period was 59%, the extreme being an 86% loss in baby iceberg lettuce. There was no significant correlation between initial ascorbate content and the proportion lost during leaf storage ($r^2 = 0.142$, n = 11; $p > 0.1$; Fig. S1).

The decline in visual quality during storage varied greatly between the different types of salad leaf. For example, fennel and pea shoots showed very little change in colour and negligible bruising (Fig. 2), and these two species were among the highest in ascorbate content. However, two other high-ascorbate species, watercress and 'wasabi' (white wall rocket; *Diplotaxis erucoides*), did show dramatic changes in colour, with obvious yellowing occurring during storage, so there was no consistent correlation between ascorbate content and visible deterioration. Equally, the other members of the Brassicaceae showed clear signs of bruising during storage, and the Amaranthaceae species show discolouration and slight bruising (Fig. 2).

3.2. Pre-harvest factors affecting ascorbate retention

Pre-harvest factors, such as growth stage, have been reported to affect the ascorbate content of salad leaves (Bergquist et al., 2006). Therefore, of the eleven salad types surveyed in Fig. 2, we selected two species to investigate this aspect. In addition, we determined the kinetics of ascorbate loss during storage rather than only the 10-d end result. The ascorbate content of excised spinach leaves (two batches harvested at juvenile growth stage and two at mature growth stage) and watercress leaves (from seedlings and from developing plants harvested 1–4 weeks after the seedlings had been planted out) was monitored throughout 10 days' storage.

The initial ascorbate concentration of the spinach samples was 35 ± 4 mg/100 g FW (mean ± SD of the four batches representing two growth stages; n = 4), corresponding to ~2.0 mM in total sap (Fig. 3a). The loss of ascorbate occurred more quickly and more extensively in mature- than in juvenile-stage spinach. After 3 d of storage, mature spinach had lost about 80% of its ascorbate, whereas juvenile spinach had lost only 25–45%. Even after 10 d, juvenile spinach leaves still contained 30% of the initial ascorbate, whereas the mature spinach had only 10% remaining (Fig. 3a). The visual quality of juvenile spinach leaves after 10 d of storage was also better than in mature spinach, which showed more bruising and discolouration (Fig. S2a, b). Thus, ascorbate loss (but not initial ascorbate concentration) correlated with visual deterioration.

Commercially, watercress is grown from seedlings planted in water-beds, and is generally harvested three weeks after planting out. The initial ascorbate concentration of the watercress samples was 1.2–2.2× higher than in spinach (Fig. 3b). Watercress plants harvested 3 and 4 weeks after planting out had a higher ascorbate concentration than the younger samples. At all ages the ascorbate content remained fairly stable for the first 3 days of storage.

Thereafter, all samples gradually lost ascorbate at about the same absolute rate (~7 mg/100 g FW per day), so the younger samples lost a higher percentage of their initial ascorbate per day. The data suggest that the current practice of harvesting watercress at three weeks after planting is the optimum in terms of ascorbate content.

In watercress, the final ascorbate content after 10 d storage was 58 (±16) percent (mean ± SD) of the initial content, whereas in spinach it was ~34% and only ~7% for juvenile and mature plants respectively (Fig. 3). Thus watercress contained more ascorbate, and lost less, than spinach. Despite this, the watercress suffered visual deterioration, especially in terms of chlorophyll loss (Fig. S2b). The mature-stage spinach showed serious bruising after 10 days' storage, whereas the juvenile-stage spinach, with its greater ascorbate retention, resisted bruising (Fig. S2a).

3.3. Post-harvest processes affecting ascorbate retention

The salad washing process at Vitacress involves a counter-current flow system with leaves loaded at the water-outflow end of a wash-tank and exiting at the fresh spring-water inlet end. The leaves are submerged in the water, and undergo air-generated turbulence to remove particles of soil and grit. The leaves are then spun surface-dry, before being packaged in polypropylene and distributed to supermarkets. We therefore tested whether this washing process could potentially be causing a loss of ascorbate.

Spinach, watercress and wild rocket were assayed for ascorbate before and after the washing process (Fig. 4a). Spinach leaves showed a significant decrease (Student's *t*-test, $p < 0.05$, $n = 6$) in ascorbate after commercial washing, whereas ascorbate in watercress and wild rocket showed no significant change (Fig. 4a).

The washed and unwashed samples were stored for up to 10 days in plastic packaging in the dark, and assayed for ascorbate at intervals. The unwashed spinach showed consistently higher ascorbate levels than the washed spinach (Fig. 4b). Interestingly, as seen in Fig. S2, ascorbate in spinach began to decrease (at ~4 mg/100 g FW per day) immediately upon storage, whereas in watercress (Fig. 4c) it remained fairly stable for 3 days, before beginning to decline. Wild rocket (Fig. 4d) resembled watercress in this respect.

It is common practice in the pre-packaged salad industry to wash leaves in chlorinated water (typically 20–60 ppm active chlorine). However, at

Vitacress Salads Ltd, the leaves are washed in single pass spring-water only. As chlorine is an oxidising agent, it is possible that washing leaves in the presence of chlorine would lead to a loss of ascorbate. Various sanitizers have been compared (Gómez-López et al., 2013), but pure water has rarely been compared with these. The effect of washing in chlorinated water on the ascorbate content of spinach compared with washing in pure water was therefore investigated (Fig. 5a). Spinach leaves shaken in chlorinated water showed a significant ($p < 0.05$, one-way ANOVA) decrease in ascorbate content compared with those incubated, stationary, for an equivalent time in air (Fig. 5a). Leaves shaken in chlorinated water appeared to suffer a larger and more significant ascorbate loss than those in pure water; however, the difference between chlorinated and pure water was not significant.

We further investigated the loss of ascorbate in spinach leaves to determine where in the washing process the ascorbate was predominantly lost. Spinach leaves submerged in turbulent water showed a significant ($p < 0.05$, one-way ANOVA) loss of ascorbate, compared with leaves at time 0 and those incubated in air (Fig. 5b). That there was no significant difference between time 0 and leaves incubated in still water suggests that mechanical stress causes a loss of ascorbate. This difference was replicated with spinach leaf discs, although the ascorbate was more readily lost in discs than in whole leaves (Fig. 5c). This may be due to ascorbate being consumed during the wound response in the excised discs, the preparation of which would be likely to elicit some wounding.

3.4. Ascorbate degradation pathways

A loss of ascorbate during washing has been previously reported in spinach (Gómez-López et al., 2013), but the fate of this lost ascorbate had not been investigated. The degradation pathways of ascorbate have not been fully elucidated, particularly not *in planta*. The first relatively stable oxidation product of ascorbate, DHA, can undergo either further oxidation followed by various hydrolytic steps (Fig. 1) or direct hydrolysis (to diketogulonate and its downstream products, not shown in Fig. 1; Green and Fry, 2005; Kärkönen *et al.*, 2017 ; Parsons *et al.*, 2011) — in either case producing irreversible degradation products. Identifying the products formed from DHA in washed leaves would enable us to determine which degradation pathway(s) the vitamin C had taken, and thus potentially provide information about the nature of the stress causing vitamin loss during washing.

We monitored the fate of ascorbate during the washing process in spinach leaves by using radiolabelled ascorbate. The nature of the experiment required the use of leaf discs rather than whole leaves. Discs were taken from a spinach leaf that had been incubated with [^{14}C]ascorbate. HVPE analysis of extracts of the discs after various severities of washing revealed oxalate to be the major ascorbate degradation product (Fig. 6). Spinach is well known to accumulate relatively high levels of oxalate (Taylor and Curhan, 2007 ; Yang and Loewus, 1975) and our finding confirms the role of ascorbate as a precursor to oxalate in spinach leaves. The yield of [^{14}C]oxalate increased over time, correlating with a loss of ascorbate and DHA. The samples incubated for 2 h showed the greatest increase in [^{14}C]oxalate (Fig. 6d). This increase in [^{14}C]oxalate was greater in the shaken samples than in the still samples at 1 h, but fairly similar at both 0.5 h and 2 h (Fig. 6). This suggests that the difference in oxalate accumulation between the still and shaken samples was minor. [^{14}C]DHA exceeded [^{14}C]AA in all extracts (Fig. 6c–e): this of itself would not alter the nutritional value of the leaves since both DHA and AA serve as vitamin C in humans, but any DHA accumulated in the leaves may be prone to irreversible loss as oxalates and DKG. Equally, the difference between ascorbate loss in leaf discs (Fig. 5c) was minimal between still and shaken samples.

Some OxT, previously reported to be a major product of DHA oxidation (Green and Fry, 2005 ; Parsons *et al.*, 2011) was also formed in the spinach leaf discs. The content of OxT did not alter significantly over time or with the severity of the washing procedure (Fig. 6b–d). The presence of OxT is notable as this is the first demonstration of OxT formation in planta, the compound previously being identified in vitro and in rose cell-suspension culture (Green & Fry, 2005). The presence of oxalate and OxT shows that the DHA has undergone oxidation. This in turn suggests that oxidative stress is causing the loss of ascorbate during washing.

The mechanical stress, caused by turbulent water, could lead to cellular wounding, which may lead to the production of various reactive oxygen species; ascorbate could then be consumed during the quenching of these reactive oxygen species, in its role as an antioxidant.

In order to avoid the loss of ascorbate during the commercial washing of pre-packaged salads, specifically spinach, it would be recommended that mechanical agitation should be minimised.

4. DISCUSSION

The post-harvest storage of salad leaves has been fairly extensively studied in terms of nutritional factors such as ascorbate, but the loss of ascorbate during the washing process itself has been much less thoroughly characterised.

All the leaves tested in this study showed some degree of ascorbate loss during post-harvest storage. The 10-day storage period used here is slightly longer than the recommended storage time (the 'best before end' date is generally 5–7 days after packaging, rather than 10 days), but this time period is often used to study the effects of post-harvest storage time (Wagstaff et al., 2007).

Spinach leaves were found to be more susceptible to the loss of ascorbate during post-harvest processing than watercress or wild rocket leaves. Although cutting has been shown to negatively impact the ascorbate content of spinach (Cocetta *et al.*, 2014 ; Lee and Kader, 2000), washing in various sanitisers has had inconclusive results (Cefola and Pace, 2015 ; Karaca and Velioglu, 2014). Many studies into post-harvest processing of spinach have involved the influence of storage temperature and storage time but the packing process had not been investigated in further detail for effects on ascorbate retention.

Our experiments showed that the mechanical stress experienced by spinach leaves during washing led to a significant loss of ascorbate, whereas mere submersion of the leaves did not significantly affect the ascorbate content. Further investigation demonstrated that ascorbate undergoes oxidation during spinach washing, producing oxalate. Spinach leaves are known to accumulate oxalate (Yang & Loewus, 1975), and oxalate can act as an anti-nutrient, as it insolubilises calcium, potentially leading to calcium deficiency and to the formation of kidney stones (Franceschi and Nakata, 2005; Libert and Franceschi, 1987 ; Noonan and Savage, 1999).

The oxidative stress within leaves is known to be increased by mechanical stress and by processes such as cutting prior to the washing procedure (Cocetta et al., 2014). The use of leaf discs rather than whole leaves in some of our experiments could have affected the oxidation state of the samples, potentially with some oxidation being due to a wound response, as well as to the washing. Interestingly, wounding was demonstrated to increase ascorbate levels in *Arabidopsis* but decrease levels in tomato leaves; this decrease in ascorbate was attributed to an increase in oxidation by H_2O_2 (Suza et al., 2010). Oxalate is known to be produced from ascorbate during the reaction with H_2O_2, although in vitro oxalate was a fairly minor product, generated in smaller amounts than OxT (Green and Fry, 2005 ; Parsons *et al.*, 2011), whereas it was the major product as observed in spinach leaves (Fig. 6). This difference is likely to be due to the presence of oxalyl esterase activity in vivo, rapidly converting initially formed OxT to oxalate (Green & Fry, 2005).

Pre-harvest factors, such as the growth stage at harvest can also affect the ascorbate content of salad leaves (Lee & Kader, 2000). This was verified in the current study, in which spinach leaves harvested at earlier growth stages showed an increased retention of ascorbate during a 10-day post-harvest storage period. Younger leaves are known to have higher ascorbate concentrations, reflecting the role of ascorbate in plant growth (Bergquist et al., 2006). Although the initial concentrations of ascorbate did not differ between juvenile and mature spinach leaves, it is possible that the younger leaves were more efficient at recycling ascorbate, and thereby less prone to irreversibly losing it.

The washing process used by Vitacress Ltd is unusual in the pre-packaged salad business as it does not use any chlorine, ozone or other anti-microbial agent; instead the leaves are washed only in single pass spring water. Ozone and chlorine have previously been shown to negatively affect the ascorbate content of salad leaves, including rocket and iceberg lettuce (Karaca and Velioglu, 2014; Kenny and O'Beirne, 2009 ; Martínez-Sánchez *et al.*, 2006). Conversely, it was reported that there was no difference in the ascorbate content of cabbage when washed in the presence of ozone or chlorine compared with pure water (da Silva et al., 2015). The literature regarding the effect of chlorine washing on spinach is not conclusive. In contrast to results reported by Karaca and Velioglu (2014), the current study shows a significant loss of ascorbate when spinach is washed in the presence of chlorine rather than in pure water.

It has recently been suggested that the levels of chlorate residue on post-harvest processed salad leaves should be minimised (Gil, Marín, Andujar, & Allende, 2016). This could be done by using more dilute chlorine solutions or using an alternative sanitiser. Alternative substances have been investigated for their efficacy as suitable washing treatments for pre-packaged salad. Citric acid was shown to preserve vitamin C content of lamb's lettuce more effectively than exogenous ascorbate treatment (Cocetta, Francini, Trivelline, & Ferrante, 2016). Equally oxalic acid was shown to improve the retention of antioxidant activity in both rocket and spinach during postharvest storage when compared to sodium hypochlorite treatment, however a water-only treatment (as in the current study) was not included (Cefola & Pace, 2015). Spinach leaves washed in calcium chloride showed greater vitamin C loss than control leaves or leaves washed in calcium lactate (Oliveira, Amaro, de Sain, & Pintado, 2016).

Although OxT concentrations did not alter with the different washing treatments, the presence of OxT is still noteworthy, as this is the first demonstration of its presence in spinach, though it has recently been detected in tomato (Truffault et al., 2017). OxT had previously been found to be produced from radiolabelled ascorbate in rose cell-suspension culture (Green and Fry, 2005 ; Parsons *et al.*, 2011).

5. CONCLUSION

Spinach leaves are more susceptible to loss of ascorbate during the washing process than wild rocket or watercress. The loss of ascorbate would seem to be primarily due to mechanical stress, which manifests as oxidative stress, causing the oxidation of ascorbate. This ultimately leads to an increase in oxalate, an anti-nutrient. These findings are of commercial importance as changes to washing procedures could result in greater retention of ascorbate during postharvest storage.

CONFLICT OF INTERESTS

The authors declare that there is no conflict of interest regarding the publication of this paper.

ACKNOWLEDGEMENTS

RAD thanks the UK Biotechnology and Biological Sciences Research Council (BBSRC) for a CASE studentship (grant BB/I015531/1).

REFERENCES

1. Airianah, O. B., Vreeburg, R. A. M., & Fry, S. C. (2016). Pectic polysaccharides are attacked by hydroxyl radicals in ripening fruit: Evidence from a fluorescent fingerprinting method. Annals of Botany, 117(3), 441–455.
2. Ames, B. N., Shigenaga, M. K., & Hagen, T. M. (1993). Oxidants, antioxidants, and the degenerative diseases of aging. Proceedings of the National Academy of Sciences of the United States of America, 90(17), 7915–7922.
3. Barry-Ryan, C., & O'Beirne, D. (1999). Ascorbic acid retention in shredded iceberg lettuce as affected by minimal processing. Journal of Food Science, 64(3), 498–500.
4. Bergquist, S. Å., Gertsson, U. E., & Olsson, M. E. (2006). Influence of growth stage and postharvest storage on ascorbic acid and carotenoid content and visual quality of baby spinach (Spinacia oleracea L.). Journal of the Science of Food and Agriculture, 86(3), 346–355.
5. Cefola, M., & Pace, B. (2015). Application of oxalic acid to preserve the overall quality of rocket and baby spinach leaves during storage. Journal of Food Processing and Preservation, 39(6), 2523–2532.
6. Cocetta, G., Baldassarre, V., Spinardi, A., & Ferrante, A. (2014). Effect of cutting on ascorbic acid oxidation and recycling in fresh-cut baby spinach (Spinacia oleracea L.) leaves. Postharvest Biology and Technology, 88, 8–16.

7. Cocetta, G., Francini, A., Trivelline, A., & Ferrante, A. (2016). Effect of washing treatments on chlorophyll a fluorescence and Vitamin C content in minimally processed lamb's lettuce during storage. Agrochimica, 60(1), 1–14.
8. da Silva, J. P., Costa, S. M., de Oliveira, L. M., Vieira, M. C. S., Vianello, F., & Lima, G. P. P. (2015). Does the use of ozonized water influence the chemical characteristics of organic cabbage (Brassica oleracea var. capitata)? Journal of Food Science and Technology, 52(11), 7026–7036.
9. DeBolt, S., Hardie, J., Tyerman, S., & Ford, C. M. (2004). Composition and synthesis of raphide crystals and druse crystals in berries of Vitis vinifera L. cv. Cabernet Sauvignon: Ascorbic acid as precursor for both oxalic and tartaric acids as revealed by radiolabelling studies. Australian Journal of Grape and Wine Research, 10(2), 134–142.
10. Foyer, C. H., & Halliwell, B. (1977). Purification and properties of dehydroascorbate reductase from spinach leaves. Phytochemistry, 16(9), 1347–1350.
11. Franceschi, V. R., & Nakata, P. A. (2005). Calcium oxalate in plants: Formation and function. Annual Review of Plant Biology, 56(1), 41–71.
12. Fry, S. C. (1998). Oxidative scission of plant cell wall polysaccharides by ascorbateinduced hydroxyl radicals. Biochemical Journal, 332, 507–515.
13. Gallie, D. R. (2013). L-Ascorbic acid: A multifunctional molecule supporting plant growth and development. Scientifica, 2013, 795964.
14. Gil, M. I., Marín, A., Andujar, S., & Allende, A. (2016). Should chlorate residues be of concern in fresh-cut salads? Food Control, 60, 416–421.
15. Gómez-López, V. M., Marín, A., Medina-Martínez, M. S., Gil, M. I., & Allende, A. (2013). Generation of trihalomethanes with chlorine-based sanitizers and impact on microbial, nutritional and sensory quality of baby spinach. Postharvest Biology and Technology, 85, 210–217.
16. Green, M. A., & Fry, S. C. (2005). Vitamin C degradation in plant cells via enzymatic hydrolysis of 4-O-oxalyl-L-threonate. Nature, 433(7021), 83–87.
17. Hancock, R. D., & Viola, R. (2005). Improving the nutritional value of crops through enhancement of L-ascorbic acid (vitamin C) content: Rationale and biotechnological opportunities. Journal of Agricultural and Food Chemistry, 53 (13), 5248–5257.
18. Hodges, D. M., & Forney, C. F. (2003). Postharvest ascorbate metabolism in two cultivars of spinach differing in their senescence rates. Journal of the American Society for Horticultural Science, 128(6), 930–935.
19. Huang, W., Wang, G. L., Li, H., Wang, F., Xu, Z. S., & Xiong, A. S. (2016). Transcriptional profiling of genes involved in ascorbic acid biosynthesis, recycling, and degradation during three leaf developmental stages in celery. Molecular Genetics and Genomics, 291(6), 2131–2143.
20. Karaca, H., & Velioglu, Y. S. (2014). Effects of ozone treatments on microbial quality and some chemical properties of lettuce, spinach, and parsley. Postharvest Biology and Technology, 88, 46–53.
21. Kärkönen, A., Dewhirst, R. A., Mackay, C. L., & Fry, S. C. (2017). Metabolites of 2,3-diketogulonate delay peroxidase action and induce non-enzymic H2O2 generation: Potential roles in the plant cell wall. Archives of Biochemistry and Biophysics, 620, 12–22.
22. Kenny, O., & O'Beirne, D. (2009). The effects of washing treatment on antioxidant retention in ready-to-use iceberg lettuce. International Journal of Food Science & Technology, 44(6), 1146–1156.
23. Koh, E., Charoenprasert, S., & Mitchell, A. E. (2012). Effect of organic and conventional cropping systems on ascorbic acid, vitamin C, flavonoids, nitrate, and oxalate in 27 varieties of spinach (Spinacia oleracea L.). Journal of Agricultural and Food Chemistry, 60(12), 3144–3150.
24. Lee, S. K., & Kader, A. A. (2000). Preharvest and postharvest factors influencing vitamin C content of horticultural crops. Postharvest Biology and Technology, 20 (3), 207–220.

25. Libert, B., & Franceschi, V. R. (1987). Oxalate in crop plants. Journal of Agricultural and Food Chemistry, 35(6), 926–938.
26. Mandl, J., Szarka, A., & Bánhegyi, G. (2009). Vitamin C: Update on physiology and pharmacology. British Journal of Pharmacology, 157(7), 1097–1110.
27. Martínez-Sánchez, A., Allende, A., Bennett, R. N., Ferreres, F., & Gil, M. I. (2006). Microbial, nutritional and sensory quality of rocket leaves as affected by different sanitizers. Postharvest Biology and Technology, 42(1), 86–97.
28. McCance, R., & Widdowson, E. (1991). McCance and Widdowson's the composition of food (5th ed.). London: Royal Society of Chemistry (Great Britain) and Ministry of Agriculture, Fisheries and Food (5th revise).
29. Noctor, G., & Foyer, C. H. (1998). Ascorbate and glutathione: Keeping active oxygen under control. Annual Review of Plant Physiology and Plant Molecular Biology, 49 (1), 249–279.
30. Noonan, S. C., & Savage, G. P. (1999). Oxalate content of foods and its effect on humans. Asia Pacific journal of clinical nutrition, 8(1), 64–74.
31. Oliveira, A. L. S., Amaro, A. L., de Sain, J., & Pintado, M. (2016). Impact of different calcium dips and solution pH on quality of ready-to-eat baby-leaf spinach. Postharvest Biology and Technology, 121, 36–42.
32. Parsons, H. T., Yasmin, T., & Fry, S. C. (2011). Alternative pathways of dehydroascorbic acid degradation in vitro and in plant cell cultures: Novel insights into vitamin C catabolism. The Biochemical Journal, 440(3), 375–383.
33. Ponting, J. D. (1943). Extraction of ascorbic acid from plant materials. Industrial and Engineering Chemistry, 15(6), 389–391.
34. Ren, J., Chen, Z., Duan, W., Song, X., Liu, T., Wang, J., ... Li, Y. (2013). Comparison of ascorbic acid biosynthesis in different tissues of three non-heading Chinese cabbage cultivars. Plant Physiology and Biochemistry, 73, 229–236.
35. Smirnoff, N., Wheeler, G. L., & Loewus, F. A. (2000). Ascorbic acid in plants: Biosynthesis and function. Critical Reviews in Biochemistry and Molecular Biology, 354(354), 291–314.
36. Suza, W. P., Avila, C. A., Carruthers, K., Kulkarni, S., Goggin, F. L., & Lorence, A. (2010). Exploring the impact of wounding and jasmonates on ascorbate metabolism. Plant Physiology and Biochemistry, 48(5), 337–350.
37. Taylor, E. N., & Curhan, G. C. (2007). Oxalate intake and the risk for nephrolithiasis. Journal of the American Society of Nephrology, 18(7), 2198–2204.
38. Truffault, V., Fry, S. C., Stevens, R. G., & Gautier, H. (2017). Ascorbate degradation in tomato leads to accumulation of oxalate, threonate and oxalyl threonate. The Plant Journal, 89(5), 996–1008.
39. Wagstaff, C., Clarkson, G. J. J., Rothwell, S. D., Page, A., Taylor, G., & Dixon, M. S. (2007). Characterisation of cell death in bagged baby salad leaves. Postharvest Biology and Technology, 46(2), 150–159.
40. Yang, J. C., & Loewus, F. A. (1975). Metabolic conversion of L-ascorbic acid to oxalic acid in oxalate-accumulating plants. Plant Physiology, 56(2), 283–285.

CHAPTER 6

Copper and Mercury in Food, Biological and Pharmaceutical Samples: Spectrophotometric Estimation as $Cu(DDTC)_2$

Mohammad Nasir Uddin*, Nur Mostaq Shah, Mohammad Amzad Hossain, Md. Muzahidul Islam

Department of Chemistry, University of Chittagong, Chittagong, Bangladesh

ABSTRACT

An alternative spectrophotometric method was optimized and validated for the estimation of mercury using diethyldithiocarbamate (DDTC), a common reagent, widely used for the preconcentration and isolation of metal ions in complex matrices followed by their estimation by varied techniques. Diethyldithiocarbamate forms yellow $Cu(DDTC)_2$ with copper and white $Hg(DDTC)_2$ with mercury (having d^{10} system) which are extracted in CCl_4. The UV-visible spectrum of $Cu(DDTC)_2$ is very stable at pH 5.0 and has a maximum absorption (λ_{max}) at 435 nm. $Hg(DDTC)_2$ is more stable than $Cu(DDTC)_2$. Estimation of mercury is based on a quantitative displacement of Cu(II) of $Cu(DDTC)_2$ with the addition of mercury followed by the measurement of reduced absorbance. Primarily, method was optimized and validated for the estimation of copper. Therefore, simultaneous determination of Cu(II) and Hg(II) in mixture is proposed fractionating the extract. The molar specific coefficient (ε) for the mercury was 1.4×10^4 $mol^{-1}\cdot L\cdot cm^{-1}$ and for copper was 3.16×10^5 $mol^{-1}\cdot L\cdot cm^{-1}$ at 435 nm. The detection limits of Cu^{2+} and Hg^{2+} were 0.023 μg·mL^{-1} and 0.029 μg·mL^{-1}, respectively. The calibration curve shows good linearity of 0.02 - 12.0 and 0.02 - 15.0 μg·mL^{-1} for the Cu^{2+} and Hg^{2+} determination, respectively. Proposed technique was applied to food, biological and pharmaceutical samples for the determination of Cu(II) and Hg(II).

Keywords: Copper, Mercury, Diethyldithiocarbamate, Spectrophotometric Method, Food, Biological, Pharmaceutical Samples

1. INTRODUCTION

A variety of metals could enter industrial wastewaters as a result of anthropogenic activity. Due to corrosion and geological factors some metal ions in complex form being environmental contaminants could also be present in drinking waters and natural water bodies. By the biological cycle, some of them through the food chain pass into plants, animals and man, thus affecting them negatively. Copper is one of the important and essential nutrients for human health as well as the growth of animals and plants [1] [2] . Although copper is an essential micronutrient and is required by the body in very small amounts, excess copper in the human body can cause stomach and intestinal distress such as nausea, vomiting, diarrhea, and stomach cramps. People with Wilson's disease, a rare genetic disorder, are more sensitive to the effects of copper [1] [2] . Mercury and its compounds are also used in medicine. Merbromin (Mercurochrome) is a topical antiseptic used for minor cuts and scrapes is still in use in some countries. It is used as a preservative in laboratory reagents and related chemicals. Mercury is used in dental amalgam. Mercury is a heavy metal which is much toxic for human health. It causes many harmful diseases when absorbed by human body. Mercury can be inhaled and absorbed through the skin and mucous membranes. The chief sources of mercury pollution are chlor-alkali plants, paper, pulp, cellulose and plastic industries, electrical, paint, pharmaceutical industries, etc. Uses of mercury as fungicides, pesticides, etc., also add mercury to the environment. Compounds of mercury consumed in fish, cereals and other food stuffs have resulted in numerous poisoning [1] -[3] .

Therefore, trace and ultra-trace determination of copper [4] [5] and mercury [6] [7] is of great importance. Atomic absorption spectroscopy (AAS) and inductively coupled plasma-atomic emission spectroscopy (ICP-AES) have been the method of choice for the elemental analysis because of their utility, sensitivity and reliability. Cold vapor atomic fluorescence spectrometry [8] as well as catalytic or electro-analytical [9] procedure has been more often used for mercury determination. These methods can rapidly determine metals in trace amounts in many types of matrices. Although these techniques are reliable and sensitive, they suffer from the limitations of being rather costly (considering instrument acquisition and maintenance), time-consuming (with respect to sample preparation), and not always readily available. Contrary, in laboratories of developing countries like us the choice of any analytical methods depends on the availability of reagents, cost effectiveness of instruments and the time required for analysis as well as safety and ease of operation. UV-visible spectroscopy is a well-established analytical technique with mature methods and equipment. UV-visible spectrophotometry is more sensitive technique widely used in inorganic trace analysis. It is commonly used in both research and science as well as in industry.

A number of spectrophotometric reagents as chloro (phenyl) glyoxime [3] , 2,5-dimercapto-1,3,4-thiadiazole [10] , thiosemicarbazone [11] , 3-methoxy-4-hydroxy benzaldehyde-4-bromophenyl hydrazine [12] and so on are reported for copper determination. Mercury being a non-transition element, a sensitive colourful reagent is required for the spectrophotometric determination which is rare available. A few have been reported for the mercury determination like diacetylmonoxime isonicotinoylhydrazone [13] , diphenylthiocarbazone [14] . One of the most common reagents, diethyldithiocarbamate (DDTC) is used for spectrophotometric determination of transition elements Cu(II), Ni(II), Mn(II), Pb(II) and V(V) [15] [16] . Hg(II) gives white insoluble complex, $Hg(DDTC)_2$which is more stable than the $Cu(DDTC)_2$. Primarily, a spectrophotometric method was aimed to be optimized and validated for the estimation of copper. Mercury can be determined by a quantitative displacement Cu(II) with the addition of additional mercury followed by the extraction of remaining of $Cu(DDTC)_2$ in a carbon tetrachloride. The reduced absorbance is measured subsequently which is equivalent to mercury. Finally, the aim of this research is to develop a rapid spectrophotometric method for the determination of mercury and copper with DDTC which has not previously been used for their simultaneous determination.

2. EXPERIMENTAL

2.1. Chemicals and Reagents

High-purity carbontetrachloride, various acids, salts and reagent grade Na-DDTC (Merck) were used. The standard stock solutions (1000 $\mu g \cdot mL^{-1}$)

were prepared by dissolving appropriate amount of each salt in water. Solutions of a large number of inorganic ions and complexing agents were prepared from their analytical grade or equivalent grade and water soluble salts. Doubly distilled deionized water, which is non-absorbent under ultraviolet radiation, was used throughout. Suitable portions of the solutions were mixed to get the desired pH. Stock solutions and environmental water samples (1000-mL each) were kept in polypropylene bottles containing 1-mL of concentrated HNO_3.

2.2. Synthesis of Complex, $[Cu(en)_2][Hg(SCN)_4]$

Complex, $[Cu(en)_2][Hg(SCN)_4]$, was synthesized according to the procedure described in literature [17] and was analyzed for the metal analysis as certified substance.

2.3. Instrumentation

A Shimadzu UV Visible UV-1800 spectrophotometer model with suitable settings equipped with 1-cm quartz cells was used for absorbance. The spectral band length was 1 nm, the wavelength accuracy was 0.5 nm with automatic wavelength correction and the recorder was a computer-controlled in the wavelength range 350 - 600 nm. A Jenway (England, U.K) (Model-30100) pH meter were used for the measurements of pH. A Varian (Australia) AAS spectrophotometer was used for comparing the results.

2.4. Preparation of Standards

A 0.1% (5.84×10^{-3} M) stock solution of Na-DDTC was prepared by dissolving 0.1 g sodium diethyldithiocarbamate reagent in approximately 80 ml water heated at 60°C. Afterwards, the volume was made up to 100 ml by adding distilled water in volumetric flask and filtered. A stock solution (100 $\mu g{\cdot}mL^{-1}$) of copper or mercury was prepared by dissolving appropriate amount of copper sulphate pentahydrate or mercuric chloride (Merck, Germany) in 250 mL of doubly distilled deionized water. The working standards were prepared by suitable dilutions of this stock solution. The buffer solutions were prepared by mixing 1 $mol{\cdot}L^{-1}$ HCl and 1 $mol{\cdot}L^{-1}$ sodium acetate (pH 1 - 3), 0.2 $mol{\cdot}L^{-1}$ acetic acid and 0.2 $mol{\cdot}L^{-1}$ sodium acetate (pH 3.2 - 6.0), 1 $mol{\cdot}L^{-1}$ sodium acetate and 0.2 $mol{\cdot}L^{-1}$ acetic acid (pH-7.0) and 2 $mol{\cdot}L^{-1}$ ammonium hydroxide and 2 $mol{\cdot}L^{-1}$ammonium chloride (pH 8.0 - 12.0). Suitable portions of these solutions were mixed to get the desired pH. A 100 ml stock solution of tartrate (0.25 $mol{\cdot}L^{-1}$) or ammonium thiocyanate (0.4 $mol{\cdot}L^{-1}$) was prepared by dissolving 7.055 g of A.C.S grade (99%) potassium sodium tartrate tetrahydrate or 3.0428 g of solid ammonium thiocyanate in deionized water. A 100 ml solution of dilute ammonium hydroxide was prepared by diluting 10 ml concentrated NH_4OH (28% - 30% A.C.S grade) to 100 ml with deionized water. A 100 ml stock solution of EDTA (4000 $\mu g \cdot mL^{-1}$) was prepared by dissolving 0.4 g of A.C.S grade (>90%) dehydrated disodium salt of ethylenediaminetetraacetic acid in 100 ml deionized water. Stock solutions were stored at 4°C, protected from light and were used within three months.

2.5. Biological Sample Collection

Blood and urine samples were kindly provided from Chittagong Medical College Hospital, Chittagong. Both urine and whole blood samples were collected into EDTA-anticoagulated polypropylene tubes. Blood specimens were refrigerated then aliquoted within two hours into cryovials immediately frozen and shipped on dry ice. All blood samples were stored frozen at or lower than −20°C until analyzed. Metal caps or metal-based containers, colored containers should be avoided.

2.6. General Procedure

To determine mercury and copper simultaneously, a mixture (1:1) 10 $\mu g \cdot mL^{-1}$ solution of (various volume) Hg(II) and Cu(II) was placed in a 25-mL separating flask along with 0.5 ml 0.05 M H_2SO_4,and 5 ml of the reagent 1000 $\mu g \cdot mL^{-1}$ DDTC solution. Its volume was made up 15 ml with deionized water. Its pH was adjusted to 5.0 using acetate buffer. The solid product, $Hg(DDTC)_2$and yellow $Cu(DDTC)_2$ complexes so formed, was extracted carefully with the addition of 4 ml (2 × 2) CCl_4, and organic phase was separated after vigorous stirring for 10 min. The solution was divided in two fractions. The absorbance of one fraction was measured by a spectrophotometer at 435 nm against a blank which is equivalent of copper content present. To another fraction known amount Hg(II) solution was added when pH value was adjusted to 10 using ammonium buffer. Copper from $Cu(DDTC)_2$ was replaced by mercury immediately forming equivalent amount $Hg(DDTC)_2$ complexes. Organic phase was further separated after vigorous stirring for 10 min. The absorbance was measured by a spectrophotometer at 435 nm against a blank. Reduced absorbance was equivalent to the mercury content present. Procedure has been followed for the analysis of food, biological and pharmaceutical samples.

2.7. Composition of the Complex

Job's method of continuous variation was applied to ascertain the stoichiometric composition of the complex according to the procedures described below [18] .

2.8. Effect of Diverse Ions

The effect of various anions and cations on the determination of Cu(II) and Hg(II) under optimal conditions was studied to find out the tolerance limits of these ions in the present method. The determination of Cu(II) and Hg(II) amount was carried out using the above described general procedure. The tolerance limit of a foreign ion is taken as the amount that caused an error in the absorbance value of ≤10%.

3. RESULTS AND DISCUSSION

3.1. Method Optimization

3.1.1. Absorption Spectra

The absorption spectra (Figure 1) of the reagent and the complexes are recorded in the wavelength range 350 - 600 nm at pH 5.0 against CCl_4 (standard) or reagent blank (real samples). $Cu(DDTC)_2$ complex solution has an absorption maximum at 435 nm, where as $Hg(DDTC)_2$ or the reagent (DDTC) shows no appreciable absorbance at this wavelength. It was found that the molar absorptivity (ε) and Sandal's sensitivity (for the absorbance of 0.001) at this wavelength are 1.41×10^4 $mol^{-1}\cdot L\cdot cm^{-1}$ and 4.184 $ng\cdot cm^{-2}$ respectively for the mercury determination. For the copper determination corresponding values are 3.16×10^5 $mol^{-1}\cdot L\cdot cm^{-1}$ and 3.165 $ng\cdot cm^{-2}$, respectively. Therefore, UV-vis spectrophotometric measurements were carried out at a wavelength of 435 nm for subsequent studies. It's determination in the different matrices based on the direct measurement of its absorption for ultraviolet light is not susceptible to potential interferences from the matrix excipients.

3.1.2. Effect of Acid

Acid effect was primarily tested for HCl, HNO_3, H_2SO_4 while last one was supposed to be suitable for $Cu(DDTC)_2$. The influence of acid concentration on the reaction was investigated by carrying out the reaction in varying acid values. Test at different acid concentration (0.02 - 1.0) × 10^{-2}mol·L^{-1}, H_2SO_4) for a constant concentration (1.0 μg·mL^{-1}) of Cu(II) produced a constant absorbance for (0.1 - 0.8) × 10^{-2} mol·L^{-1} H_2SO_4. Therefore, 0.1 × 10^{-2} mol·L^{-1}, H_2SO_4 was selected as optimized concentration, and all measurements have been peformed at this concentration level.

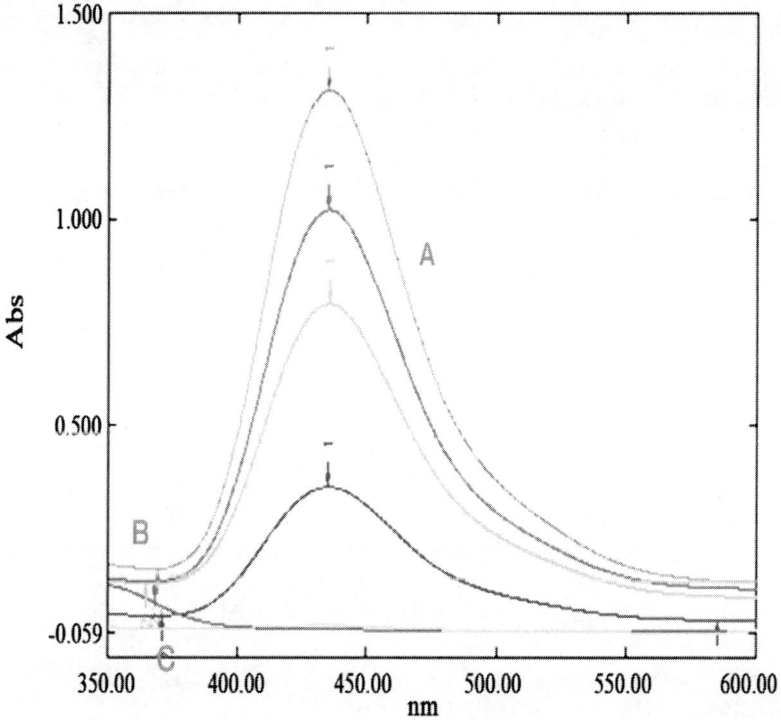

Figure 1. The typical UV-vis spectra of (A) $Cu(DDTC)_2$, gradually decreasing due to the addition of mercury at varying conc., (B) $Hg(DDTC)_2$ in CCl_4 and (C) reagent blank.

3.1.3. Effect of Reagent Concentration

Study of the effect of DDTC concentration on its reaction with copper revealed that the reaction was dependent on the DDTC concentration as the absorbance increased with the increase in the reagent concentration. The absorbance of the complex solution at 435 nm is measured according to the standard procedure at different molar excesses (1:1 - 1:40) of sodium diethyldithiocarbamate keeping Cu(II) concentration (1.0 μg·mL^{-1}) constant at optimized acid concentration. Copper metal (optical path length 1 cm) and the reagent molar ratios of 1:10 to 1:40 produced a constant absorbance. A greater excess of the reagent was not studied. Therefore, a 15 fold molar excess of DDTC was optimized for constant color development and was used in all the subsequent experiments. Excess of the reagent has no effect on the absorbance of the complex.

3.1.4. Effect of Temperature and Time

The effect of temperature on the reaction was not studied due to the lake of instrumental facilities. Therefore, further experiments were carried out at room temperature (25°C ± 2°C). In order to determine the optimum time that is required for completion the reaction, it was allowed to proceed at room temperature for varying periods of time. The reaction goes to almost completion within 1 min. However, for higher precision of readings the reaction was allowed to proceed for quite longer time. Reactions in all the subsequent experiments were carried out for 10 min.

3.1.5. Stability of the Cu(II)-DDTC Complex

The effect of time on the stability of the Cu(II)-DDTC complex was studied by following the absorption intensity of the reaction solution (after dilution) at different time intervals. The absorbance of the complex remains stable for at least 72 h. This allowed the process of large batches of samples, and their comfortable measurements with convenience. This gives the high throughput property to the proposed method when applied for analysis of large number of samples in quality control/analytical laboratories.

3.1.6. Effect of the Aqueous Phase Volume

The volume of the aqueous phase is an important factor for the extraction of metal ions as Cu(II)-DDTC is sparingly soluble in aqueous phase. The effect of volume of the aqueous phase was also studied. The solutions containing 1 $\mu g \cdot mL^{-1}$ of Cu(II) were diluted in the volume range from 5 ml to 30 ml with deionized water. The recovery of Cu(II) from aqueous phase was plotted against the aqueous phase volume. The absorbance of Cu(II) from aqueous phase decreases when the aqueous phase volume increases. Rapid decrease in absorbance was occurred for aqueous volume beyond 15 ml. Hence, total aqueous volume was always confined to maximum volume 15 ml throughout all experiments for convenient of operation with confidence.

3.1.7. Optimum Extraction Period

The efficiency of $Cu(DDTC)_2$ extraction in organic phase depends on the extraction period. Optimum extraction period was determined according to the procedure described previously by extracting copper from the solutions containing 1 μg·mL^{-1} of Cu(II) for various periods of time. The amount of Cu(II) dissolved in CCl_4 was determined on the basis of measurement of the absorbance using a UV-vis spectrophotometer. After 10 min constant absorbance was obtained and it was constant up to studied period of 30 min. Hence, reaction mixture was extracted at least 10 min for getting maximum recovery.

3.1.8. Composition of the Complex

Under the optimum conditions (Table 1), the stoichiometry of the reaction between Cu(II) and DDTC was investigated by Job's method [18]. Experimental data has been presented graphically in Figure 2. The stoichiometry was found to be 1:2 (Copper:Ligand).

3.1.9. Effect of pH on the Extraction

The effect of pH on the color intensity is studied in the pH range 1 - 10. Experimental results showed that as the pH increased the absorbance was increased rapidly. The optimum pH value for the reaction of Cu(II) and DDTC is attained at pH 4 and remains constant up to 8. At higher pH, sharp decrease in the readings occurred. The pH value should not be lower than 4 due to the fast decomposition of dithiocarbamates. On the other hand, pH value over 8 would accelerate the $Cu(OH)_2$ precipitation. Hence, pH 5.0 is chosen for further studies as convenient. Acetic acid-acetate buffer solution (pH = 5.0) was chosen for the subsequent studies.

The influence of pH on the metal replacement of $Cu(DDTC)_2$ and Hg(II) was investigated. Percentage recovery of $Hg(DDTC)_2$ complex showed the maximum at pH 10.0. $Hg(DDTC)_2$ complex was formed instantaneously by the quantitative replacement of Cu from $Cu(DDTC)_2$, and it was well dissolved in CCl_4 media at pH 10.0. It is assumed that the reaction to form this complex could have competed against hydroxide precipitation above pH 10.0 and at acidic pH, as the sulfur atom in the chelating site of DDTC has more affinity power with proton at a higher concentration of protons.

Table 1 summarizes the optimum values of optimization parameters for the proposed method.

Table 1. Summary for the optimization and Validation variables of the proposed spectrophotometric method.

Optimization Variables			Validation Parameters		
Variables	**Studied range**	**Selected**	**Parameters**	(Mercury)	(Copper)
Acid concentration/M	(0.02 - 1.0) × 10^{-2}	0.1 × 10^{-2}	Measurement wavelength (nm)	435	
Reagent molar fold excess (M:L)	1:1 - 1:40	15	Linear range ($\mu g{\cdot}mL^{-1}$)	0.02 - 15	0.02 - 12
Temperature	Ambient	Ambient	Linearity equation	Y = −0.065x + 1.140	Y = 0.203x + 0.113
Time/min	1 - 20	10	Standard deviation of the slope	0.004	0.003
Stability	1 min-72 hrs.		Correlation coefficient (r^2)	0.999	0.999
Wavelength (nm)	350 - 600	435	Relative standard deviation (%)	3.13	3.92
Extraction variables			Limit of detection, LOD ($\mu g{\cdot}mL^{-1}$)	0.031	0.024
Extraction period/min	1 - 30	10	Limit of quantification, LOQ ($\mu g{\cdot}mL^{-1}$)	0.116	0.075
Aqueous phase volume/mL	5 - 30	15	Molar absorptivity, ε ($L{\cdot}mol^{-1}{\cdot}cm^{-1}$)	1.41 × 10^4	3.16 × 10^5
pH	1 - 10	5	Sandall's sensitivity ($ng{\cdot}cm^{-2}$)	4.169	3.165
pH of Cu replacement	3 - 12	10	Precision (%) of λ_{max} (n = 10)	0.105	

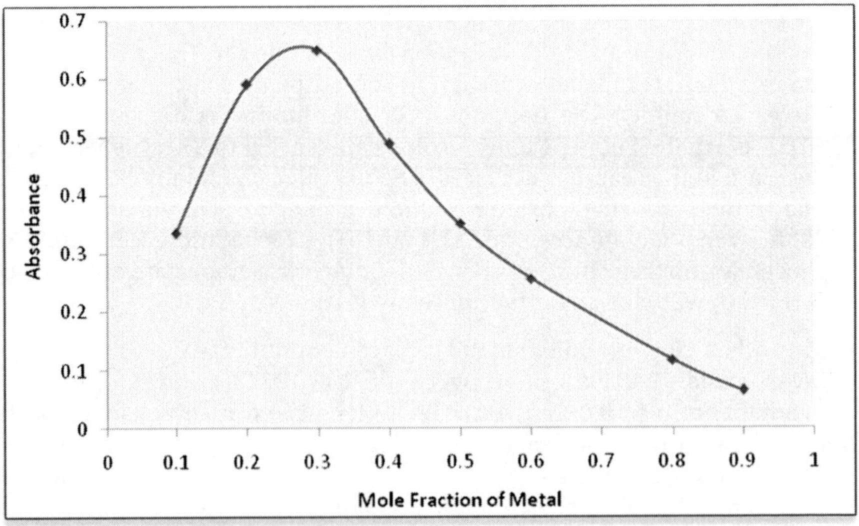

Figure 2. Job's method for determining the composition of complex.

3.2. Method Validation

3.2.1. Calibration Curves

The calibration curve was constructed by plotting absorbance against corresponding concentrations for ten standard solutions containing 0.01-12.0 μg·mL^{-1} of copper and 0.01 - 15.0 μg·mL^{-1} of mercury according to the general procedure. The calibration curves are shown in Figure 3 for copper determination and Figure 4 for mercury determination. The linearity range, regression equation and correlation coefficient were obtained by the method of least squares. The liner plot between the absorbance and the amount of Cu(II) ion is drawn, and the straight line obeyed the equation $y = 0.203x + 0.113$ for copper having regression coefficient of (r^2) 0.999. Li nearity range for copper according to the Beer's law was obtained of 0.02 - 12 μg·mL^{-1}. The liner plot between the absorbance obtained after addition of mercury and the amount of Hg(II) is drawn, and the straight line obeyed the equation $y = -0.065x + 1.140$ for mercury having regression coefficient of (r^2) 0.999. Linearity range for mercury according to the Beer's law corresponded to the Cu(II) concentrations was obtained of 0.02 - 15 μg·mL^{-1}. The analytical sensitivity, the calibration sensitivity, which is the slope of the analytical curve, the limit of detection, and the limit of quantitation as well as other analytical characteristics of the procedure are calculated and summarized in Table 1.

3.2.2. Accuracy

The accuracy of the proposed method was evaluated by the recovery studies for three different concentrations (1, 5, 8 μg·mL^{-1}) of standards. The recovery values were 94.7% - 102.0% for both intra and inter day analyses indicating the good accuracy of the proposed method. For an accuracy check of the method, certified reference substances (alloy, amalgam, synthetic compound) were analyzed. The recovery values calculated for the certified reference materials were 94.2% - 97.6% (Table 2). The results indicate that the certified and found values are very concordant.

3.2.3. Precision

The intra-assay precision of the proposed method was determined on samples of standard mixture solutions at varying concentration levels by analyzing six replicates of each sample. The inter assay precision was determined by analyzing the same samples as duplicates in three consecutive days (n = 6). The relative standard deviations (RSD) did not exceed 4.43% proving the high precision of the proposed method for the routine application in the analysis of mercury.

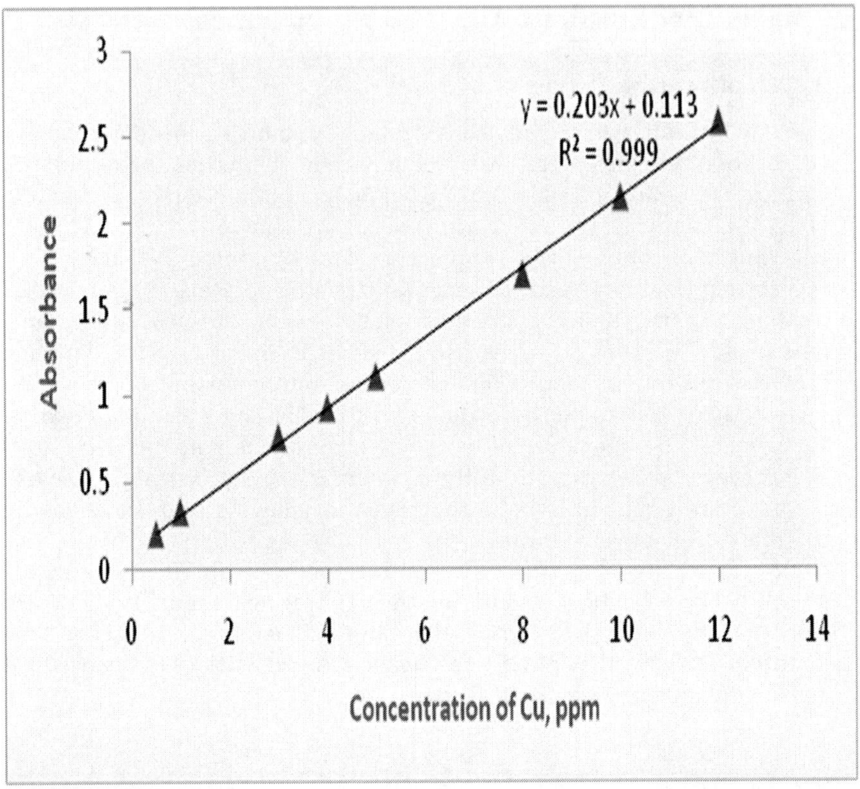

Figure 3. Calibration curve constructed by plotting absorbance against corresponding concentrations (μg·mL^{-1}) of copper.

Table 2. Analysis of certified substances and synthetic mixtures.

Certified substance	Mercury (μg·mL^{-1}), n = 3				Copper (μg·mL^{-1}), n = 3			
	Cert. values	Expt. Values	RSD (%)	R (%)	Cert. values	Expt. values	RSD (%)	R (%)
Alloy	6.20	6.17 ± 0.03	0.49	99.10				
Amalgam	4.10	4.07 ± 0.02	0.49	99.30	6.00	6.06 ± 0.02	0.33	100.90
$[Cu(en)_2][Hg(SCN)_4]$	2.00	2.04 ± 0.01	0.49	102.30	4.00	3.88 ± 0.01	0.26	97.10

Synthetic mixtures Ions	Conc. of Mercury (μg mL^{-1})	Exp. Conc. (μg mL^{-1})	% Er	RSD (%) n = 3
Hg^{2+}	3.00	2.95 ± 0.05	−2.96	1.70
Hg^{2+} + Ag^{+} + citrate		2.87 ± 0.03	−5.83	1.10
Hg^{2+} + Ag^{+} + K^{+} + citrate		2.96 ± 0.02	−2.62	0.68
Hg^{2+} + Ag^{+} + K^{+} + Sr^{2+} + citrate		2.85 ± 0.04	−5.47	1.40
Hg^{2+} + Ag^{+} + K^{+} + Sr^{2+} + Al^{3+} + citrate		3.12 ± 0.01	+3.31	0.19
Hg^{2+} + Ag^{+} + K^{+} + Sr^{2+} + Al^{3+} + Mg^{2+} + citrate		3.02 ± 0.01	+1.12	0.17

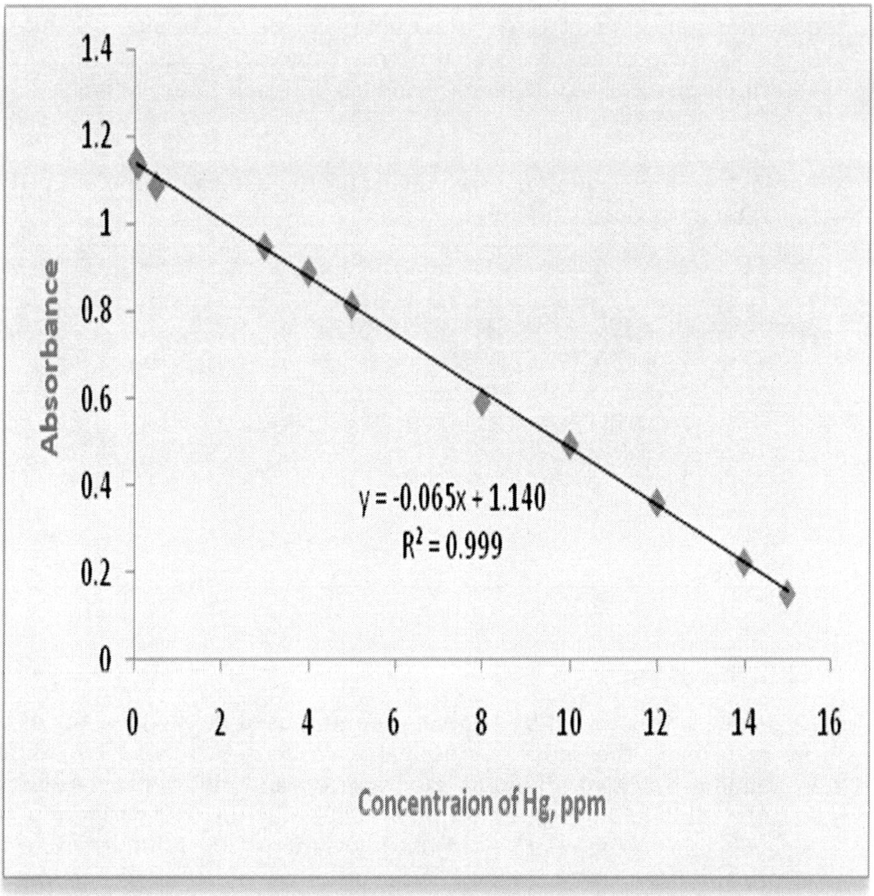

Figure 4. Calibration curve constructed by plotting absorbance against corresponding concentrations ($\mu g \cdot mL^{-1}$) of mercury.

3.2.4. Sensitivity

The analytical sensitivity, the calibration sensitivity, which is the slope of the analytical curve, the limit of detection, and the limit of quantitation as well as other analytical characteristics are calculated from the data obtained for constructing calibration curve. The calculation method is based on the standard deviation of the response (S_{xy}) and the slope of the calibration curve (a). The limit of detection were calculated from calibration graph by the formula; LOD = $3 \cdot S_{xy}/a$, and the limit of quantification; LOQ = $10 \cdot S_{xy}/a$. The standard deviation of the method for ten determinations is found to be not more than 0.066. The lower detection limit and quantification limit of Hg(II) were found to be 0.031 μg·mL^{-1} and 0.116 μg·mL^{-1}, respectively. And the corresponding values for copper determination were 0.024 μg·mL^{-1} and 0.075 μg·mL^{-1}, respectively.

Table 3 .Tolerance limit of foreign ions for the determination of Cu(II) and Hg(II) by the proposed method.

Foreign ions	Tolerance limit (μg·mL^{-1})
NaF, NaCl, NaBr, acetate, ascorbic acid tartrate, citrate, oxalate, EDTA, sulfate, chloride, nitrate, phosphate, Iodide,	2000
Ca^{2+}, Cr^{2+}, Na^{+}, K^{+}, Ba^{2+}, Mg^{2+}, Al^{3+}, Li^{+}	1500
Urea, thiourea	750
Bromate, thiosulphate, Ru^{4+}, Pt^{4+}, Cr^{3+}, Zr^{4+},	650
Bi^{3+}, Mn^{2+}, Ru^{3+}, Pt^{4+}	500
Ti^{4+}, Bi^{3+}, Sr^{2+}	300
Fe^{3+}, Cd^{2+}, Fe^{2+}, V^{5+}, Mo^{6+}, Pb^{2+}, Mn^{2+}, Te^{4+}, U^{6+}, Th^{4+}, W^{6+}, Ce^{6+},	250
Ti^{4+}, Co^{2+}, Ni^{2+}, Cu^{2+}, Zn^{2+}	100 (100 μg·mL^{-1} of EDTA)
Ag+, Pd^{2+}	<50 (100 μg·mL^{-1} of citrate)

3.2.5. Ruggedness

Ruggedness was tested by applying the proposed methods to the assay of mercury using the same operational conditions but using two different instruments at two different laboratories and different elapsed time. Results obtained from lab-to-lab and day-to-day variations were reproducible, as the relative standard deviations (RSD) did not exceed 4%. A pre-validated AAS method was used as a reference method for determination of copper in whole blood and urine samples as to get inter method variation. The result obtained in proposed method for the determination of copper in whole blood and urine samples was compared with that obtained from the reference method by statistical analysis with respect to the accuracy (by t-test) and precision (by F-test). No significant differences were found between the calculated and theoretical values of tand F-tests at 95% confidence level proving similar accuracy and precision in the determination of mercury by both methods.

3.2.6. System Suitability

For standard at five different concentration levels within 1.0 - 10.0 μg·mL^{-1} maximum wavelength (435.52 ± 0.33) of absorption was checked and their relative standard deviations were calculated was to be 0.077%. Molar absorptivity calculated for the four different concentrations gave straight line parallel to concentration axis (x-axis) when plotted. Both indicate the excellent system suitability for the proposed method.

3.3. Effect of Diverse Ions

The tolerance limit of various anions and cations on the determination of Cu(II) or Hg(II) under optimal conditions in the present method is verified. The tolerance limit of a foreign ion is taken as the amount that caused an error in the absorbance value of ≤10%. Large amounts of commonly associated cations and anions do not interfere in the present method. Among the various ions studied, all the anions and the cations Pb^{2+}, Te^{4+}, U^{6+}, Na^{+}, K^{+}, Li^{+}, Ca^{2+}, Mg^{2+}, Bi^{3+}, Th^{4+}, W^{6+}, Ce^{6+}, Ti^{4+}, Al^{3+} do not interfere even when present in more than 100 fold excess. Ag^{+} and Pd^{2+} interfered seriously at all proportions. It (up to 50 µg) can be masked by 100 $\mu g \cdot mL^{-1}$ of citrate. Interference from Ti^{4+}, Co^{2+}, Ni^{2+}, Cu^{2+}, Zn^{2+} and be masked up to 100 µg by the use 100 $\mu g \cdot mL^{-1}$ of EDTA. Therefore, citrate has been selected to be suitable masking agent in the present method for the simultaneous determination of copper and mercury, and also for the determination of mercury alone. Tolerance limit of foreign ions for the determination of Cu(II) and Hg(II) by the proposed method has been given in Table3

4. APPLICATION: STANDARD AND REAL SAMPLE

The present method was successfully applied for the determination of mercury in series of synthetic mixtures of various compositions. The method was also extended to the determination of mercury and copper in a number of environmental water and soil samples, biological, pharmaceutical, fertilizer, food samples. The results of biological analyses by spectrophotometric method were found to be in excellent agreement with those obtained by AAS. The results of pharmaceutical, fertilizer, food and soil samples analyses by the spectrophotometric method shown in tables gave the satisfactory RSD within analytical agreement.

4.1. Reference Materials

A 0.1 g Certified reference materials (alloy, amalgam, synthetic compounds) containing different composition of metals was accurately weighed and placed in a 50 ml Erlenmeyer flask. It was decomposed as the acid digestion method. The solution was then cooled and neutralized with a dilute NH_4OH

buffer solution. A suitable aliquot (1 ml) of the decomposed solution each was taken into a calibrated flask and the metal content was determined as described under general procedure. Based on five replicate analyses, the average mercury or copper concentration determined was in good agreement with the certified values. The average percentage recovery of mercury or copper obtained in certified reference substances was quantitative as shown in Table2RSD value obtained indicates precision of the measurements reasonable.

4.2. Synthetic Mixtures

Several synthetic mixtures of varying compositions (1:1, 5:5, 8:8) containing mercury and diverse ions of known concentrations were determined by the present method using citrate as a masking agent. The results shown in Table 2 were found to be highly reproducible. The accurate recoveries were achieved in all solutions with maximum RSD value of 1.7% for triplicate measurements. It is indicated that mercury can be measured eliminating possible interferences from diverse ions.

4.3. Food Samples

5 g amount of fish samples, 10 g amount of radish and 25 g amount of cabbage samples were accurately weighed was subjected to acid digestion. The solution was neutralized by dilute ammonia and made up to the mark into 50 ml volumetric flask with de ionized water to form stock solution. A suitable aliquot (1 ml or 2 ml) of the above solution was taken into a calibrated flask and the metal content was determined under general procedure as described. Based on five replicate analyses, the average metal content determined by spectrophotome tric method was in good agreement with the leveled values. RSD values for all measurements (n = 3) were less than 5%. Table 4 presents the results of determination of mercury and copper in food samples by the proposed method. Results of triplicate measurements are given in table indicate RSD well enough.

4.4. Biological Samples

Human blood 2 ml or urine 25 ml was collected in polyethane bottles from the affected persons. The samples were taken into beaker. Acid decomposition was performed. The content of the beaker was filtered and neutralized with dilute ammonia. The resultant solution was then transferred quantitatively into a 20-ml calibrated flask and made up to the mark with de-ionised water. An aliquot (1-ml) of this digested biological sample was pipetted into a calibrated flask and the metal content was determined under the general procedure as described. The results of the biological sample analyses by the spectrophotometric method were found to be in excellent agreement with those obtained by AAS. The results are given in Table 5 indicating the method reproducible.

4.5. Mercury in Pharmaceutical Samples

1 g of tablet or powder or 1 ml of homeopathic medicine was dissolved in hot water and made 100 ml solution to form experimental solution. An aliquot amount was analyzed according to the general procedure. Results are summarized in Table6 RSD obtained for the triplicate measurements was not exceeding 3%.

Table 4. Determination of mercury and copper in food samples by the proposed method.

Samples	Source		Mercury		Copper	
			Concentration (μg·mL^{-1})	RSD (%)	Concentration (μg·mL^{-1})	RSD (%)
Food	Fish	Laitta	0.70 ± 0.03	4.3	0.25 ± 0.005	2.1
		Shrimp	0.31 ± 0.02	6.4	0.43 ± 0.02	4.6
	Vegetables	Radish	0.08 ± 0.005	5.9	0.20 ± 0.01	4.9
		Cabbage	0.07 ± 0.003	3.9	0.44 ± 0.008	1.8

Table 5. Biological sample analysis for the determination of mercury and copper by proposed method.

Biological Sample	Patients	Mercury		Copper			
		Conc. of Mercury (μg·mL^{-1})	RSD (%)	Conc. of Copper (μg·mL^{-1})	RSD (%)	AAS (μg·mL^{-1})	R (%)
Blood	Lung	1.845 ± 0.01	0.5	0.716 ± 0.03	4.1	0.698	96.5
	Kidney	1.808 ± 0.02	1.1	0.224 ± 0.01	4.4	0.221	102.5
	Stroke	1.831 ± 0.01	0.6	0.327 ± 0.51	1.5	0.326	103.1
Urine	Lung	0.251 ± 0.03	1.2	1.671 ± 0.40	2.4	1.650	101.1
	Kidney	1.878 ± 0.01	0.4	0.150 ± 0.21	3.5	0.149	97.1
	Stroke	1.088 ± 0.02	3.9	0.414 ± 0.22	4.8	0.422	99.6

Table 6. Pharmaceutical sample analysis for the determination of mercury and copper by proposed method.

Pharm. Samples	Name and Power	Brand Name	Claimed value (μg·mL^{-1})	Expt. Value (μg·mL^{-1})	RSD (%)	R (%)
Tablet	Merc-sol (3X)	Max pair	1.00	1.03 ± 0.02	1.9	102.7
Powder Medicine	Hepar sulph Mercury detox (2X)	B & T	0.50	0.49 ± 0.01	2.0	99.1
Anticeptic	Per chloride of Mercury (1:4000) sol^{n}	Momtaz Chemicals	0.50	0.53 ± 0.01	1.9	105.0
Potentized Medicine	Merc-cur (200 m)	B & T	1.00	0.99 ± 0.02	2.1	99.7
	Syphilinum(50 m)	B & T	1.00	1.03 ± 0.03	2.9	102.8

Table 7. Comparison on sensitivity and application of present method with other reported methods.

Sl. No.	Reagent	Media	λ_{max} nm	ε, $L \cdot mol^{-1} \cdot cm^{-1}$	Linearity $\mu g\ mL^{-1}$	LOD $\mu g \cdot mL^{-1}$	Applications	Ref. Vol. page, year
1	6-hdroxy-3-(2-oxoindolin-3-lideneamino)-2-thioxo-2 H-1,3-thiazin-4(3H)-one	Britton-Robinson (B-R) buffer (pH 4 - 6)	505	4×10^4	0.3 - 3.0	0.026	Dental unit wastewater, fertilizer samples	J. Hazardous Mat. 178, 287, 2010
2	Anthrone phenylhydrazone	pH 7	367	1.267×10^4	0.815 - 8.146	0.0939	spiked water samples	Analele Uni. Bucuresti 20 (1), 57, 2011
3	5-5(p-aminobenzylidene) -rhodanine	SPE, eluent-DMF	540	1.16×10^5	0.01 - 3.0	N.R.	Tobacco additives	J. Chinese Chem. Soc. 51, 297, 2004
4	2-(2-thiazolylazo)-p-cresol	Cetylpyridinium chloride, at pH 9.5	548	4.69×10^4	0.02 - 0.112	0.006	Tap water, dental wastewater	Turk J Chem 36, 159, 2012
5	Diacetyl monoxime isonicotinoyl hydrazone	acidic buffer pH 5.5	351	2.23×10^4	1.00 - 12.03	N.R.	Synthetic samples of alloy	Int. J. Chem. 3(2), 227, 2011
6	O-carboxyphenyl diazoamino p-azobenzene	alkaline medium	540	2.22×10^5	0.08 - 0.8	N.R.	Air, water, soil, fungicide samples	Talanta 57, 461, 2002
7	Diphenylthiocarbazone (dithizone)	50% aqueous 1,4-dioxane	488	2.5×10^4	0.1 - 25	20	Biological, soil, plant samples	Spectroscopy 17, 45, 2003
8	1,5-diphenylthiocarbazone	Sodium dodecyl sulphate	490	5.02×10^4	0.5 - 10	1.0	Biological, food, pharm. samples	Anal. Sci. 21, 507, 2005
9	Isonitriso p-isopropyl Acetophenone Phenyl Hydrazone	Methyl isobutyl ketone	395	2.678×10^{-3}	1.0 - 20		sewage waste, spiked water	Int. J. Life Sci. Pharm. Res. 1(1), 75, 2011
10	Iodide and Rhodamine B	dichloromethane benzene	565	17.68×10^4	0.25 - 2.7	0.10		Sains Malaysiana 41(2), 213, 2012
11	*Cold vapor AAS	ultrasonic assisted extraction			0.721 - 1.41 mg/kg	0.133	muscle tissues of marine fish	Pak. J. Anal. Environ. Chem. 11(2), 12, 2010
12	*Flow injection-cold vapor-ICPOES	Cloud point extraction				$4\ ng \cdot L^{-1}$	tap water	Spectrochimica Acta Part B 57, 365, 2002
13	*Electrothermal-atomic absorption spectrometry	Ultrasound separation, back-extraction-dithizone and cyclohexane				$0.27\ \mu g \cdot L^{-1}$	human urine	Talanta, 64, 217, 2004
14	Hydentoin, 5-amino1, 3,4-thiadizole-2-thiol	sodium acetate-acetic acid buffer	490	6.45×10^4	2.2		Beef and sheep liver and kidney	Int J Curr Pharm Res, 3(2), 102, 2011
15	2-mercaptobenzo thiazole	0.001% CTAB at pH 10.0			1.0×10^{-7} - 1.0×10^{-5} $M \cdot L^{-1}$	$6.2\ ng \cdot mL^{-1}$	Waste water	Bull. Korean Chem. Soc. 25(12), 1877, 2004
16	*Thio-Michler's Ketone	cloud point extraction using the nonionic surfactant Triton X-114	570		5.0 - 80.0 $ng \cdot mL^{-1}$	$0.83\ ng \cdot mL^{-1}$	Tap, river, well, waste water	J. Hazardous Materials 165, 1200, 2009
17	Ion-associate of HgL_4^{2-} and ferroin aqueous and n-heptane interface at pH 5	CH_2Cl_2		6.53×10^6	8.0×10^{-9} - 1.6×10^{-7} $M \cdot L^{-1}$	5.0×10^{-10} M	Environmental, waste water	Talanta 67, 555, 2005
18	2,4-dihydroxy propio-phenone benzoic acid hydrazone	pH 3, ferrocyanide solution	430	N.R.	0.108 - 0.977	0.041	Environmental water	Chem. Sci. Journal CSJ-14, 1, 2010
19	4-hydroxy-3,5-dimethoxy-benz-aldehyde-4-hydroxy benzoylhydrazone	Basic surfactant SDBS (5%)	420	4.56×10^4	0.300 - 3.09	N.R.	Environmental water, biological samples	Int. J. Green Chem. Bioprocess 1(1), 10, 2011
20	Sodium diethyl-dithiocarbamate	CCl_4	435	1.5×10^5	0.02- 15	0.29	Biological, soil, water, vegetable and fish samples	P. M.

In radish (vegetable sample) mercury was found 0.07 $\mu g \cdot mL^{-1}$ but in laitta fish sample it was found 0.70 $\mu g \cdot mL^{-1}$, which is greater than vegetables. On the other hand in all blood samples (Lung, Kidney and Storke patient) mercury was found around the 1.8 $\mu g \cdot mL^{-1}$. But in urine samples of kidney patient has higher than lung and storke patient. Lung patient contains higher amount of copper than the others. The biological tolerance limits for mercury concentration are 50 μg·mL^{-1} in urine and 10 μg·mL^{-1} in blood.

Diethylthiocarbamate is proved to be more suitable than other complexing reagents cited in literature in respect to molar absorptivity and sensitivity values and because of their selectivity towards Hg(II). Using suitable masking agent the method shows wide application for mercury determination. Both copper and mercury were simultaneously determined in all real samples. All determinations provided high accuracy and excellent precision. Method is applicable avoiding interferences as most of the alkaline and alkaline earth metals form colourless complexes with the DDTC.

Table 7 shows the comparison on sensitivity and application of present method with other methods recently published available in literature (supplementary data). It shows that there are a number of methods though sensitive (Sl. no. 11-13, 16) involve complicated procedures employing sophisticated equipment, expensive and tedious separation and pre-concentration steps or they have limited application.

5. CONCLUSION

A spectrophotometric procedure for the simultaneous determination of mercury(II) and copper using diethyldithiocarbamate (DDTC) as ligand is described. Primarily Cu(II) complex was extracted with CCl_4 and the absorbance was measured at 435 nm. Mercury determination was based on the quantitative displacement of diethyldithiocarbamate (DDTC) from copper complex with Hg(II), and subsequent measurement of reduced absorbance. Therefore, the method described herein has many advantages: it is simple and rapid; it has high accuracy and sensitivity, use of inexpensive reagents available in any analytical laboratory. Therefore, the method is practical and valuable for its wide application. The method was applied to a number of environmental water and soil samples, and biological, pharmaceutical and food samples. The results show a good agreement with certified values in alloy and amalgam, theoretical values in synthetic compounds, pharmaceutical preparations and with the results obtained by AAS method for biological samples. Moreover, $Hg(DDTC)_2$ is more stable in basic aqueous media than $Cu(DDTC)_2$ as to perform the mercury determination by the quantitative replacement. Also, great advantage that is interesting of this method is its selective application to mercury determination individually or copper and mercury determination simultaneously as required.

CONFLICTS OF INTEREST

Mohammad Nasir Uddin, Nur Mostaq Shah, Mohammad Amzad Hossain, Md. Muzahidul Islam The authors declared no conflicts of interest.

REFERENCES

1. Fergusson, J.E. (1990) The Heavy Elements: Chemistry, Environmental Impact and Health Effects. Pergamon Press, Oxford, 85-547.
2. Venugopal, B. and Lucky, T.D. (1978) Metal Toxicity in Mammals. Vol. 2, Plenum Press, New York, 86-99.
3. Turkoglu, O. and Soylak, M. (2005) Spectrophotometric Determination of Copper in Natural Waters and Pharmaceutical Samples with Chloro(phenyl) glyoxime. Journal of the Chinese Chemical Society, 52, 575-579.
4. Moreno-Cid, A. and Yebra, M.C. (2002) Flow Injection Determination of Copper in Mussels by Flame Atomic Absorption Spectrometry after On-Line Continuous Ultrasound-Assisted Extraction. Spectrochimica Acta Part B: Atomic Spectroscopy, 57, 967-974.
5. Cindric, I.J., Krizman, I., Zeiner, M., Kampic, S., Medunic, G. and Stingeder, G. (2012) ICP-AES Determination of Minorand Major Elements in Apples after Microwave Assisted Digestion. Food Chemistry, 135, 2675-2680.
6. Alexiu, V., Chirtop, E., Vladescu, L. and Simion, M. (2004) Determination of Mercury in Pharmaceuticals by Graphite Furnace Atomic Absorption Spectrometry with Chemical Modifier. Acta Chimica Slovenica, 51, 361-372.
7. Han, F.X., Patterson, W.D., Xia, Y., Maruthi, B.B. and Su, Y. (2006) Rapid Determination of Mercury in Plant and Soil Samples Using Inductively Coupled Plasma Atomic Emission Spectroscopy, a Comparative Study. Water, Air, and Soil Pollution, 170, 161-171.
8. da Silva, M.J., Paim, A.P., Pimentel, M.F., de la Cervera M.L. and Guardia, M. (2010) Determination of Mercury in Rice by Cold Vapor Atomic Fluorescence Spectrometry after Microwave-Assisted Digestion. Analytica Chimica Acta, 667, 43-48.
9. Rajawat, D.S., Srivastava, S. and Satsangee, S.P. (2012) Electrochemical Determination of Mercury at Trace Levels Using Eichhornia Crassipes Modified Carbon Paste Electrode. International Journal of Electrochemical Science, 7, 11456-11469.
10. Ahmed, M.J., Jahan, I. and Banoo, S. (2002) A Simple Spectrophotometric Method for the Determination of Copper in Industrial, Environmental, Biological and Soil Samples Using 2,5-Dimercapto-1,3,4-Thiadiazole. Analytical Sciences, 18, 805-810.
11. Ramanjaneyulu, G., Reddy, P.R., Reddy, V.K. and Reddy, T.S. (2008) Direct and Derivative Spectrophotometric Determination of Copper(II) with 5-Bromosalicylaldehyde Thiosemicarbazone. The Open Analytical Chemistry Journal, 2, 78-82.
12. Rekha, D., Suvardhan, K., Kumar, K.S., Reddy, P., Jayaraj, B. and Chiranjeevi, P. (2007) Extractive Spectrophotometric Determination of Copper(II) in Water and Alloy Samples with 3-methoxy-4-hydroxy Benzaldehyde-4-bromophenyl Hydrazone. Journal of the Serbian Chemical Society, 72, 299-310.
13. Reddy, G.C., Devanna, N. and Chandrasekhar, K.B. (2011) Derivative Spectrophotometric Determination of Mercury (ii) Using Diacetyl Monoxime Isonicotinoyl Hydrazone. International Journal of Chemistry, 3, 227-232.

14. Khan, H., Ahmed, M.J. and Bhanger, M.I. (2005) A Simple Spectrophotometric Determination of Trace Level Mercury Using 1,5-diphenylthiocarbazone Solubilized in Micelle. Analytical Sciences, 21, 507-512.
15. Cesur, H. (2007) Selective Solid-Phase Extraction of Cu(II) Using Freshly Precipitated Lead Diethyldithiocarbamate and Its Spectrophotometric Determination. Chemical Papers, 61, 342-347.
16. Uddin, M.N., Salam, M.A. and Hossain, M.A. (2012) Spectrophotometric Measurement of Cu(DDTC)2 for the Simultaneous Determination of Copper and Zinc. Chemosphere, 90, 366-373.
17. Uddin, M.N., Chowdhury, D.A. and Islam, J. (2013) Synthesis, Characterization and Antibacterial Evaluation of Some Mixed-Metal Mixed-Ligand Complexes. Chiang Mai Journal of Science, 40, 625-635.
18. Skoog, D.A., West, D.M. and Holler, F.J. (1988) Fundamentals of Analytical Chemistry. 5th Edition, Saunders, New York.

CHAPTER 7

Maillard Reaction in Milk – Effect of Heat Treatment

Tomoko Shimamura and Hiroyuki Ukeda

[1] *Faculty of Agriculture, Kochi University, Nankoku,, Japan*

1. INTRODUCTION

Milk is usually subjected to heat treatment to ensure microbiological safety before retail and consumption. There are three types of heat treatment; (1) low temperature long time (LTLT) pasteurization, (2) high temperature short time (HTST) pasteurization, and (3) ultra-high temperature (UHT) treatment. In all types of heat treatment, the Maillard reaction occurs in milk.

The Maillard reaction (nonenzymatic glycation) is a chemical reaction between amino group and carbonyl group; it is the extremely complex reaction that usually takes place during food processing or storage. In the case of milk, lactose reacts with the free amino acid side chains of milk proteins (mainly ε-amino group of lysine residue) to proceed to early, intermediate, and advanced stages of Maillard reaction and forms enormous kinds of Maillard reaction products. The reactions of lactose and milk proteins have been frequently investigated and the formations of various Maillard reaction products in milk during heat treatment have been demonstrated [1]. In the general Maillard reaction, firstly an Amadori product is generated, and it progresses to the 3-deoxyosone or 1-deoxyosone route depending on the reaction pH. In the case of the Maillard reaction of disaccharides such as lactose, there is a third reaction route. It is the 4-deoxyosone route. A main carbohydrate in milk is lactose. Thus, the Maillard reaction in milk progresses via the above described three routes. Finally, the Maillard reaction results in the formation of melanoidins (browning compounds).

2. EFFECT OF THE MAILLARD REACTION ON MILK PROTEINS

The Maillard reaction shows various effects on milk proteins such as bioavailability, solubility, forming property, emulsifying property, and heating stability [1-4]. In addition, the formation of flavor compounds and browning compounds is caused as the consequences of the Maillard reaction between lactose and milk proteins [1, 5].

As for the effect of the Maillard reaction on the bioavailability of milk proteins, various studies were performed. Generally, in the Maillard reaction in milk, lactose mainly reacts with ε-amino group of lysine residue of milk proteins. Thus, the lysine loss by the Maillard reaction increases with a severity of heat treatment. The modified lysine cannot be available as a nutrient any more. For example, steam injection process (direct heating) generated 3.6% (120°C for 400 sec) and 6.8% (130°C for 290 sec) of the blocked lysine in whole milk. The indirect heating at 115°C for 10 to 40 min increased the modified lysine from 11.0 to 13.0% [6]. In addition, it was revealed that the lysine residues in skim milk powder were more susceptible to heating than those in skim milk [7].

Le *et al.* [3] recently suggested that the Maillard reaction was responsible for the solubility loss in milk protein concentrate powder. It was also reported that the glycated β-lactoglobulin was more stable at acidic pH and more stable against heating. The glycation of β-lactoglobulin, moreover, could improve its forming and emulsifying properties [4]. These results suggested the usefulness of the Maillard reaction for enabling milk proteins to have different properties.

3. MONITORING OF THE MAILLARD REACTION OF MILK USING XTT ASSAY

The Maillard reaction has a lot of effects on the function of milk proteins and sensory property of milk and dairy products as described above. Particularly, in the manufacturing of milk, the excess progress of Maillard reaction and the formation of melanoidins are undesired, because a commercial value of milk is drastically decreased by them. Therefore, the detection of the Maillard reaction products is important for the quality control of milk. So far, several heat-induced markers have been proposed to control and check the heat treatment given to milk and dairy products. For example, furosine, hydroxymethylfurfural (HMF), and lactulose concentrations have been recognized to be the most promising indicators, since these concentrations increase with the heat treatment [1]. In Japan, protein reducing substance value (PRS) obtained by a ferricyanide assay is also widely-used conventional indicator [8]. It is based on the detection of reducing substances such as sulfhydryl group which are generated by heating in the fraction of acid-precipitated milk protein. These method, however, are generally time-consuming and complicated.

We proposed an assay method for determining the ability of milk to reduce 3'-[1-[(phenylamino)-carbonyl]-3,4-tetrazolium]-bis(4-methoxy-6-nitro)benzensulfonic acid hydrate (XTT: Figure 1) as a method of evaluating the extent of the Maillard reaction [8-11]. The tetrazolium salt XTT is reduced to water-soluble formazan which is suitable for the spectrophotometric measurement. Taking account into an economical view and a rapidity of assay, we tried to develop a microplate assay. The assay conditions were as follows: XTT concentration, 0.5 mM; reaction pH, 7.0; menadione concentration, saturation level (ca. 0.55 mM); reaction temperature, room temperature; volume of XTT solution, 60 μL; volume of sample solution, 40 μL; detection wavelength, 492 nm; reference wavelength, 600 nm; reaction time, 20 min. The procedure of the XTT assay was as follows: the XTT solution was added into each well in a microplate. Afterward, the sample solution was added to the well. After mixing on a microplate shaker at 500 rpm for 15 sec, a difference in the absorbance between 492 and 600 nm was read on a microplate reader as the absorbance at 0 min. After 20 min, the absorbance difference was read again. An increase in the absorbance for 20 min was recorded as the ability of the sample to reduce XTT (XTT reducibility).

Using the XTT assay described above, the XTT reducibility of milk was examined (Figure 2). In this test, an LTLT milk (Milk A: 65°C for 30 min) and two kinds of UHT milk (Milk B: 130°C for 2 sec, Milk C: 140°C for 3 sec) were used. When the milk was mixed with XTT solution, intense orange color which was derived from the XTT formazan was recognized. As shown in Figure 2, Milk C showed the highest XTT reducibility and the order of XTT reducibility was Milk C, B, and A. At the same time, the HMF content in Milk B and C was also determined as the conventional indicator of the Maillard reaction, because it was reported that the HMF content clearly increased with the severity of the heating treatment of milk [12, 13]. As a result, the content of HMF in Milk C (12.6 μM) was higher than that of Milk B (8.95 μM). From these results, it was revealed that the XTT assay could estimate the degree of thermal stress delivered to the milk as well as the HMF value.

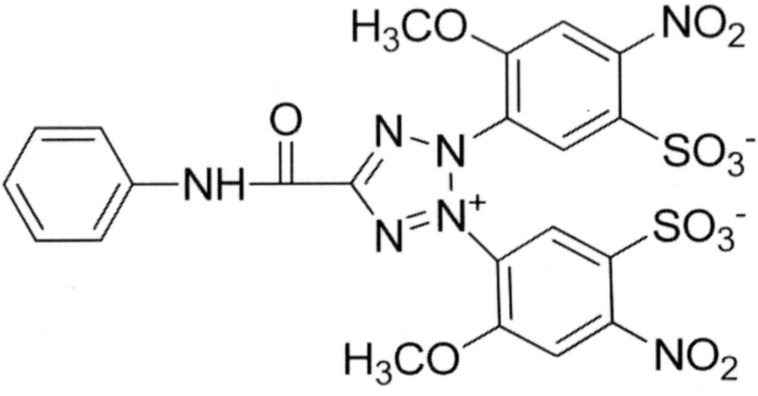

FIGURE 1. Structure of tetrazolium salt XTT.

In addition, the changes in the XTT reducibility during storage of UHT milks (Milk B and C) were investigated. For purpose of comparison, the PRS of Milk B and C were also examined by the ferricyanide assay [8]. The UHT milks were stored at 4°C and 37°C for about 4 weeks (Figure 3, 4). The XTT reducibility of Milk B and C gradually decreased depending on the storage period and the rate of decrease in the XTT reducibility was clearly larger at higher storage temperature (Figure 3). This result strongly suggested that, when the XTT assay is applied to the milk heated under a given condition, the result can serve to estimate not only the heating condition but also the storage period after heat treatment if the storage temperature is known or if storage period is known.

On the other hand, the PRS of Milk B and C were almost constant at all storage conditions (Figure 4). This result showed that the ferricyanide-reducing substances might be stable unlike the XTT-reducing substance. In addition, we found that the HMF value of Milk C did not significantly change during its storage for 60 days at 5°C [11]. It was in accordance with the result of Fink and Kessler about HMF [13]. They also reported that the lactulose concentration of UHT milk was constant throughout the storage period for 70 days at room temperature [13]. It could be concluded, therefore, that the XTT assay is applicable for the estimation of storage conditions which was impossible by the conventional method.

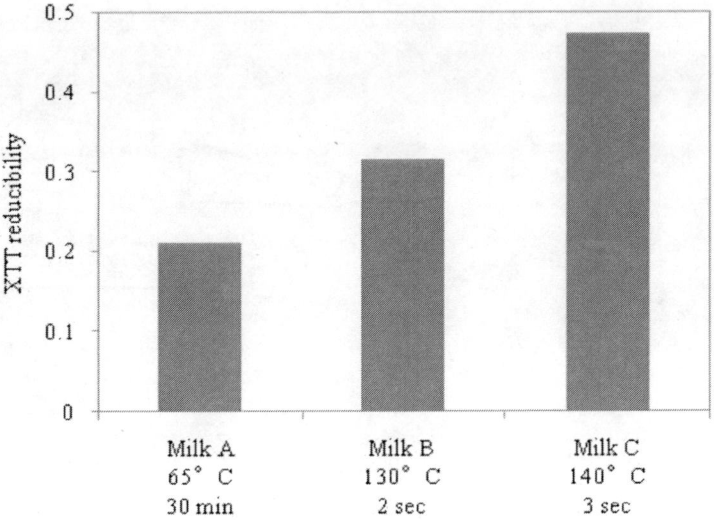

FIGURE 2. Comparison of XTT reducibility of LTLT milk (Milk A) and UHT milks (Milk B and C).

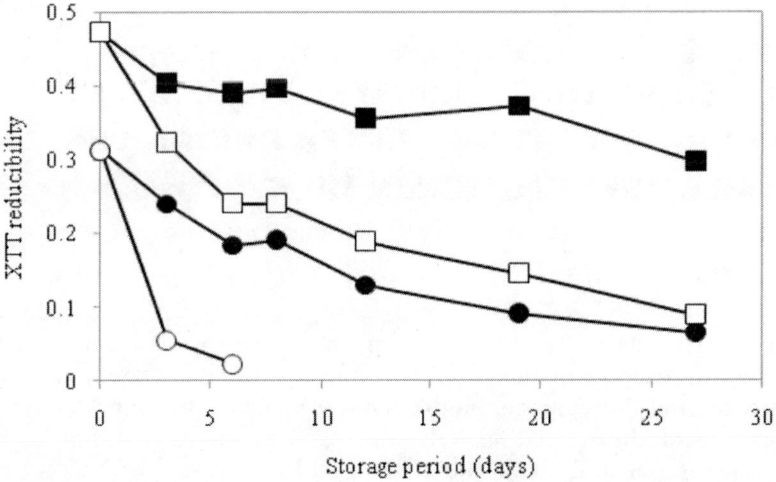

Milks were stored at 4°C (●, Milk B; ■, Milk C) and 37°C (○, Milk B; □, Milk C).

FIGURE 3. Changes in XTT reducibility during storage.

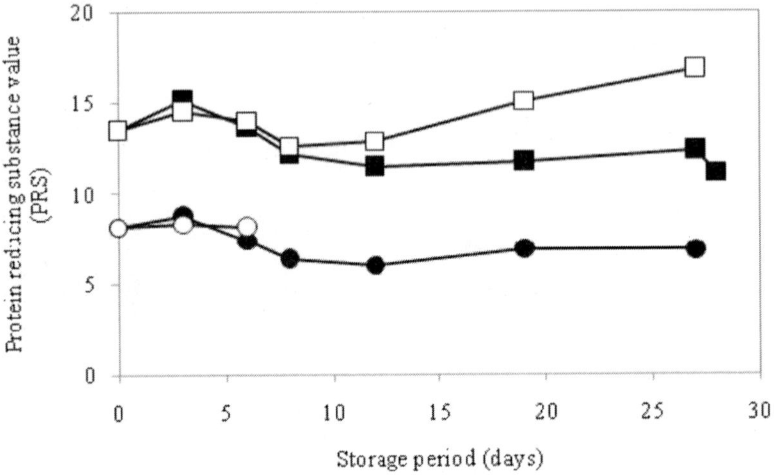

Milks were stored at 4°C (●, Milk B; ■, Milk C) and 37°C (○, Milk B; □, Milk C).

FIGURE 4. Changes in PRS during storage.

4. DEMONSTRATION OF THE PRESENCE OF AMINOREDUCTONE FORMED DURING THE MAILLARD REACTION IN MILK

4.1. In The Model System of Lactose and Butylamine [9]

In order to clarify the XTT-reducing substance that is formed during the Maillard reaction in milk, we firstly used a model system consisting of lactose and butylamine, and then performed the spectrophotometric analysis of the heated model solution. The model solution of lactose-butylamine heated at 80-100°C for 0-30 min showed a characteristic UV absorption maximum at 320 nm. During the heating at 80-100°C for 30 min, the changes in the absorbance at 320 nm (Figure 5) and the XTT reducibility (Figure 6) were investigated. As a result, both indices increased gradually in accordance with the rise in temperature and heating time. This result indicated that the compound with the absorption maximum at 320 nm was formed by heating. Moreover, the behavior of increase in absorbance at 320 nm was similar to that of the XTT reducibility. Since the similar trend was recognized in the time course of both indices, the XTT reducibility was plotted against the absorbance at 320 nm. In consequence, a significant relationship between them was recognized with a correlation coefficient of 0.967 (n = 19, $p < 0.001$). From these results it was found that the compound with the absorption maximum at 320 nm might be responsible for the reduction of XTT.

The ^{13}C- and ^{1}H-NMR analyses of the compound with the absorption maximum at 320 nm which was extracted from lactose-butylamine model solution heated at 100°C for 15 min were performed. The signals of the ^{13}C- and ^{1}H-NMR could be assigned the compound as the aminoreductone, 1-(butylamino)-1,2-dehydro-1,4-dideoxy 3-hexulose (Figure 7). This compound was reported as the Maillard reaction product formed in the 4-deoxyosone route. It was also reported as the characteristic compound in the Maillard reaction of disaccharides [14]. In addition, we demonstrate a linear relationship between the XTT reducibility and the amount of aminoreductone which was determined more specifically by HPLC [10]. These results strongly indicated that the aminoreductone formed during Maillard reaction of lactose was mainly responsible for the reduction of XTT.

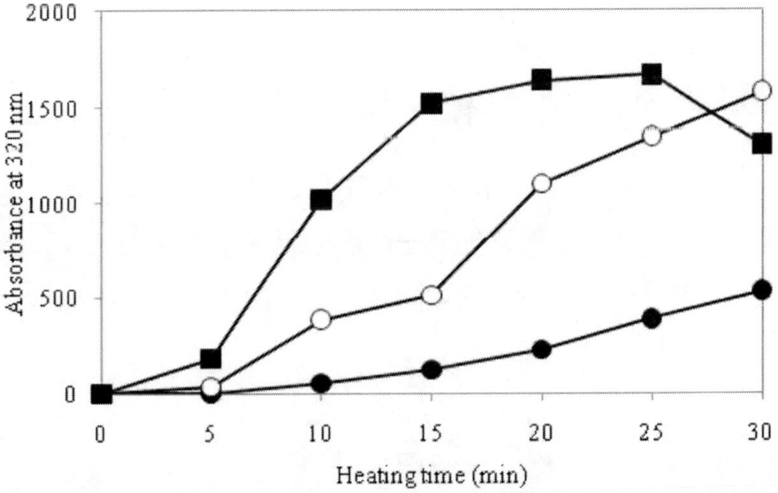

Lactose (262 mM) and butylamine (1.16 M) in 1.28 M phosphate buffer (pH 7.0) were heated at 80°C (●), 90°C (○), and 100°C (■).

FIGURE 5. Effect of heating temperature and time on the absorbance at 320 nm.

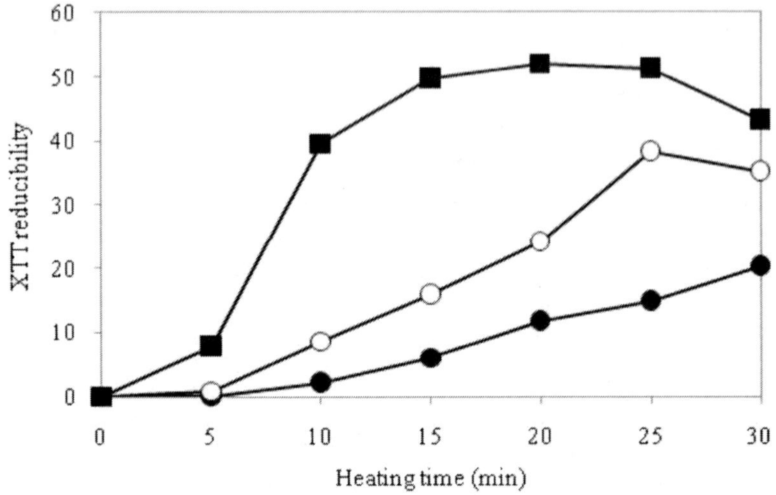

Lactose (262 mM) and butylamine (1.16 M) in 1.28 M phosphate buffer (pH 7.0) were heated at 80°C (●), 90°C (○), and 100°C (■).

FIGURE 6. Effect of heating temperature and time on the XTT reducibility.

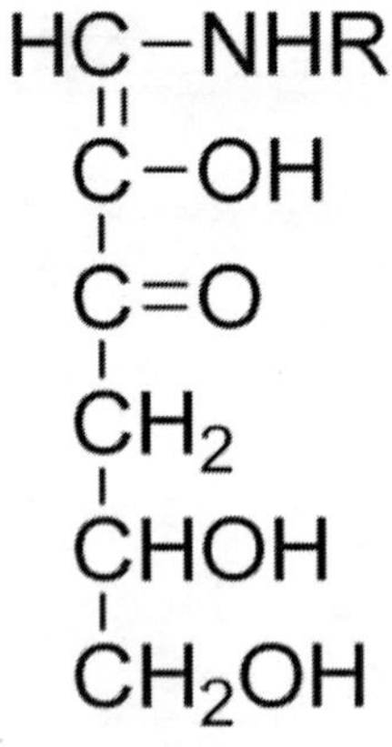

FIGURE 7. Structure of aminoreductone generated by the Maillard reaction of lactose and butylamine.(R = butyl group)

From these results, we presumed that the aminoreductone is formed by the Maillard reaction between lactose and ε-amino groups of milk proteins, and then it is responsible for the reduction of XTT. However, at that time, there was no report to prove the presence of aminoreductone in milk. Thus, we tried to demonstrate it using model system consisting of lactose and milk proteins and UHT milk.

4.2. In The Model System of Lactose and Milk Proteins

As a model system of milk, the solution consisting of lactose (4.6%) and casein (2.6%), α-lactalbumin (0.12%), or β-lactoglobulin (0.32%) was used and heated at 130°C for 15 min. After heating, the characteristic absorption maximum or shoulder at 320 nm was recognized. In addition, the changes of the absorbance at 320 nm (Figure 8) and the XTT reducibility (Figure 9) were investigated. In all model systems, the increases in the absorbance at 320 nm and the XTT reducibility depended on the heating time. Because similar tendencies were observed between two indices in all model systems, correlations were examined. Consequently, there were significant linearities as follows: casein ($r = 0.993$, $n = 6$, $p < 0.001$), α-lactalbumin ($r = 0.996$, $n = 6$, $p < 0.001$), and β-lactoglobulin ($r = 0.975$, $n = 6$, $p < 0.001$). From these results, it was suggested that aminoreductone is generated in the Maillard reaction between lactose and milk proteins and it is responsible for the reduction of XTT.

4.3. IN MILK [15]

As described above, a possibility of the formation of aminoreductone on the milk proteins during the Maillard reaction with lactose was clearly shown. However, direct demonstration of the presence of aminoreductone in milk had not been accomplished because of a difficulty in isolation of an intact aminoreductone from milk proteins. For instance, aminoreductone is labile and hence not suitable for enzyme hydrolysis and multiple extraction steps. To achieve the practical application of the XTT assay in food industries including dairy products, it was essential to demonstrate the presence of aminoreductone in milk. Because of this background, we attempted to isolate aminoreductone from milk proteins using 2,4-dinitrophenylhydrazine (DNP), a common labeling reagent for the carbonyl group, and Cu^{2+} [16]. A mechanism of derivatization of aminoreductone in milk is shown in Figure 10. In this derivatization step, Cu^{2+} plays as an oxidizing agent against aminoreductone, and the oxidized aminoreductone (OAR) has two or three carbonyl groups. Finally, it was assumed that two or three carbonyl groups in OAR are derivatized by DNP (OAR-DNP).

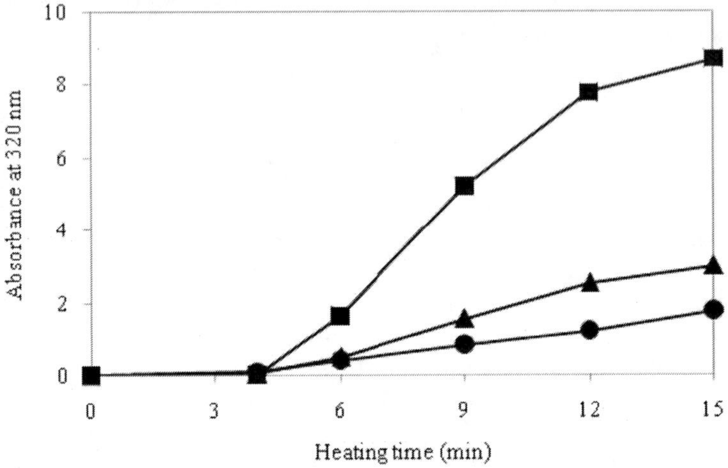

Lactose (4.6%) and casein (■: 2.6%), α-lactalbumin (●: 0.12%), or β-lactoglobulin (▲: 0.32%) in 20 mM phosphate buffer (pH 6.7) were heated at 130°C.

FIGURE 8. Effect of heating time on the absorbance at 320 nm.

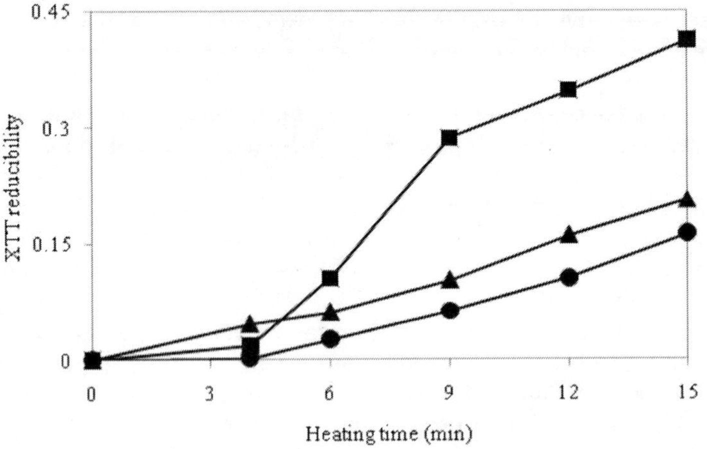

Lactose (4.6%) and casein (■: 2.6%), α-lactalbumin (●: 0.12%), or β-lactoglobulin (▲: 0.32%) in 20 mM phosphate buffer (pH 6.7) were heated at 130°C.

FIGURE 9. Effect of heating time on the XTT reducibility.

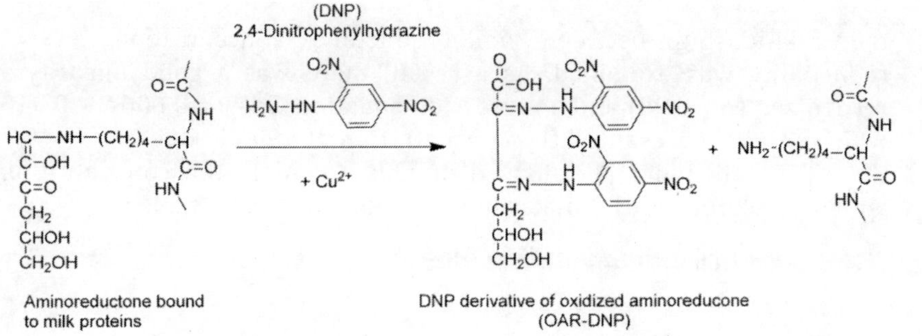

FIGURE 10. Derivatization mechanism of aminoreductone in milk using DNP and Cu^{2+}.

The derivatization using DNP and Cu^{2+} was applied to aminoreductone in UHT milk (140°C, 3 sec). In this study, the UHT milk was reheated at 130°C for 15 min in order to increase the content of aminoreductone, because the original content of aminoreductone in the commercially available UHT milk was not so high. As a result, the reheating of UHT milk could increase the content of DNP derivative by 40 times. The DNP derivative which was thought to be corresponding to OAR-DNP in the reheated UHT milk was purified by preparative normal-phase HPLC and preparative reversed-phase HPLC. Finally, the purified compound (4.2 mg) was obtained from 980 mL of UHT milk and analyzed by ^{13}C- and ^{1}H-NMR. The NMR signals of the DNP derivative from UHT milk could be assigned to the structure of OAR-DNP shown in Figure 10. In addition, the NMR signals of DNP derivative from UHT milk were nearly the same as those of the OAR-DNP from lactose-butylamine model system. These results demonstrated that aminoreductone was formed by the Maillard reaction on the milk proteins and present in milk.

Therefore, considering the above, the principle of the present XTT assay can be concluded as follows (Figure 11): (1) Lactose and ε-amino groups of lysine residue in milk proteins react non-enzymatically to form the Amadori product by the heating process. (2) Aminoreductone structure is formed on the milk proteins after elimination of galactose moiety from lactose through 4-deoxyosone pathway. (3) Aminoreductone is oxidized by XTT, whereas XTT is simultaneously reduced to the corresponding water-soluble formazan.

It would be thought that the above mentioned steps (1) and (2) progresses depending on the time and temperature of heating process in milk production, so the XTT assay can differentiate the extent of heat treatment of milk. Based on the study using model system of lactose and butylamine, the relationship between aminoreductone concentration and XTT reducibility was examined. As a result, there was a good linearity was recognized (r = 0.98) and a regression equation was $y = 0.606x + 0.046$, in which x and y represented the concentration of aminoreductone (mM) and the XTT reducibility [17]. Based on this equation, the concentration of aminoreductone in UHT milk could be estimated as 0.44 mM.

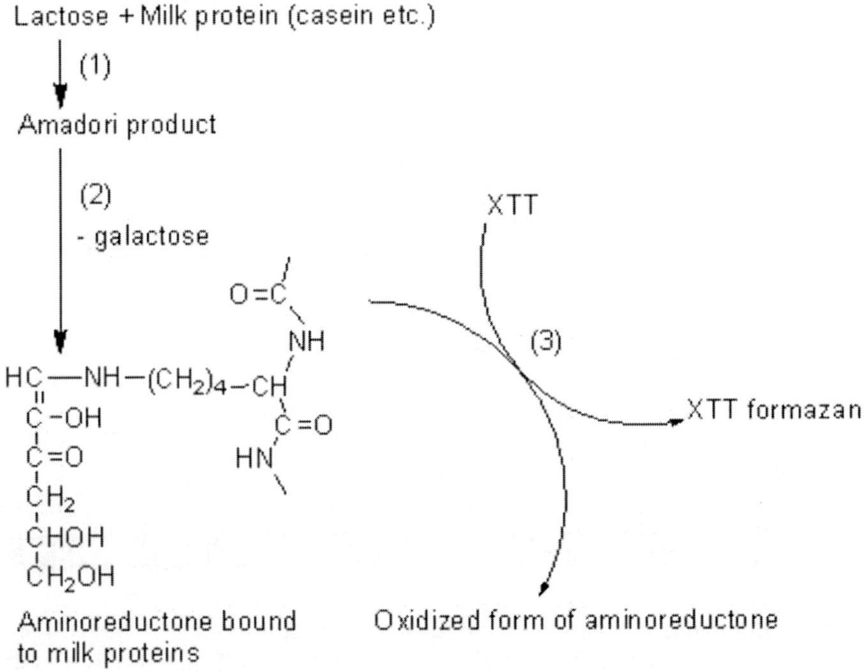

FIGURE 11. Principle of XTT assay.

As described above, during the course of elucidation of the XTT-reducing substance, Cu^{2+} was used as the oxidizing agent (Figure 10), because we empirically knew that it was easily decomposed by the addition of Cu^{2+} [18]. In fact, the aminoreductone extracted from the heated model solution of lactose and butylamine was rapidly oxidized by the addition of Cu^{2+}, and simultaneously the XTT reducibility was also lost. On the other hand, in our previous work, it was revealed that the commercially available UHT milk contains Cu^{2+} at the concentration of 30 μg/L [18]. Thus, it was easily presumed that the aminoreductone formed by the heating process was gradually oxidized by endogenous Cu^{2+} in milk during storage period. This was the reason why that the XTT reducibility decreased depending on the storage period (Figure 3). The detailed investigation about the relationship between the aminoreductone concentration, Cu^{2+} concentration, and storage stability of milk is now in progress.

5. FUNCTIONALITY OF AMINOREDUCTONE

The functionalities of aminoreductone have attracted interest and, so far, some studies were performed. Trang *et al.* [17] reported a protective effect of aminoreductone against riboflavin (vitamin B_2) photolysis. It is well known that the milk is important source of riboflavin and its content is 1.5 mg/L. It is stable to heat and oxidation, but is rapidly photo-degraded. In experimental condition at 7000 lux light intensity, the riboflavin (1.5 mg/L) was almost completely degraded for 150 min. On the other hand, the addition of aminoreductone (0.22 mM: half concentration in UHT milk) could extend the half-life period of riboflavin. The protective effect of aminoreductone against riboflavin photolysis was higher than that of ascorbic acid which was famous antioxidant. In addition, the antioxidative activity of aminoreductone was reported [19]. From these results, it was suggested that the aminoreductone formed by the Maillard reaction would contribute to keep the nutritional value and sensory quality of milk.

Furthermore, aminoreductone showed antimicrobial activity against *Helicobacter pylori* (*H. pylori*). *In vitro* it effectively inhibited the growth of 24 kinds of *H. pylori* strains including antibiotic-resistant strains and had a bactericidal activity [20, 21]. The Killing ability was observed even in acidic condition. In addition, aminoreductone also had the antimicrobial activity against methicillin-resistant *Staphylococcus aureus* (MRSA) [21, 22]. These results indicated that foods containing aminoreductone, such as milk and dairy products, have a potential health benefits in medical practice.

6. CONCLUSION

In this chapter, the demonstration of the presence of aminoreductone formed by the Maillard reaction in milk and the specific assay method for aminoreductone were focused and introduced. Since the novel functionality of aminoreductone have come out one after another, the information of aminoreductone obtained by the XTT assay would gain importance in the quality control in milk and dairy products.

REFERENCE

1. M. A. J. S. van Boekel, 1998 Effect of heating on Maillard reactions in milk. Food Chem. 62 403414 .
2. F. Evangelisti, C. Calcagno, S. Nardi, P. Zumin, 1999 Deterioration of protein fraction by Maillard reaction in dietetic milks. J. Dairy Res. 66 237243 .
3. T. T. Le , B. Bhandari, H. C. Deeth, 2011 Chemical and physical changes in milk protein concentrate (MPC80) powder during storage. J. Agric. Food Chem. 59 54655473 .
4. M. A. Augustin, P. Udabage, 2007 Influence of processing on functionality of milk and dairy proteins. In: Taylor S. L., editor. Advances in Food and Nutrition Research. Amsterdam: Elsevier. 138 .
5. M. Freedman, 1996 Food browning and its prevention: an overview. J. Agric. Food Chem. 44 631653 .
6. P. A. Finot, R. Deutsch, E. Bujard, 1981 The extent of the Maillard reaction during the processing of milk. Prog. Fd. Nutr. Sci. 5 345355 .
7. T. Henle, H. Walter, H. Klostermeyer, 1991 Evaluation of the extent of the early Maillard-reaction in milk products by direct measurement of the Amadori-product lactuloselysine. Z. Lebensm. Unters. Forsch. 193 199122 .
8. T. Shimamura, H. Ukeda, M. Sawamura, 2001 Quality evaluation of milk using XTT assay-comparison with ferricyanide assay. Nippon Shokuhin Kagaku Kogaku Kaishi (in Japanese) 48 840843 .
9. T. Shimamura, H. Ukeda, M. Sawamura, 2000 Reduction of tetrazolium salt XTT by aminoreductone formed during the Maillard reaction of lactose. J. Agric. Food Chem. 48 62276229 .
10. T. Shimamura, H. Ukeda, M. Sawamura, 2004 Relationship between the XTT reducibility and aminoreductone formed during the Maillard reaction of lactose: the detection of aminoreductone by HPLC. Food Sci. Technol. Res. 10 69 .
11. H. Ukeda, Y. Goto, M. Sawamura, H. Kusunose, H. Kamikado, T. Kamei, 1995 Reduction of tetrazolium salt XTT with UHT-treated milk: its relationship with the extent of heat treatment and storage conditions. Food Sci. Technol. Int. 1 5257 .
12. H. G. Kessler, R. Fink, 1986 Changes in heated milk and stored milk an interpretation by reaction kinetics. J. Food Sci. 51 11051111 .
13. R. Fink, H. G. Kessler, 1988 Comparison of methods for distinguishing UHT treatment and sterilization of milk. Milchwissenschaft 43 275280 .
14. M. Pischetsrieder, C. Schoetter, T. Severin, 1998 Formation of an aminoreductone during the Maillard reaction of lactose and Nα-acetyllysine or proteins. J. Agric. Food. Chem. 46 928931 .
15. T. Shimamura, Y. Kurogi, S. Katsuno, T. Kashiwagi, H. Ukeda, 2011 Demonstration of the presence of aminoreductone formed during the Maillard reaction in milk. Food Chem. 129 10881092 .
16. K. Shinohara, Y. Fukumoto, Y. Tseng, K. , Y. Inoue, H. Omura, 1974 Action of amino reductones on nucleic acids. Nippon Nogei Kagaku Kaishi (in Japanese) 48 499506 .
17. V. T. Trang, Y. Kurogi, S. Katsuno, T. Shimamura, H. Ukeda, 2008 Protective effect of on the photo-degradation of riboflavin. Int. Dairy J. 18 344348 .
18. H. Ukeda, T. Shimamura, T. Hosokawa, Y. Goto, M. Sawamura, 1998 Monitoring of the Maillard reaction based on the reduction of tetrazolium salt XTT. Food Sci. Technol. Int. Tokyo 4 258263 .
19. M. Pischetsrieder, F. Rinaldi, U. Gross, T. Severin, 1998 Assessment of the antioxidative and prooxidative activities of two aminoreductones formed during the Maillard reaction: effects on the oxidation of β-carotene, Nα-acetylhistidine, and cis-alkenes. J. Agric. Food Chem. 46 29452950 .

20. V. T. Trang, H. Takeuchi, H. Kudo, A. Aoki, S. Katsuno, T. Shimamura, T. Sugiura, H. Ukeda, 2009 Antimicrobial activity of aminoreductone against Helicobacter pylori. J. Agric. Food Chem. 57 1134311348 .
21. H. Takeuchi, T. Shimamura, H. Kudo, H. Ukeda, T. Sugiura, 2012 Antimicrobial activity of natural products and food components. Current Research in Agriculture and Food Chemistry 1 112 .
22. V. T. Trang, H. Takeuchi, H. Kudo, S. Katsuno, T. Shimamura, T. Kashiwagi, V. H. Son, T. Sugiura, H. Ukeda, 2011 In vitro antimicrobial activity of aminoreductone against the pathogenic bacteria methicillin-resistant Staphylococcus aureus (MRSA). J. Agric. Food Chem. 59 89538960 .

CHAPTER 8

Carotenoid Retention in Biofortified Maize using Different Post-Harvest Storage and Packaging Methods

Víctor Taleon [a], Luke Mugode [b], Luisa Cabrera-Soto [c], Natalia Palacios-Rojas [c]

[a] *HarvestPlus, c/o IFPRI 2033 K Street, NW, Washington, DC 20006-1002, USA*

[b]*Department of Food Science, Postharvest Technology Research Laboratory, Stellenbosch University, Private Bag X1, Stellenbosch 7602, South Africa*

[c]*Global Maize Program, International Maize and Wheat Improvement Center (CIMMYT), CIMMYT Research Station, Km. 45 Carretera Mexico-Veracruz, El Batan, Texcoco CP 56237, Edo. de México, Mexico*

ABSTRACT

Orange maize is being promoted as a source of provitamin A carotenoids (pVAC) in Zambia. Carotenoid retention in orange maize grains stored in metal silos, multilayer polyethylene and common woven bags, and maize meal packaged in single and multilayer polyethylene bags was evaluated. Significant differences in total pVAC retention were found between grain storage methods (48.1–57.2%) after 6 months of storage. Total pVAC retention in hammer meal (73.1–73.5%) was higher than in breakfast meal

(64.3–69.3%) after 4 months of storage; however, no differences in pVAC retention were found between meal types when stored in single and multilayer polyethylene bags. In general, β-cryptoxanthin (βCX) had higher retention than β-carotene (βC). Potential contribution of stored orange maize to the estimated average requirement of children and women was 26.5% and 24.3%, respectively. Orange maize meal can provide significant amounts of provitamin A to diets of Zambians even after 4 months of storage.

Keywords: Biofortified maize; Carotenoid degradation; Hermetic storage

1. INTRODUCTION

About 54% of children 0.5–5 years old in Zambia are vitamin A deficient (WHO, 2009). Maize is the most important crop for human consumption in Zambia, where annual production is 2.8 million metric tons and per capita intake is 287 g per day among women and 172 g per day among children 2–5 years old (FAOSTAT, 2016 ; Hotz *et al.*, 2011). Maize consumed in Zambia as well as in other Southern and Eastern African countries is mostly white-grained, and the consumption of yellow grains is minor (Smale et al., 2015). White maize has low levels of total carotenoids and does not have provitamin A carotenoids (pVAC), while common yellow maize has up to 66 $\mu g\,g^{-1}$ of total carotenoids but typically has less than 2 $\mu g\,g^{-1}$ of pVAC, making it a poor source of provitamin A (Pixley et al., 2013). The first group of biofortified orange maize varieties, which has been conventionally bred as an alternative to improve vitamin A status of populations, contains 6–9 $\mu g\,g^{-1}$ of pVAC. Biofortified orange maize varieties have been grown commercially in Zambia, Nigeria and Ghana since 2013 (Alamu *et al.*, 2015; Pixley *et al.*, 2013 ; Smale *et al.*, 2015). Malawi, Zimbabwe and Tanzania have also released biofortified orange maize recently (Hwang *et al.*, 2015 ; International Maize and Wheat Improvement Center (CIMMYT), 2016).

The efficacy of biofortified orange maize as an alternative for improving vitamin A status in children has been shown by two studies conducted in Zambia (Gannon *et al.*, 2014 ; Palmer *et al.*, 2016). The current target for maize biofortification efforts is 15 µg pVAC per g of maize, which should be sufficient for children and pregnant women to reach 50% of the estimated average requirement (EAR), assuming several factors including up to 50% pVAC losses during grain storage and cooking (Lividini and Fiedler, 2015 ; Saltzman *et al.*, 2013). However, recent modeling studies on the adoption of pVAC maize and its benefits in addressing vitamin A deficiency (VAD) in

Zambia suggest that positive impact may be limited by both the low retention of carotenoid during storage and postharvest grain losses (Lividini & Fiedler, 2015).

Postharvest grain losses of up to 30% occur in African countries when maize is stored using common storage methods, such as artisanal silos or woven bags. Losses are mainly attributed to infestation with insects, rodents or fungi and accumulation of mycotoxins (Tefera et al., 2011). Improved methods for preventing losses during storage have been promoted in African countries, including the use of multilayer polyethylene bags (PICS) and metal silos hermetically sealed with or without a candle (Tefera *et al.*, 2011 ; Yakubu *et al.*, 2011). Respiration of insects and fungi reduces oxygen (O_2) and increases carbon dioxide (CO_2) available in hermetic storage systems, creating a modified atmosphere that prevents migration of moisture and growth of such (micro) organisms (Baoua *et al.*, 2012 ; Garcia-Lara *et al.*, 2013).

In Zambia, maize is consumed mostly as *nshima*, a stiff porridge prepared with maize meal and boiling water (Smale et al., 2015). Different types of maize meal are used to prepare *nshima*, namely hammer meal, breakfast meal and roller meal. Hammer and breakfast meals are the most consumed throughout the country. Hammer meal is made of whole maize grain ground in a hammer or stone mill, whereas breakfast meal is produced by grinding the maize endosperm after removing most of the pericarp and germ. Typically, hammer and breakfast meals are not consumed on the same day they are milled but are stored for 4–6 months in woven bags or single layer polyethylene bags. The shelf life of hammer meal is shorter than that of breakfast meal due to the high fat content in the germ that leads to rancidity (Fiedler *et al.*, 2013 ; Gwirtz and Garcia-Casal, 2014).

It is known that carotenoids in different food matrices are highly degraded by oxygen, temperature and light during storage (De Moura *et al.*, 2015 ; Rodriguez-Amaya, 2015). In matrixes with high exposure to oxygen, such as flour, it was found that the oxygen level had greater influence on carotenoid degradation than temperature (Alves *et al.*, 2012 ; Bechoff *et al.*, 2010). In general, pVAC degradation in biofortified maize grain may reach 60% when grain is stored for 6 months using traditional methods, depending on the genotype (De Moura *et al.*, 2015 ; Ortiz *et al.*, 2016). Also, previous studies reported 78–100% pVAC retention when cooking freshly ground biofortified orange maize meal to prepare *nshima* (Gannon *et al.*, 2014 ; Mugode *et al.*, 2014). However, to our knowledge, there is no information on carotenoid retention during storage of breakfast and hammer meal produced with biofortified orange maize.

The objectives of this study were to: 1) determine the rate of carotenoid degradation during kernel storage under different storage conditions; 2) assess pVAC retention in hammer and breakfast meal stored for 4 months using different packaging methods; and 3) estimate the percent EAR of vitamin A provided by *nshima* made with biofortified orange maize.

2. MATERIALS AND METHODS

2.1. Maize grain for grain storage methods

Maize grains of biofortified orange maize variety GV664A collected from three commercial farms in Central Province, Zambia, in June 2015, were used to evaluate different grain storage methods. Ears were harvested at <14% grain moisture, dried for three days in shaded and ambient temperature conditions and then shelled using a mechanical sheller. Grains were immediately placed in different containers, as described in the grain storage methods section.

2.2. Grain storage methods

Maize grains from Central Province were stored for 6 months under ambient conditions with a daily average temperature of 22 °C. Average daily temperature during the first 60 days of storage was 17 °C, with a weekly variation between 15 and 19 °C. Grains were stored in traditional 50-kg woven bags, 100-kg PICS bags, 100-kg metal silos with candle and 100-kg metal silos without candle. Fifty kilograms of grain were used for each storage condition, in triplicate. PICS bags were filled and hermetically closed, as described by Baoua et al. (2012). Metal silos, with and without candle, were hermetically sealed with rubber bands, as described by Tefera et al. (2011). In addition, 50 kg of unshelled grain in woven bags were stored under the same conditions. Grains stored at the same temperature and for the same time period but in 3.75-kg vacuum-sealed aluminum Mylar bags were used as controls. Oxygen absorbers were placed inside the aluminum Mylar bags and immediately heat-sealed to create the vacuum. Total available O_2 at the beginning of the experiment was 27.9 L for metal silos and 5.2 L for woven and PICS bags. Grain samples for carotenoid analysis were taken at 0, 30, 90 and 180 days of storage from the metal silos and PICS, whereas for the woven bags, samples were taken at 0, 5, 10, 15, 30, 60, 90, 135 and 180 days of storage. For carotenoid analysis, samples from each repetition were collected in triplicate. Immediately after sampling, samples were placed in aluminum Mylar bags with oxygen absorbers and stored at −20 °C until shipment and analysis.

2.3. Evaluation of milling and packaging types

To evaluate different packaging types for maize meal, orange maize grains of variety GV664A were collected from Lusaka Province, Zambia, (June 2015) but not from the Central Province, due to logistical reasons. Samples from three different commercial farms were used. After shelling the ears, grains were stored for 3 months in 50-kg woven bags following common storage practices used by households in this region of Zambia. After storage, grains were processed either as breakfast meal or hammer meal. Breakfast meal was obtained by removing about 10% of the germ and pericarp from 150 kg of maize grains (90% extraction rate) using a de-huller and then milling the remaining endosperm into coarse flour using a commercial hammer mill and a sieve with 0.6-mm openings. Hammer meal was obtained by grinding 150 kg of whole maize kernels using the same commercial hammer mill and sieve.

2.4. Packaging types for maize meal

Hammer and breakfast meal were stored in 100-kg multilayer PICS and in 3.75-kg polyethylene and aluminum bags for 4 months under ambient conditions. Daily average temperature during the storage period was 24 °C, with a monthly variation between 23 and 25 °C. The amount of maize meal used was 25 kg for PICS bags and 2.5 kg for polyethylene and aluminum bags, in triplicate. Total available O_2 at the beginning of hammer meal storage was 2.92 L for PICS bags and 0.28 L for polyethylene bags. For breakfast meal, total available oxygen was 3.58 L for PICS bags and 0.34 L for polyethylene bags. Availability of O_2 in aluminum bags with oxygen absorbers for hammer and breakfast meal was <0.02 L. Maize meal samples for carotenoid analysis were taken at 0, 30, 60, 90 and 120 days of storage. Samples were stored at −20 °C in aluminum Mylar bags with oxygen absorbers until shipping and analysis. Sampling for carotenoid analysis was done in triplicate for each repetition.

2.5. Oxygen measurement

The O_2 and CO_2 levels were monitored with a portable gas analyzer Checkmate 3 (PBI Dansensor, Ringstead, Denmark) by inserting a needle through a rubber septum placed in each storage unit. Before the experiments, small 5-mm holes were opened in the metal silos and double septa were placed in the hole to measure oxygen over time without interfering with the atmosphere generated inside the silos during the storage period. Double septa were also placed in PICS bags for the same purpose. For grains, O_2 and CO_2 were measured daily for the first 90 days in modified metal silos, every week in PICS bags and just before sample collection in woven bags. For maize meal, O_2 and CO_2 were measured before sample collection.

2.6. Physical characterization of maize grain and meal

Bulk density of maize kernels and flour was determined using AACC Method 55-10 (AACC, 2000). Kernel dimensions were measured using a digital caliper (Salinas & Vazquez, 2006). The true density of grain and maize meal was determined using a liquid displacement method and particle size was measured as described by Salinas and Vazquez (2006).

2.7. Carotenoid analysis

All reagents and chemicals used were of HPLC and analytical grade. All *trans*-β-carotene (βC), 9-*cis*-β-carotene (9-*cis*-βC), 13-*cis*-β-carotene (13-*cis*-βC), β-cryptoxanthin (βCX), lutein (LUT) and zeaxanthin (ZEA) were purchased from CaroteneNature (Ostermundigen, Switzerland). An internal standard of β-apo-8′-carotenal was purchased from Sigma-Aldrich (St. Louis, MO). Methyl *tert*-butyl ether, tetrahydrofuran, dichloroethane, methanol, acetonitrile, ethyl alcohol and potassium hydroxide were obtained from Merck (Darmstadt, Germany). Carotenoid extraction and analysis was done as described in Rosales, Agama-Acevedo, Bello-Perez, Gutiérrez-Dorado, and Palacios-Rojas (2016). Total pVAC was computed as βC + (1/2)(13-*cis*-βC) + (1/2)(9-*cis*-βC) + (1/2)(βCX), which also represents total β-carotene equivalents (βCE).

2.8. Rate of carotenoid degradation and calculation of carotenoid retention

Rate of carotenoid degradation (k) during storage was calculated as:

$$Ln\frac{x}{x_0} = -kt$$

where x is the concentration of carotenoid at time t (day), x_0 is the concentration of carotenoid at the beginning of storage, and k is the constant rate of carotenoid degradation (day^{-1}). Since the experiment was carried out under field conditions, the temperatures were not constant. However, in order to use this rate of degradation (k) model, which assumes constant temperatures, a permissible temperature variation of 4 °C (weekly variation of daily average) for grains and 3 °C (monthly variation of daily average) for maize meal were used. With such restrictions, the storage time used for calculating the carotenoid degradation rate in grains was 60 days, and 120 days in maize meal.

The true retention (%TR) was calculated as:

$$\%TR = \frac{\text{pVAC per g of maize after storage (dry basis)} * 100}{\text{pVAC per g of maize before storage(dry basis)}}$$

2.9. Percent vitamin A EAR from orange maize nshima

To estimate the potential contribution of orange maize *nshima* to% vitamin A EAR for children 4–6 years old and women in Zambia (under typical grain and maize meal storage conditions), two scenarios of maize intake were considered based on assumptions from modeling studies by Lividini and Fiedler (2015). The first scenario assumed that biofortified maize fully substituted consumption of white maize and that the daily maize intake was 287 g for women and 172 g for children 4–6 years old. The second scenario assumed partial adoption of biofortified maize; therefore, consumption was 117 g for women and 71 g for children 4–6 years old. To calculate total true retention during storage of maize grain and meal, a maximum of 8 months of storage as grain and 4 months of storage either as hammer or breakfast meal was assumed. These values were based on the common shelf life of hammer meal which is 4 months (Gwirtz & Garcia-Casal, 2014). The *k* values obtained (day^{-1}) for maize grains and meal using traditional storage and packaging methods were used to calculate the monthly carotenoid retention for grains and hammer meal and combined to generate the total retention for a season in Zambia. Vitamin A EAR of 500 μg for women and 275 μg for children 4–6 years was used (IOM, 2001). Retention during cooking of *nshima* was considered to be 78%, as reported by Gannon et al. (2014).

2.10. Statistical analysis

Data were expressed as mean ± standard deviation (SD). For the experiments with grains, the effects of storage methods and time were evaluated using a 2-factor ANOVA model. ANOVA and linear regression analysis were performed using the MIXED and REG procedures in SAS 9.4. Means separation was calculated using Duncan's Multiple Range Test. Differences between means were considered significant at $p < 0.05$.

3. RESULTS AND DISCUSSION

3.1. Physical characteristics of GV664A maize grains and meal

Grain samples from Central Province had 11% moisture content before milling. The thousand-kernel weight was 224 ± 2 g. Average kernel dimensions were 8.91 ± 0.25 mm length, 8.14 ± 0.14 mm width and 4.08 ± 0.21 mm thickness. True density was 1.288 ± 0.030 kg L^{-1} and bulk density was 0.794 ± 0.010 kg L^{-1}. Bulk density was lower than values reported by Ortiz et al. (2016) for flint type orange maize (0.887–0.984 kg L^{-1}at 8–16% moisture), but much higher than common white or yellow maize (0.682–0.772 kg L^{-1} at 14% moisture) (Sahasrabudhe, 2015). High bulk density compared to the values reported for common yellow maize varieties was possibly due to the high true density of grains as well as small grain dimensions. A higher bulk density could be important because when the storage unit is hermetically sealed, less O_2 will be available inside to react with carotenoids, compared to most common commercial white and yellow maize.

Bulk density of breakfast and hammer meal was 0.842 ± 0.005 and 0.764 ± 0.003 kg L^{-1}, respectively. True density for both meal types was 1.50 ± 0.01 kg L^{-1}. Particle size of breakfast meal was finer than hammer meal, with 25% and 46% of particles bigger than 0.50 mm and 14% and 5% smaller than 0.25 mm, respectively. Significantly higher bulk density in breakfast meal could be attributed to factors such as different particle size distribution and less germ content compared to hammer meal.

3.2. Oxygen content in storage units

The gas composition in the woven and PICS bags at the beginning of the experiment was that of normal air (20.6% O_2 + 0.03% CO_2), whereas for metal silos with lit candles, the O_2 level was 16.5%. Initial O_2 concentration in metal silos with lit candles was lower than in the other storage methods, due to the O_2 consumed by the candles inside the silos. To our knowledge, no reports of the minimum O_2 level in a hermetically sealed metal silo with a candle inside are available. The average O_2 level observed in the aluminum silos during the 90 days of monitoring in grain was 18.5%, whereas in PICS bags it was 18.7%. The oxygen level in both hermetic storage methods did not drop to less than 1–5% as suggested by other studies that monitored O_2 levels in hermetic jars in the laboratory and PICS bags in the field. This difference might be explained because in such studies accelerated O_2 consumption inside the storage methods used was due to the growth of inoculated insects (Moreno-Martinez *et al.*, 2000; Mutungi *et al.*, 2015; Tefera *et al.*, 2011 ; Yakubu *et al.*, 2011). Natural O_2 reduction due to insect respiration in hermetically sealed commercial metal silos filled with grain has not been clearly documented; however, in field experiments insect infestation was reduced to only 4% when using small commercial metal silos (De Groote et al., 2013).

3.3. Carotenoid composition of maize grains and hammer meal

Total carotenoid content of orange maize grains from the Central Province before being stored as grain was 20.4 ± 0.2 $\mu g\,g^{-1}$ in dry basis. Individual pVAC levels were between 1.2 ± 0.1 and 4.9 ± 0.4 $\mu g\,g^{-1}$, representing 55.1% of the total carotenoids quantified and 7.6 ± 0.1 $\mu g\,g^{-1}$ pVAC (Table 1). The most predominant pVAC found was βCX with 24.1%, followed by βC with 19.1%. The 9-*cis*-βC and 13-*cis*-βC isomers were found in lower proportions, with 6.1% and 5.8%, respectively. The proportions of βCX and βC were within the ranges observed in other orange maize genotypes, 19.0–28.0% for βCX and 12–23% for βC (Mugode *et al.*, 2014; Muzhingi *et al.*, 2016; Ortiz *et al.*, 2016 ; Pillay *et al.*, 2014). Non pVAC ZEA and LUT represented 31.2% and 12.2% of the total quantified carotenoids, respectively (Table 1).

Table 1. Carotenoid content ($\mu g\,g^{-1}$ dry weight)[a] of orange maize grain, meal and bran variety GV664A grown in Zambia.

Grain source	Grain fraction or type	LUT	ZEA	βCX	13-*cis*-βC	βC	9-*cis*-βC	pVAC
Central Province[§]	Whole grain	2.6 ± 0.2[a]	6.5 ± 0.1[a]	4.9 ± 0.4[a]	1.2 ± 0.0[a]	3.9 ± 0.2[a]	1.2 ± 0.1[a]	7.6 ± 0.1[a]
Lusaka Province[€]	Whole grain	1.7 ± 0.1[b]	4.1 ± 0.2[b]	3.4 ± 0.3[b]	0.7 ± 0.0[c]	2.1 ± 0.1[c]	0.9 ± 0.0[b]	4.6 ± 0.3[c]
Lusaka Province[€]	Breakfast meal	2.0 ± 0.3[b]	4.9 ± 0.8[b]	3.6 ± 0.5[b]	0.9 ± 0.0[b]	2.6 ± 0.2[b]	1.1 ± 0.1[a]	5.4 ± 0.1[b]
Lusaka Province[€]	Bran + germ	0.3 ± 0.2[c]	0.8 ± 0.4[c]	0.7 ± 0.2[c]	0.3 ± 0.1[d]	0.6 ± 0.2[d]	0.3 ± 0.1[c]	1.3 ± 0.4[d]

[a] Different lower case letters indicate significant differences within columns ($p < 0.05$). Values are average ± standard deviation (three farms). LUT = lutein; ZEA = zeaxanthin; βCX = β-cryptoxanthin; 13-*cis*-βC = 13-*cis*-β-carotene; βC = all *trans*-β-carotene; 9-*cis*-βC = 9-*cis*-β-carotene; pVAC = provitamin A carotenoids (expressed as β-carotene equivalents). Whole grain = hammer meal.

[§] Freshly harvested grains from Central Province.

[€] Obtained from grain stored for 3 months in farms.

Considering that maize grain used to produce either hammer meal or breakfast meal is usually not fresh, and to understand carotenoid degradation in maize meal stored under common storage conditions in Zambia, orange maize harvested in Lusaka Province was stored for three months and then milled to produce these two meal types. Total carotenoid content in the stored grains used to produce both meal types was 12.8 ± 0.7 $\mu g\,g^{-1}$, whereas their pVAC was 4.6 ± 0.3 $\mu g\,g^{-1}$, representing 55.0% of the total carotenoids identified (Table 1). The most predominant pVAC was βCX with 26.4%, whereas βC, 9-*cis*-βC, 13-*cis*-βC represented 16.0%, 6.9% and 5.7%, respectively. These stored grains from Lusaka

Province had 37% less pVAC than fresh grains from Central Province. Taking into consideration the storage time and environmental effect, the difference in pVAC between maize grains from the two provinces was reasonable and representative of what could be expected in typical farms in Zambia. The carotenoid content in hammer meal was the same as the carotenoid content found in its unprocessed stored grains, since the particle size reduction required to analyze carotenoids in maize grains resembles the production of hammer meal.

3.4. Carotenoid profile of breakfast meal

In freshly milled breakfast meal at 90% extraction rate obtained with the stored grain from Lusaka Province, the total carotenoid content and pVAC was 14.8 ± 1.1 $\mu g\ g^{-1}$ and 5.4 ± 0.1 $\mu g\ g^{-1}$, respectively. The pVAC present in the highest proportion was βCX (24.4%), followed by βC (17.6%), 9-*cis*-βC (6.9%) and 13-*cis*-βC (5.8%), whereas ZEA and LUT represented 32.1% and 13.3% of total quantified carotenoids. In general, the proportions of carotenoids found in breakfast meal were similar to what was found in their grains. Levels of LUT, ZEA and βCX were not significantly different between freshly milled breakfast meal and hammer meal; however, βC, 9-*cis*-βC, 13-*cis*-βC and pVAC were significantly higher in breakfast meal (Table 1).

The bran fraction obtained when producing breakfast meal had the lowest total carotenoid and pVAC with 3.0 ± 1.2 and 1.3 ± 0.4 $\mu g\ g^{-1}$, respectively (Table 1). Proportions of βCX, βC, 9-*cis*-βC and 13-*cis*-βC were 23.6%, 21.2%, 11.7% and 11.2%, respectively. The lower carotenoid content found in the bran is in agreement with reported values for other maize genotypes (Burt *et al.*, 2010; Ndolo and Beta, 2013). The low carotenoid content in bran removed from grain when producing breakfast meal may explain the higher carotenoids levels observed in freshly milled breakfast meal compared to hammer meal. However, the total proportion of pVAC removed from the bran represented only 3% of the pVAC present in the kernel, whereas the proportion of bran removed was 10%. This is an important aspect to consider when calculating per capita pVAC intake based on total maize grain consumed at the household level, because elimination of germ and pericarp from breakfast meal (mostly endosperm) likely results in greater carotenoid intake compared to hammer meal.

3.5. Carotenoid degradation in grain under different storage conditions

The pVAC contents in maize grains after 180 days of storage were 8.0 ± 0.4, 4.3 ± 0.2, 4.3 ± 0.3, 3.9 ± 0.2, 3.8 ± 0.2 and 3.6 ± 0.1 $\mu g\ g^{-1}$, representing true retention of 104.3%, 57.2%, 57.1%, 51.4%, 50.6% and 48.1% for aluminum bags, PICS bags, silo with candle, woven bags, silo without candle and ears in woven bags, respectively (Fig. 1). Carotenoid retention in maize stored in aluminum bags with oxygen absorbers was the highest, followed by silos with candles and PICS bags, whereas grains in woven bags, silos without candles and ears in woven bags had the lowest carotenoid retention ($p < 0.05$). The minimum retention values found in the different storage methods used were slightly higher than the 40.0% retention limit reported by Mugode et al. (2014) for orange maize stored at 20 °C (average temperature) using traditional methods in Zambia. However, in that study, fast degradation was observed during the first two weeks of storage and no information on the kinetics of degradation was provided.

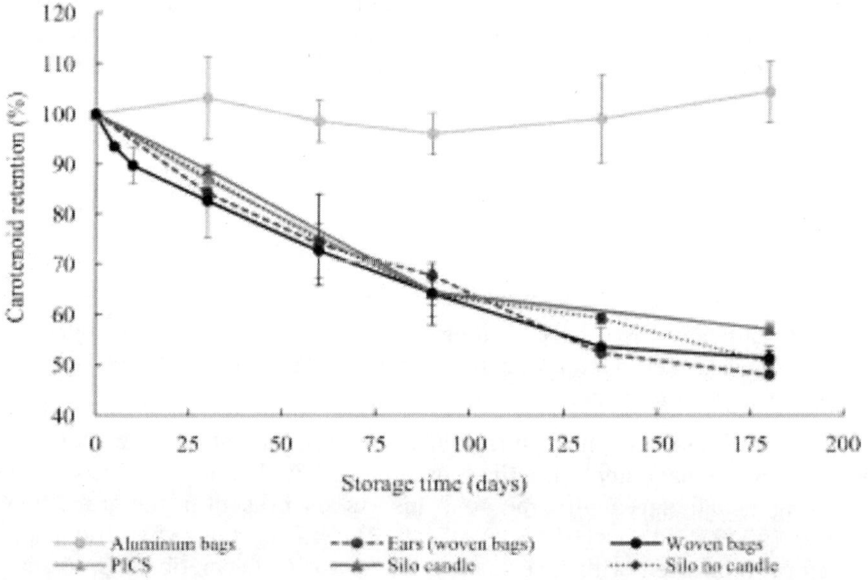

Fig. 1. Effect of storage method on carotenoid pVAC retention in orange maize grain variety GV664A stored using different methods for 180 days under ambient conditions in Zambia.

Degradation rate of βCX, 13-*cis*-βC, βC and 9-*cis*-βC in orange maize grains stored in woven bags was calculated for the first 60 days of storage. Based on the coefficient of determination (r^2) observed when obtaining the degradation rate constant values, all pVAC followed first-order degradation kinetics. First-order degradation has been also reported for pVAC in orange maize grains (Ortiz et al., 2016) and for βC in sweet potato and cassava (Bechoff *et al.*, 2010; Bechoff *et al.*, 2015). When comparing *k* values of each carotenoid, βC had the highest *k* values (0.00777 ± 0.00060), followed by 13-*cis*-βC (0.00511 ± 0.00051) and βCX (0.00381 ± 0.00058), whereas 9-*cis*-βC (0.00276 ± 0.00043) had the lowest *k* values ($p < 0.05$). The *k* values for βCX and βC were within the ranges reported for different orange maize genotypes by Ortiz et al. (2016), who also reported higher *k* values for βC than βCX in maize stored at 22 °C under non-hermetic conditions. The lower *k* values (51% lower) observed for βCX compared to βC are relevant because maize grain bred with higher proportions of βCX, as opposed to βC, could have better pVAC retention during storage and consequently higher impact on vitamin A deficient populations. Higher βCX stability during storage together with higher βCX bioavailability compared to βC, and βCX's similar bioconversion and bioefficacy to βC (Schmaelzle et al., 2014), adds strong support to the βCX breeding strategy proposed by Dhliwayo, Palacios-Rojas, Crossa, and Pixley (2014), which is based on genetically controlling pVAC accumulation in maize grains. Large genetic diversity for βCX content in maize is already documented and makes the strategy feasible (Swarno, Palacios-Rojas, Kaeppler, & Babu, 2015).

In general, these results suggest that improved storage methods that reduce grain losses (such as hermetic silos with low O_2 and PICS bags) could also contribute to reduce pVAC degradation in orange maize grains. The lower degradation in these storage systems may be due to the lower oxygen content inside them. Despite the very high carotenoid retention in aluminum bags with oxygen absorbers, this method is too expensive and not feasible for use in rural households; however, results showed that oxygen is a key factor when storing orange maize grains at room temperature in Zambia.

Grains to be stored must be free of insect contamination; however, given the poor management practices in the field and during harvesting and drying, usually larvae or even adult insects are present in the grain. Baoua et al. (2012) and Garcia-Lara et al. (2013) reported that when the grain is naturally infested with insects before storage in hermetic silos, available oxygen inside the silos decreases to 5% due to insect respiration, resulting in minimum grain damage at the end of the storage period. If under such conditions grain integrity is well preserved, there is a possibility that the carotenoid degradation rate during storage of initially infested grains will be lower than in non-infested grains.

3.6. Degradation kinetics of carotenoids in maize meal

Carotenoid levels in maize meal after storage were significantly lower than in the freshly milled maize for all the packaging methods used except aluminum bags. The pVAC levels at the end of the storage period were 4.8 ± 0.5, 3.3 ± 0.2, and 3.3 ± 0.2 $\mu g\ g^{-1}$ for hammer meal in aluminum, PICS and polyethylene bags, whereas for breakfast meal these values were

4.9 ± 0.3, 3.8 ± 0.3 and 3.5 ± 0.1 μg g^{-1}, respectively. The retention of pVAC in hammer meal after 120 days of storage was 104.8%, 73.5%, 73.1% for aluminum bags, polyethylene bags and PICS, whereas for breakfast meal it was 89.8%, 64.3% and 69.3%, respectively (Fig. 2). As observed in the grain, carotenoid retention in maize meal stored in aluminum bags was the highest among all storage methods after 120 days of storage ($p < 0.05$).

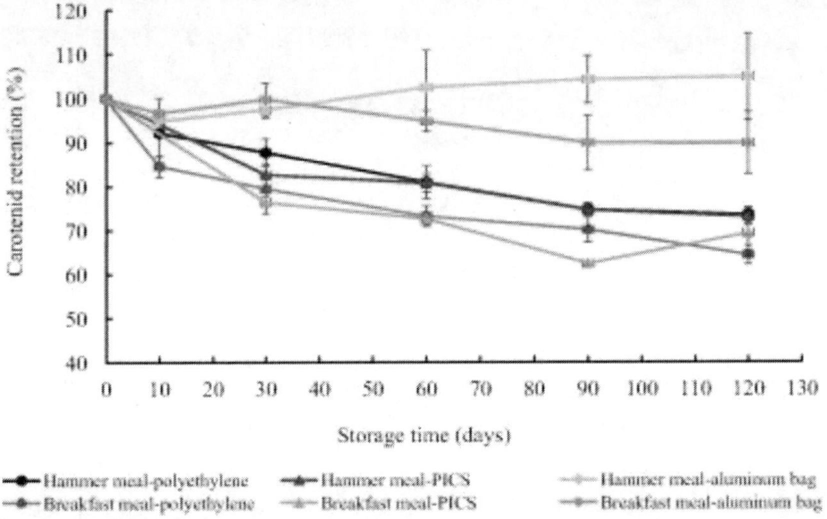

Fig. 2. Effect of storage method on pVAC retention in breakfast meal and hammer meal made from orange maize variety GV664A and stored using different packaging methods for 120 days under ambient conditions in Zambia.

Degradation constant rates for pVAC were determined for breakfast and hammer meal stored in different packaging methods. All pVAC followed first-order degradation kinetics for all packaging methods based on the high coefficient of determination ($r^2 > 0.8$), except for the control (aluminum foil with oxygen absorbers), which had low r^2 values (Table 2). Hammer meal in single layer polyethylene bags had lower k values than breakfast meal in the same packaging type for 13-*cis*-βC, 9-*cis*-βC, βC and pVAC, whereas hammer meal in PICS bags had lower k values than breakfast meal in PICS bags for 13-*cis*-βC, βC and pVAC (Table 2).

Table 2. Degradation constant rate (k) [a] expressed as day^{-1} for pVAC in orange maize meal stored for 120 days in different packaging methods.

Type of meal	Package	Carotenoid	k	r^2	p
BF meal	Polyethylene	βCX	0.00327^a	0.92	<0.01
BF meal	PICS		0.00349^a	0.82	<0.01
BF meal	Aluminum		0.00088^b	0.26	0.16
HM meal	Polyethylene		0.00283^a	0.93	<0.01
HM meal	PICS		0.00292^a	0.92	<0.01
HM meal	Aluminum		0.00023^b	−0.05	0.99
BF meal	Polyethylene	13-*cis*-βC	0.00345^a	0.89	<0.01
BF meal	PICS		0.00323^a	0.89	<0.01
BF meal	Aluminum		0.00107^c	0.77	<0.01
HM meal	Polyethylene		0.00215^b	0.97	<0.01
HM meal	PICS		0.00215^b	0.95	<0.01
HM meal	Aluminum		0.00059^c	0.67	<0.01
BF meal	Polyethylene	βC	0.00526^a	0.88	<0.01
BF meal	PICS		0.00508^a	0.88	<0.01
BF meal	Aluminum		0.00141^c	0.70	<0.01
HM meal	Polyethylene		0.00344^b	0.95	<0.01
HM meal	PICS		0.00344^b	0.95	<0.01
HM meal	Aluminum		0.00072^c	0.37	<0.01
BF meal	Polyethylene	9-*cis*-βC	0.00327^a	0.92	<0.01
BF meal	PICS		0.00300^{ab}	0.91	<0.01
BF meal	Aluminum		0.00095^c	0.75	<0.01
HM meal	Polyethylene		0.00221^b	0.97	<0.01
HM meal	PICS		0.00232^b	0.96	<0.01
HM meal	Aluminum		0.00043^c	0.36	0.43
BF meal	Polyethylene	pVAC	0.00416^a	0.91	<0.01
BF meal	PICS		0.00416^a	0.88	<0.01
BF meal	Aluminum		0.00095^c	0.52	<0.01
HM meal	Polyethylene		0.00296^b	0.95	<0.01
HM meal	PICS		0.00302^b	0.94	<0.01
HM meal	Aluminum		0.00035^c	0.14	0.15

[a] Different lower case letters indicate significant differences within columns ($p < 0.05$). Values of k are average of three farms. BF = breakfast meal; HM = hammer meal; βCX = β-cryptoxanthin; 13-cis-βC = 13-*cis*-β-carotene; βC = all *trans*-β-carotene; 9-*cis*-βC = 9-*cis*-β-carotene; pVAC = provitamin A carotenoids (expressed as β-carotene equivalents).

Significant differences in k values for all carotenoids were observed between packaging types after 120 days of storage. Similar to the results on grains, the k values of βC (0.00344–0.00526) in maize meal were significantly higher ($p < 0.05$) than for all other pVAC (0.00215–0.00349), suggesting that βC is more unstable than all other pVAC in maize regardless of the physical properties of the matrix. These k values for βC are lower than values reported in sweet potato flour (0.00930–0.01930) when stored at 20–30 °C (Bechoff et al., 2010), and dry cassava granules stored at 19–33 °C ($k = 0.0083–0.02710$) (Bechoff et al., 2015). Considering the large effect of oxygen observed in grains and maize meal, it could be hypothesized that an important reason why k values obtained for maize meal were lower than reported values for sweet potato at slightly lower temperature (20 °C) is that maize meal granules are bigger than sweet potato flour granules. The dense matrix of maize meal granules could also have restricted the reaction of oxygen from the environment with the carotenoids compared to cassava granules (at 20 °C) which have light granules. Similarly, the higher k values of breakfast meal compared to hammer meal may be due in part to the smaller particle size of breakfast meal which exposes it to oxygen, but also, to the presence of tocochromanols (mainly tocopherols) in hammer meal. The germ contained in hammer meal could have prevented degradation of its carotenoids, considering that tocopherols are present mostly in the corn germ, and tocopherols reduce the degradation of carotenoids due to their antioxidant capacity (Muzhingi et al., 2016). The integrity of the cellular compartments may be crucial for reducing the oxygen exposure of carotenoids, thereby their stability. Similar observations were reported by Rosales et al. (2016) when comparing pVAC retention during traditional and extruded lime-cooking process. Retention of pVAC was higher in traditional methods where grains are milled after being cooked, as compared to methods where milling occurs before extrusion.

In general, based on the calculated *k* values for maize grain and hammer meal of variety GV664A, and assuming a maximum storage period of 8 months for grain and 4 months for meal, the average retention during storage using traditional methods was 46.7% for hammer meal and 43.8% for breakfast meal. Therefore, biofortified orange maize should be commercialized as grain and milled shortly before consumption, preferably as hammer meal. Also, farmers and consumers should be advised to consume biofortified orange maize first, before extensive carotenoid degradation occurs, and consume white maize when orange maize is no longer available. For urban commercialization of milled orange maize, the time it is kept on store shelves should be minimized.

3.7. Contribution of orange maize nshima to daily intake of vitamin A

Assuming total replacement of white maize with GV664A or similar cultivars of biofortified orange maize (containing about 7.6 $\mu g\ g^{-1}$ pVAC) and 78% retention during cooking, *nshima* prepared with orange maize hammer meal made from freshly ground maize provides 56.8% and 52.1% EAR, while maize stored for up to 4 months provides 26.5% and 24.3% EAR, for children 4–6 years old and women, respectively. Biofortified orange maize provides more than the target 50% EAR if grain and flour are processed fresh, with no storage period. New orange maize genotypes currently in the variety release pipeline contain 15 or more $\mu g\ g^{-1}$ pVAC, and the% EAR that they provide will increase correspondingly (www.cimmyt.org). Given the genotypic differences in pVAC retention, further research is needed to verify that new cultivars with 15 or more $\mu g\ g^{-1}$ indeed provide the predicted 112.6% and 103.4% EAR when consuming fresh maize and 52.6% and 48.2% EAR for stored maize, for children 4–6 years of age and women, respectively.

Considering Lividini and Fiedler's (2015) estimates that orange maize will be only partially adopted, with consumption levels reaching 71 g for children 4–6 years of age and 117 g for women, then the expected contribution of orange maize to% EAR would be 21.7% and 19.7%, respectively, when storing grain and meal of varieties with 15 $\mu g\ g^{-1}$pVAC. These values are nutritionally important although they are lower than the target 50% EAR expected for biofortified maize in Zambia. Under this scenario, alternatives to ensure higher retention would be important, as well as continued breeding for higher content of pVAC, including higher βCX content. Breeding for increased content of antioxidant compounds (i.e., tocochromanols) mainly in the grain endosperm could also enhance resistance to oxidative degradation of pVAC (Muzhingi et al., 2016). To reduce vitamin A deficiency in Zambia, nutritional education, diet diversification, sanitation and disease prevention efforts, as well as creating awareness and demand for biofortified orange maize should be continued.

4. CONCLUSION

Carotenoid degradation was similar for ears and shelled orange maize. Improved maize storage methods, such as PICS bags and metal silos with low O_2, enhance carotenoid retention, compared to traditional storage methods. More emphasis should be given to increase βCX in new orange maize genotypes because βCX is more stable than βC in all storage and packaging methods for either grains or maize meal. Also, pVAC carotenoids are more stable in hammer meal compared to breakfast meal, mainly due to the larger particle size and smaller surface exposed to oxygen for hammer meal relative to breakfast meal. The potential scenario that had the lowest pVAC in cooked nshima, considering the current pVAC levels of orange maize being cultivated in Zambia, provided 26.5% and 24.3% of vitamin A% EAR for children 4–6 years old and women. The time during which orange maize meal sits on market shelves should be minimized so that its contribution to the vitamin A status of Zambians is maximized.

ACKNOWLEDGMENTS

Authors wish to thank the HarvestPlus team in Zambia, Howard Tembo, Ivor Mukuka, and the Storage Unit at the Zambian Agricultural Research Institute (ZARI) for their support in sample storage. We also thank Erick Boy and Kevin Pixley for editing and reviewing the manuscript.

*Financial support:*HarvestPlus and CRP-MAIZE.

REFERENCES

1. AACC (2000). Approved methods of analysis. Method 55-10 (11th ed.). St. Paul, Minn., USA: American Association of Cereal Chemist.

2. Alamu, E. O., Maziya-Dixon, B., Menkir, A., & Olaofe, O. (2015). Effects of husk and harvesting on provitamin A activity and sensory properties of boiled fresh orange maize hybrids. Journal of Food Quality, 38, 387–395.
3. Alves, R. M. V., Ito, D., Carvalho, J. L. V., Melo, W. F., & Godoy, R. L. O. (2012). Stability of biofortified sweet potato flour. Brazilian Journal of Food Technology, 15, 59–71.
4. Baoua, I. B., Margam, V., Amadou, L., & Murdock, L. L. (2012). Performance of triple bagging hermetic technology for postharvest storage of cowpea grain in Niger. Journal of Stored Products Research, 51, 81–85.
5. Bechoff, A., Chijioke, U., Tomlins, K., Govinden, P., Ilona, P., & Westby, A. (2015). Carotenoid stability during storage of yellow gari made from biofortified cassava or with palm oil. Journal of Food Composition and Analysis, 44, 36–44.
6. Bechoff, A., Dhuique-Mayer, C., Dornier, M., Tomlins, K. I., Boulanger, R., Dufour, D., & Westby, A. (2010). Relationship between the kinetics of b-carotene degradation and formation of norisoprenoids in the storage of dried sweet potato chips. Food Chemistry, 121, 348–357.
7. Burt, A. J., Grainger, C. M., Young, C., Shelp, B. J., & Lee, E. A. (2010). Impact of postharvest handling on carotenoid concentration and composition in highcarotenoid maize (Zea mays L.) kernels. Journal of Agricultural and Food Chemistry, 58(14), 8286–8292.
8. De Groote, H., Kimenju, S. C., Likhayo, P., Kanampiu, F., Tefera, T., & Hellin, J. (2013). Effectiveness of hermetic systems in controlling maize storage pests in Kenya. Journal of Stored Products Research, 53, 27–36.
9. De Moura, F. F., Miloff, A., & Boy, E. (2015). Retention of provitamin A carotenoids in staple crops targeted for biofortification in Africa: Cassava, maize and sweet potato. Critical Reviews in Food Science and Nutrition, 55, 1246–1269.
10. Dhliwayo, T., Palacios-Rojas, N., Crossa, J., & Pixley, K. V. (2014). Effects of S1 recurrent selection for provitamin-A carotenoid content for three openpollinated maize cultivars. Crop Science, 54, 2449–2460.
11. FAOSTAT (2016). Food and Agriculture Organization of the United Nations Available at: [Accessed 07.09.2015] http://faostat.fao.org.
12. Fiedler, J. L., Lividini, K., Zulu, R., Kabaghe, G., Tehinse, J., & Bermudez, O. I. (2013). Identifying Zambia's industrial fortification options: Toward overcoming the food nutrition information gap-induced impasse. Food and Nutrition Bulletin, 34, 480–500.
13. Gannon, B., Kaliwile, C., Arscot, S. A., Schmaelzle, S., Chileshe, J., Kalungwana, N., ... Tanumiardjo, S. A. (2014). Biofortified orange maize is as efficacious as a vitamin A supplement in Zambian children even in the presence of high liver reserves of vitamin A: A community-based, randomized placebo-controlled trial. American Journal of Clinical Nutrition, 100, 1541–1550.
14. Garcia-Lara, S., Ortiz-Islas, Sofia., & Villers, P. (2013). Portable hermetic storage bags resistant to Prostephanus truncatus, Rhyzopertha dominica, and Callosobruchus maculatus. Journal of Stored Products Research, 54, 23–25.
15. Gwirtz, J. A., & Garcia-Casal, M. N. (2014). Processing maize flour and corn meal food products. Annals of the New York Academy of Sciences, 1312, 66–75.
16. Hotz, C., Palaniappan, U., Chileshe, J., Kafwembe, E. M., & Siamusantu, W. (2011). Nutrition survey in central and eastern provinces, Zambia 2009: Focus on vitamin A and maize intakes, and vitamin A status among women and children Washington (DC), National Food and Nutrition Commission of Zambia, Lusaka, Zambia; Tropical Diseases Research Centre, Ndola, Zambia; HarvestPlus, CGIAR, IFPRI, Washington, D. C..
17. Hwang, T., Ndolo, V. U., Katundu, M., Nyirenda, B., Bezner-Kerr, R., Arntfield, S., & Beta, T. (2015). Provitamin A potential of landrace orange maize variety (Zea mays L.) grown in different geographical locations of central Malawi. Food Chemistry, 196, 1315–1324.

18. Institute of Medicine (IOM) (2001). Dietary reference intakes for vitamin a, vitamin k, arsenic, boron, chromium, copper, iodine, iron, manganese, molybdenum, nickel, silicon, vanadium, and zinc. Washington, D. C.: National Academy Press.

19. International Maize and Wheat Improvement Center (CIMMYT) (2016). Biofortification to fight "hidden hunger" in Zimbabwe Available at: [Accessed 30.10.2016] http://www.cimmyt.org/biofortification-to-fight-hidden-hungerin-zimbabwe/.

20. Lividini, K., & Fiedler, J. (2015). Assessing the promise of biofortification: A case study of high provitamin A maize in Zambia. Food Policy, 54, 65–77.

21. Moreno-Martinez, E., Jimenez, S., & Vazquez, M. E. (2000). Effect of Sitophilus zeamais and Aspergillus chevalieri on the oxygen level in maize stored hermetically. Journal of Stored Products Research, 36, 25–36.

22. Mugode, L., Ha, B., Kaunda, A., Sikombe, T., Phiri, S., Mutale, R., ... De Moura, F. F. (2014). Carotenoid retention of biofortified provitamin A maize (Zea mays L.) after Zambian traditional methods of milling, cooking and storage. Journal of Agricultural and Food Chemistry, 62, 6317–6325.

23. Mutungi, C., Affognon, H. D., Njoroge, A. W., Manono, J., Baributsa, D., & Murdock, L. L. (2015). Triple-layer plastic bags protect dry common beans (Phaseolus vulgaris) against damage by Acanthoscelides obtectus (Coleoptera: Chrysomelidae) during storage. Journal of Economic Entomology, 108, 2479–2488.

24. Muzhingi, T., Palacios-Rojas, N., Miranda, A., Cabrera, M. L., Yeum, K., & Tang, G. (2016). Genetic variation of carotenoids, vitamin E and phenolic compounds in provitamin A enriched maize. Journal of the Science of Food and Agriculture.

25. Ndolo, V., & Beta, T. (2013). Distribution of carotenoids in endosperm, germ, and aleurone fractions of cereal kernels. Food Chemistry, 139, 663–671.

26. Ortiz, D., Rocheford, T., & Ferruzi, M. G. (2016). Influence of temperature and humidity on the stability of carotenoids in biofortified maize (Zea mays L.) genotypes during controlled post-harvest storage. Journal of Agricultural and Food Chemistry, 64, 2727–2736.

27. Palmer, A. C., Chileshe, J., Hall, A. G., Barffour, M. A., Molobeka, N., West, K. P., & Marjorie, J. H. (2016). Short-term daily consumption of provitamin A carotenoid-biofortified maize has limited impact on breast milk retinol concentrations in Zambian women enrolled in a randomized controlled feeding trial. The Journal of Nutrition.

28. Pillay, K., Siwela, M., Derera, J., & Veldman, F. J. (2014). Provitamin A carotenoids in biofortified maize and their retention during processing and preparation of South African maize foods. Journal of Food Science and Technology, 51, 634–644.

29. Pixley, K., Palacios-Rojas, N., Babu, R., Mutale, R., Surles, R., & Simpungwe, E. (2013). Biofortification of maize with provitamin A carotenoids. In S. A. Tanumihardjo (Ed.), Carotenoids and human health (pp. 271–292). New York: Springer Science.

30. Rodriguez-Amaya, D. B. (2015). Food carotenoids: Chemistry, biology, and technology (p. 309). UK: John Wiley & Sons.

31. Rosales, A., Agama-Acevedo, E., Bello-Perez, L., Gutiérrez-Dorado, R., & PalaciosRojas, N. (2016). Effect of traditional and extrusion nixtamalization on carotenoid retention in tortillas made from provitamin A-enriched maize (Zea mays L.). Journal of Agricultural and Food Chemistry, 64(4), 8289–8295.

32. Sahasrabudhe, S. N. (2015). Corn characterization and development of a convenient laboratory scale alkaline cooking process [Master thesis]. University of NebraskaLincoln. Salinas, M. Y., & Vazquez, C. G. (2006). Metodologías de análisis de la calidad nixtamalera-tortillera en maíz, Folleto técnico No. 24 (p. 98).

33. Mexico: INIFAP. Saltzman, A., Birol, E., Bouis, H., Boy, E., De Moura, F. F., Islam, Y., & Pfeiffer, W. H. (2013). Biofortification: Progress toward a more nourishing future. Global Food Security, 2(2013), 9–17. Schmaelzle, S., Gannon, B., Crawford, S., Arscott, S., Goltz, S., Palacios-Rojas, N., ... Tanumihardjo, S. (2014). In vitro micellarization does not predict the relationship in vivo for provitamin A

carotenoids bioefficacy from staple and mixed foods in Mongolian gerbils. Journal of Agriculture and Food Chemistry, 62, 136–143.
34. Smale, M., Simpungwe, E., Biro, E., Kassie, G. T., de Groote, H., & Mutale, R. (2015). The changing structure of the maize seed industry in Zambia: Prospects for orange maize. Agribusiness, 31, 132–146.
35. Swarno, W., Palacios-Rojas, N., Kaeppler, S., & Babu, R. (2015). Genome-wide association analysis reveals new targets for carotenoid biofortification in maize. Theoretical and Applied Genetics, 128, 851–864.
36. Tefera, T., Kanampiu, F., De Groote, H., Hellin, J., Mugo, S., Kimenju, S., ... Banziger, M. (2011). The metal silo: An effective grain storage technology for reducing postharvest insect and pathogen losses in maize while improving smallholder farmers' food security in developing countries. Crop Protection, 30, 240–245.
37. World Health Organization (WHO) (2009). Global prevalence of vitamin a deficiency in populations at risk 1995–2005.
38. WHO global database on vitamin A deficiency. Geneva: World Health Organization. Yakubu, A., Bern, C. J., Coats, J. R., & Bailey, T. B. (2011). Hermetic off-farm storage for maize weevil control in East Africa. African Journal of Agricultural Research, 6, 3311–3319.

CHAPTER 9

Green and Efficient Extraction Method to Determine Polyphenols in Cocoa and Cocoa Products

Merichel Plaza[1,2] & Diana Oliveira[3] & Anne Nilsson[3] & Charlotta Turner[1]

[1]*Department of Chemistry, Centre for Analysis and Synthesis, Lund University, P.O. Box 124, -221 00 Lund, SE, Sweden*
[2] *Department of Analytical Chemistry, Physical Chemistry and Chemical Engineering, Faculty of Biology, Environmental Sciences and Chemistry, Universidad de Alcalá, Ctra. Madrid-Barcelona Km. 33.600, Alcalá de Henares, Madrid, Spain*
[3] *Food for Health Science Centre, Lund University, Lund, Sweden*

ABSTRACT

Cocoa and cocoa products have broad and well-proven health benefits, most of which are associated with the high antioxidant capacity of cocoa flavonoids. Most extraction methods for flavonoids in cocoa products use several steps including a defatting step and large amount of organic solvents. Moreover, they are labor-intensive and time-consuming. In this work, a pressurized hot water extraction (PHWE) method has been compared to conventional sonication extraction (CSE) method. The contribution of individual compounds to the total antioxidant capacity of the extracts was evaluated by developing an analytical technique consisting of high-performance liquid chromatography (HPLC) with photodiode array coupled to electrochemical and to charged aerosol detectors and HPLC-mass spectrometry for the confirmation of the identity of compounds present. Additionally, procyanidins were analyzed by fluorescence detection. PHWE turned out to be more efficient in extracting phenolics and methylxanthines, as compared to the conventional method, in addition to being more "green" in terms of using less organic solvents.

Keywords: Cocoa Pressurized hot water extraction Flavanols Antioxidant capacity Electrochemical detector

INTRODUCTION

Polyphenols have become an intense focus of research interest due to their potent antioxidant properties and their perceived health-beneficial effects in the treatment and prevention of cancer, cardiovascular disease, and other pathologies (Chen and Chen 2013; Iriti and Varoni 2013; Martin et al. 2013).

Cocoa beans are one of the best-known sources of dietary polyphenols; the total polyphenol content of the dried fat free-mass of fresh cocoa beans is around 15–20% and of fermented beans approximately 5% (Wollgast and Anklam 2000). The polyphenol content in cocoa beans depends on their origin and processing, as they are subject to a combination of fermentation and drying treatments, followed by alkalisation and roasting, which affects the polyphenol content and consequently the end product quality (Hii et al. 2009). However, the content of polyphenols in chocolate, which is the most commonly consumed cocoa product, is lower (Adamson et al. 1999). This is not only because chocolate contains processed cocoa beans but also added sugars, milk, solids, and cocoa butter. In addition, the conditions during cocoa bean processing and chocolate making may result in polyphenol degradation (Wollgast and Anklam 2000).

Cocoa polyphenols have been reported in many studies as bioactive compounds, with antioxidant, antiradical, antihypertensive, and anticarcinogenic properties (Andújar et al. 2012; Martin et al. 2013). The main polyphenols found in cocoa are flavanols, mainly catechin, epicatechin, and procyanidins (see Fig. 1).

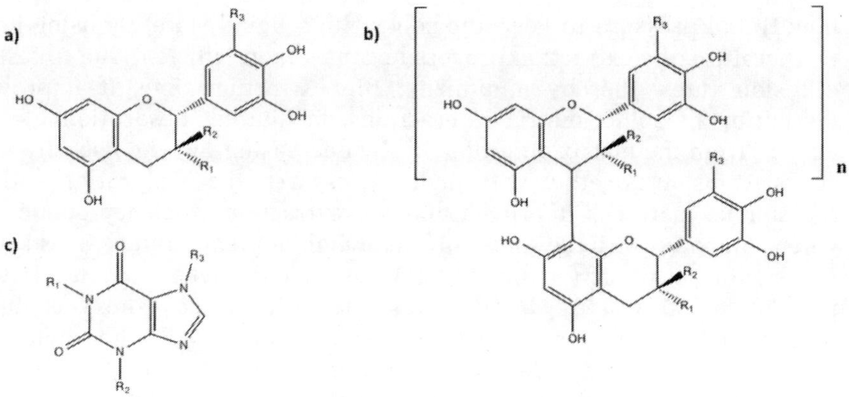

Fig. 1. Chemical structure of the main phenolic compounds, flavan-3-ol for **a** monomer units and **b** procyanidins (where catechin (R1 = H, R2 = OH, and R3 = H) and epicatechin (R1 = OH, R2 = H, and R3 = H)), and **c** methylxanthines (where theobromine (R_1 = H, R_2 = R_3 = CH_3) and caffeine (R_1 = R_2 = R_3 = CH_3)) found in cocoa

Even though most of the studies indicate that the health benefits of cocoa products are attributed to polyphenols, it should be observed that cocoa and cocoa products also contain high amounts of methylxanthines, namely theobromine and caffeine, which represent about 2.4–3.6% of defatted dry cocoa composition (Fig. 1) (Matissek 1997). Theobromine is known to be a brain stimulant, to have diurctic action, and being a potential for blood pressure reduction (Cheng et al. 2009). Besides having their own health benefits, methylxanthines may also be involved in synergistic interactions with polyphenols, although possible health effects are unclear and further studies are needed.

Common methods to extract polyphenols from cocoa products use several steps, including defatting before the extraction of the polyphenols itself (see Fig. 2) (Adamson et al. 1999; Belščak et al. 2009; Wollgast et al. 2001). In addition, large amounts of organic solvents are used and methods are often based on manual and are time consuming and highly dependent on the operator, i.e., not easily reproduced. The development of an efficient method to extract flavonoids from cocoa is a challenging task due to the structural diversity of phenolic compounds, their presence in a complex matrix and interaction with other compounds. Hence, an important aspect to consider is the development of fast, cost-effective, and environmentally adaptable extraction methods that are able to isolate the polyphenols of interest from cocoa and cocoa products. For instance, pressurized liquid extraction is an extraction technique that uses solvent subjected to elevated temperature (usually above the boiling point of the liquid) and pressure to keep the solvent in a liquid state. By using high temperature, the extraction process becomes more efficient due to faster diffusion rates, thereby minimizing the extraction time. It improves disruption of solute-matrix interactions, facilitating desorption of the analyte from the matrix. Additionally, by decreasing solvent viscosity and surface tension, together with higher pressures, the penetration within the sample matrix is improved and the extraction efficiency enhanced, when compared with the same solvent at ambient temperature. This leads to a significant reduction on extraction times and solvent volumes. In the present work, water was used as the solvent, i.e., the technique pressurized hot water extraction (PHWE). In PHWE, the dielectric constant of the water decreases with increasing temperature, which implies that the solvent water polarizability can be tuned by changing temperature. The dielectric constant is around 90 at 25 °C but is around 65 at 100 °C and around 35 at 250 °C. That is, the dielectric constant of water at 250 °C is the same as that for methanol at ambient conditions (Plaza and Turner 2015). Thus, water at high temperatures is an interesting alternative extraction solvent and could potentially replace many of organic solvents conventionally used in extraction.

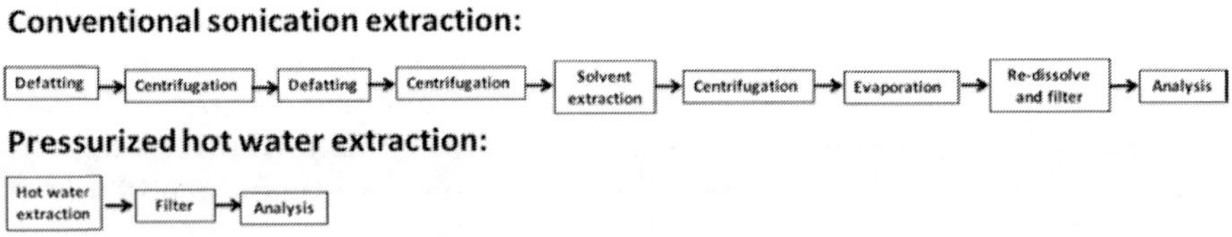

Fig. 2. Main steps employed to carry out conventional sonication extraction and pressurized hot water extraction

Therefore, our hypothesis is that PHWE can be used as a green technique to extract polyphenols from cocoa and cocoa products, avoiding the defatting step and the use of organic solvents for the extraction (Fig. 2). In this work, PHWE was compared with a conventional sonication extraction (CSE) (Wollgast et al. 2001) to extract phenolic compounds and methylxanthines from cocoa and cocoa products. In addition, a full characterization of cocoa extracts and the contribution of individual compounds to the total antioxidant capacity of the extracts was evaluated for testing both extraction techniques, PHWE and CSE. The analytical techniques used were HPLC coupled to a photodiode array detector (DAD), an electrochemical detector (ECD), and a charged aerosol detector (CAD) and HPLC with mass spectrometry detector (MS). The advantage of combining all these detectors (DAD, ECD, CAD, and MS) is that information about oxidizability can be obtained from ECD, in addition to light adsorbing properties of the compounds with DAD, as well as quantitative analysis with CAD and chemical structure from the MS detector. Moreover, the different degrees of polymerization of procyanidins were analyzed by fluorescence detection (HPLC-FLD).

MATERIALS AND METHODS

Chemicals and Reagents

All the chemicals were of analytical grade. Heptane, acetone, and acetonitrile were purchased from Honeywell Burdick and Jackson (Muskegon, MI, USA). Ammonium formate, acetic acid, theobromine, and sodium carbonate were supplied by Sigma-Aldrich (Steinheim, Germany). Formic acid, caffeine, and Folin-Ciocalteu reagent were from Merck (Darmstadt, Germany). Methanol, HPLC-MS grade, was provided by Scharlau (Barcelona, Spain). The polyphenols standards (catechin, epicatechin, and procyanidin B2) were purchased from Extrasynthese (Lyon, France). Deagglomerated alumina (α-Al_2O_3) suspensions with grain sizes 0.1 μm was provided by Struers (Ballerup, Denmark). The ultrapure water used was obtained from a Milli-Q instrument (Millipore, Billerica, MA, USA).

Samples

All samples, cocoa powders (A and B), chocolates (B, C, and D), and nibs and nibs powder were commercially available and the chocolate breads were manufactured *in house*, following the recipe below, and using the cocoa powders A and B. The cocoa nibs are cocoa beans that have been roasted, separated from their husk, and broken in smaller pieces. Before extraction, all samples were grinded.

Bread recipe: 400 g of white flour (Kungsörnen AB, Järna, Sweden), 150 g cocoa powder, 360 g water, 4.8 g dry yeast, 40 g sugar, and 4.8 g salt. The bread was baked according to a standardized procedure in a home baking machine (Severin model n° BM 3983). The menu choice was programme 2 (white bread, 1 Kg, quick (time: 2:37)). The water content of the cocoa bread was calculated by subtracting dry weight from the total weight. The fresh cocoa bread was dried in an oven at 60 °C for 56 h. This process was carried out in triplicate. The percentage of water in the bread with cocoa A and cocoa B was 40.9 ± 1.2 and 41.5 ± 0.9%, respectively.

Conventional Sonication Extraction Technique

A conventional sonication extraction technique previously described by Wollgast et al. (2001), Gu et al. (2006), and Brcanovic et al. (2013) was employed, with some modifications, to extract the polyphenols from cocoa and cocoa products. One gram of the cocoa and cocoa products was defatted twice with 10 mL heptane for 5 min using an Elmasonic S30H ultrasonic bath (Singen, Germany) with ultrasonic vibration at frequency of 37 kHz, ultrasonic power of 80 W, and a temperature of 30 °C and was subsequently centrifuged for 10 min at 3000*g*. Polyphenols were extracted from the pellet with 10 mL of a mixture of acetone-water-acetic acid (70:29.8:0.2, *v*/*v*/*v*) for 10 min at 30 °C in the ultrasonic bath at the same conditions as that in the defatting step. The sample was centrifuged for 10 min at 3000*g* and dried with a Reacti-Vap™ Evaporators (Thermo Fisher, Germering, Germany). Solvent evaporation was carried out using a nitrogen flow and magnet stirring at 30 °C until total sample dryness. The dry extracts were re-dissolved in water and filtered through a 0.45 μm PTFE filters (VWR International, West Chester, PA, USA) before HPLC separation. Samples were prepared in triplicate.

Pressurized Hot Water Extraction

Extractions were performed on a Dionex ASE 200 system (Thermo Fisher, Germering, Germany). The water was sonicated for 40 min for removing the dissolved oxygen. Extractions were performed at 125 °C for 3 min based on an experimental method used in a previous study (Plaza et al. 2013). Prior to each experiment, the extraction cell was heated-up for 6 min. Likewise, all extractions were performed in 11-mL extraction cells, containing 1 g of fresh sample. Samples were prepared in triplicate and stored protected from light at −20 °C until analysis. The volume of the extracts was adjusted to 25 mL with MilliQ water and filtered through a 0.45-μm PTFE filters (VWR International, West Chester, PA, USA) prior to analysis.

Total Phenolic Compounds Determination (Folin-Ciocalteu)

The Folin-Ciocalteu reducing capacity was estimated as epicatechin equivalents (EE), expressed as milligram epicatechin per gram of sample based on a previous protocol (Koşar et al. 2005). The total volume of the reaction mixture was miniaturized to 1 mL. 790 μL of water and 10 μL of sample were mixed, to which 50 μL of undiluted Folin-Ciocalteu reagent was subsequently added. After 1 min, 150 μL of 2% (*w*/*v*) Na_2CO_3 was added. After 2 h of incubation at 25 °C, 300 μL of the mixture was transferred into a well of the microplate, the absorbance was measured at 760 nm in a microplate spectrophotometer reader (Multiskan GO, Thermo Fisher, Germering, Germany) and compared to a epicatechin calibration curve (0.025–2.000 mg/mL) prepared equally. The data was presented as the average of triplicate analyses for each extract.

Analysis of Polyphenols by HPLC with Diode Array, Electrochemical, and Charged Aerosol Detection (HPLC-DAD-ECD-CAD)

The HPLC-DAD-ECD-CAD analysis of phenolic compounds from cocoa and cocoa products extracts was performed according to Plaza et al. (2014b). An UltiMate-3000® HPLC system from Dionex (Thermo Fisher, Germering, Germany) with an online degasser, a dual-gradient pump with degasser, an autosampler with cooler, column oven, and photodiode array detector, all controlled by a Chromeleon 6.80 (Thermo Fisher) software, was used. An ECD (Bioanalytical Systems Inc., West Lafayette, IN) was attached just after the DAD. The ECD redox potential was measured at +0.6 V vs. Ag/AgCl. A battery-powered potentiostat (Palmsens, Palm Instrument BV, Houten, The Netherlands) was connected to the flow cell and was controlled by HP iPAQ Pocket PC (HP 2200 series, Hewlett-Packard Company). PalmTime software (vs. 2.3.0.0, Palmsens) was used to collect data. A Corona CAD instrument from ESA Biosciences Inc. (a part of Thermo Fisher, Germering, Germany) was placed in series after the ECD detector. Data processing was carried out with Chromeleon 6.8 software (Thermo Fisher).

Separation was accomplished with porous-shell fused-core Ascentis Express C18 analytical column (150 × 2.1 mm, particle size 2.7 μm) from Supelco (Bellefonte, PA, USA). The column temperature was 50 °C and the flow rate 300 μL/min. Two microliters of extract was injected from a sample tray held at 4 °C. The mobile phases consisted of (A) ammonium formate buffer (pH 3.0) 60 mM in water and (B) methanol with 0.5% of formic acid in a gradient elution analysis programmed as follows: 0 min, 5% (B); 0–5 min, 5% (B); 5–35 min, 5–40% (B); 35–40 min, 40% (B), with 10 min of post-time. To circumvent the organic solvent elicited impact on the CAD response (de Villiers et al. 2007; Plaza et al. 2014b), an inverse methanol gradient (make-up gradient) was delivered to CAD with 0.3-min delay with respect to the elution gradient (de Villiers et al. 2007). All solvents were purged with nitrogen to remove oxygen.

Phenolic compounds were quantified with the CAD detector. The calibration curve of epicatechin was selected to quantify all phenolic compounds and methylxanthines, because it is normally found in cocoa (Gu et al. 2006; Langer et al. 2011). The epicatechin standard solution was injected in triplicate at six concentrations levels (1–100 µg/mL). The calibration curve of epicatechin was obtained by plotting peak area as function of concentration (in micrograms/milliliter). Responses obtained in the examined ranges were expressed by a linear equation with high r^2 determination coefficient value (0.999).

Identification of Polyphenols by HPLC with DAD and Mass Spectrometry

The structural elucidation of phenolic compounds was carried out by a Waters Acquity UPLC chromatographic system (Waters Corp., Manchester, UK) equipped with a DAD and a quadrupole and orthogonal acceleration time-of-flight tandem mass spectrometers Xevo G2 QTOF with electrospray ionization source (ESI) (Waters MS Technologies, Manchester, UK). The HPLC instrument was equipped with a binary solvent, an auto-sampler, and a column heater compartment. The system was controlled by Waters® Empower™ Chromatography software, while MassLynx™ (V 4.1, SCN 779, Waters Corp., Manchester, UK) was used for MS data acquisition and treatment. The separation conditions were the same as the previous (see section "Analysis of polyphenols by HPLC with diode array, electrochemical, and charged aerosol detection (HPLC-DAD-ECD-CAD)"). UV-vis spectra were recorded in the range of 200–500 nm. The ESI interface was operated in both positive and negative modes, and full-scan HPLC-QTOF-MS spectra were obtained by scanning from 50 to 1000 m/z. The mass spectrometer was calibrated using a solution of sodium formate. Data were collected in continuum mode and mass was corrected during acquisition using an external reference (Lock-Spray™) comprising a 10-µL/min solution of leucine-enkephalin (2 ng/µL) via a lock-spray interface. The capillary and cone voltage were set at 3 kV and 30 V, respectively. Nitrogen was used as both cone gas (50 L/h) and desolvation gas (1000 L/h). The source and desolvation temperature were set at 120 and 400 °C, respectively. Simultaneous acquisition of exact mass at high and low collision energy, MS^E (where E represents collision energy), was used to obtain full scan accurate mass fragment, precursor ion, and neutral loss information. The collision energy in function 1 (low energy) was off while in function 2 (high energy), the collision energy ranged between 15 and 60 V.

Analysis of Procyanidins by HPLC with Fluorescence Detector

The separation of flavanol oligomers was performed as previously described (Robbins et al. 2009) using an Agilent 1100 series HPLC system (Agilent Technologies, Palo Alto, CA, USA) equipped with an autosampler, solvent degasser, quaternary pump, column heater, and DAD and fluorescence detectors. A Develosil Diol 100-Å column (250 × 4.6 mm; 100-Å pore size; 5-µm particle size; 35 °C) from Phenomenex (Torrance, CA, USA) was used for separation. The mobile phase consisted of (A) acetonitrile/acetic acid (98:2, *v*/*v*) and (B) methanol/water/acetic acid (95:3:2, *v*/*v*/*v*), and the following gradient system was used: 0 min, 7% (B), 0–3 min, 7% (B), 3–60 min, 37.6% (B), 60–63 min, 100% (B), 63–70 min, 100% (B), and 70–76 min, 7% (B), with a flow rate of 1 mL/min and the injection volume was 5 µL. The autosampler was set to and held at 5 °C. UV data were collected at 280 nm, and fluorescence detection was conducted with an excitation wavelength of 230 nm and emission at 321 nm.

Quantification of procyanidins with different degrees of polymerization (DP) was carried out according to Robbins et al. (2012). They have developed a reliable, robust, and readily transferable method for the analysis and quantification of procyanidins with different DP in cocoa and cocoa products, allowing its broader implementation for the analysis of procyanidins in cocoa. The method was based on the relative response factors (RRFs) of epicatechin specifics for each DP to quantify DP of 2 to 10.

Monomer content was based on the external calibration standard (−)-epicatechin. The concentration of DP 2 to 10 was calculated using the various RRFs according to Robbins et al. (Robbins et al. 2012) by the following equation:

Concentration of flavanols inmg/g=(((AreaDPn−b)/(mepic×RRFDPn)))/(Sample (g))×DFConcentration of flavanols inmg/g=(((AreaDPn−b)/(mepic×RRFDPn)))/(Sample (g))×DF

Where DPn is the degree of polymerization indicating the oligomer, m is the slope of the epicatechin calibration curve, and DF is the appropriate dilution factor. Only a single calibration curve was needed to quantify DP 1–10 since the concentration is based on the RRF to epicatechin. The epicatechin standard solution was injected in triplicate at six concentration levels (0.1–50 µg/mL). The calibration curve of epicatechin was obtained by plotting peak area as function of concentration (in micrograms/milliliter). Responses obtained in the examined ranges were expressed by a linear equation with a good r^2 determination coefficient value (0.999). The total procyanidin concentration was determined by summing the individual DP 1–10 concentrations.

Statistical Analysis

All the analyses were conducted in triplicate. The data generated were subjected to one-way analysis of variance (ANOVA) using R i386 version 3.3.1 (R foundation for statistical computing, Vienna, Austria). Tukey's paired comparison test was used to determine statistical significant differences ($p < 0.05$) between mean values for different samples at 95% confidence level. Results are expressed as mean value and standard deviation; differences are pointed out with superscripted letters. Differences in the tested variables were considered statistically significant, when $p \leq 0.05$, highly significant when $p \leq 0.01$, and with a trend when $p \leq 0.1$.

RESULTS AND DISCUSSION

Two different extraction methods, CSE and PHWE, were employed and compared at an analytical scale to test our hypothesis. As far as our knowledge goes, this is the first time that PHWE is used in such samples. The extracts obtained by CSE and PHWE from cocoa and cocoa products were fully characterized by Folin-Ciocalteu assay, HPLC-DAD-CAD-ECD, HPLC-QTOF/MS, and HPLC-FLD methods.

Total Phenolic Compounds

Figure 3 shows the total phenolic compounds (TPC) measured by Folin-Ciocalteu method for all the different cocoas and cocoa products extracts, obtained by PHWE and CSE. The amounts of TPC extracted by PHWE and CSE were significantly different ($p \leq 0.05$) in cocoas (A and B) and in nibs (nibs and nibs powder) samples being 2.3, 3.1, 5.6, and 4.7 times, respectively, higher in PHWE. This could be due to the fact that the solvation power of a liquid increases with increasing temperature. Additionally, liquid water at elevated temperature and pressure is a solvent of lower polarizability/polarity and density that enables faster mass transfer and improved wetting of the sample due to higher diffusivity and lower viscosity and surface tension (Plaza and Turner 2015). These liquid water properties at elevated temperature could explain why the TPC was significantly higher when PHWE was used to extract antioxidants from these samples.

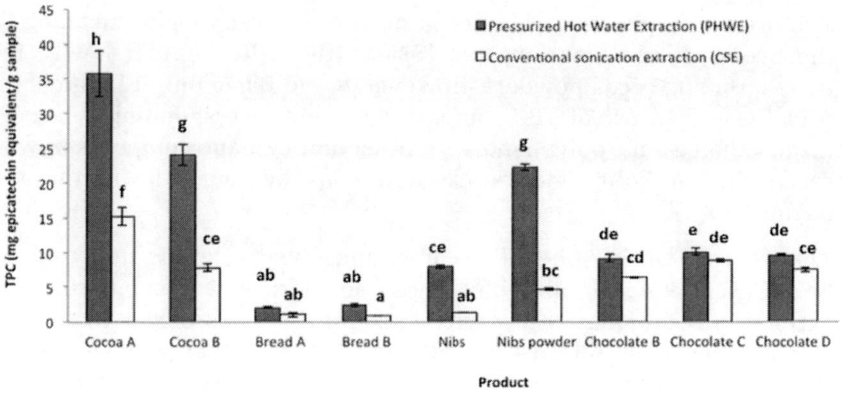

Fig. 3. Total Phenolic Coumpounds (TPC) (milligram of epicatechin equivalent per gram of sample) measured by the Folin-Ciocalteau method for all different cocoas and cocoa products, extracted by PHWE and CSE. *Different letters* represent significant differences ($p \leq 0.05$)

However, the TPC extracted using PHWE and CSE were not significantly different ($p \geq 0.05$) in chocolate (B, C, and D), possibly due to the porosity of the particles. This physical characteristic can affect the extraction of TPC. In the chocolate samples, the particles lack porosity which means that the advantages of PHWE about mass transfer could be low. Also, the bread samples (A and B) did not show significant difference ($p \geq 0.05$).

It can be observed that TPC was higher for almost all PHWE extracts of cocoa products, although the highest differences could be observed among samples of different nature (Fig. 3). The highest phenolic content (mg EE/g) was found in cocoas (A and B, 35.7 ± 3.3 and 24.1 ± 1.5, respectively) and nibs powder (22.3 ± 0.5) in PHWE extracts, while in CSE, was the cocoa A (15.3 ± 1.3) followed by chocolate C (8.8 ± 0.2), and cocoa B (7.9 ± 0.6). The lowest phenolic content in PHWE was found in breads A and B, followed by nibs and chocolates B, D, and C. For CSE extracts, the lowest values were found in breads, nibs, nibs powder, and chocolate B (Fig. 3).

The nature of the sample and its physical characteristics can also affect the capacity of the phenolic extraction. For instance, the cocoa, which is the basic ingredient of the chocolate, depending on how the cocoa beans are cultivated and how the cocoa powder is produced, can affect the concentration of phenolic compounds present in the cocoa. Therefore, cocoa A presented higher concentration of TPC than cocoa B. Also, the TPC in chocolates depends on the cocoa origin, chocolate production, and percentage of cocoa's ingredient used. Thus, chocolate C had the highest concentration followed by chocolates D and B (Fig. 3).

The bread samples had 15% cocoa powder by wet weight; meaning that the bread should have around 15% of the TPC compared with their corresponding cocoa powders. However, bread B had only 11% and bread A had less than 6% of TPC compared to their corresponding cocoa. This means than some polyphenols or other antioxidant compounds, which react with the Folin-Ciocalteu reagent, were degraded during the bread baking.

The extraction of phenolic compounds was significantly different ($p \leq 0.05$) between nibs and nibs powder, when comparing both extraction techniques. The particle size of nibs was bigger than that of nibs powder. Thus, nibs presented lower TPC. However, when the particle size of the cocoa nibs was reduced using a grinder, the extraction of phenolic compounds increased. As expected, the particle size affected the mass transfer between the nibs and the solvent, because a larger surface area resulted in greater accessibility of the extraction solvent to the analyte and the migration rate of the analyte through the pores of the solid matrix increases with decreasing particle size. Also, the mechanical treatment used to decrease particle size can result in breakage of the cell walls and cell structure, which may enhance the diffusion of the analytes.

Anyhow, the Folin-Ciocalteu assay is not specific for phenolic compounds and it has been proposed for the measurement of total reducing capacity of samples (Magalhães et al. 2008). In this work, further advance analytical techniques were used to show the chemical composition obtained in extracts produced by PHWE and CSE specifically considering phenolic compounds from cocoa and cocoa products.

Identification of the Extracted Phenolic Compounds and Methylxanthines

Separation of phenolic acids, flavanols, flavonols, and methylxanthines of all extracts obtained by PHWE and CSE was carried out using HPLC with DAD coupled to ECD and CAD in analysis time of around 40 min. Figure 4 shows the chromatograms and amperogram of cocoa A obtained by PHWE.

A HPLC-MS^E method was set up, aiming to separate and identify the phenolic compounds present in the extracts. Table 1 shows that the profile of phenolic compounds obtained when using both extraction techniques in the different cocoa products was similar. Nibs powder extracted by PHWE was chosen as representative for being one with high content in phenols (see Fig. 3 and Table 1). The careful analysis of the separated compounds, using the information provided by the DAD detector as well as the MS detector installed in series, together with the information that could be found in the literature and from commercial standards (when available), allowed the identification of 12 phenolic compounds and 2 methylxanthines in the extracts. The identified compounds as well as their chemical characteristics are shown in Table 2. All phenolic compounds were detected using an electrospray interface in negative, while methylxanthines were detected in positive ionization mode.

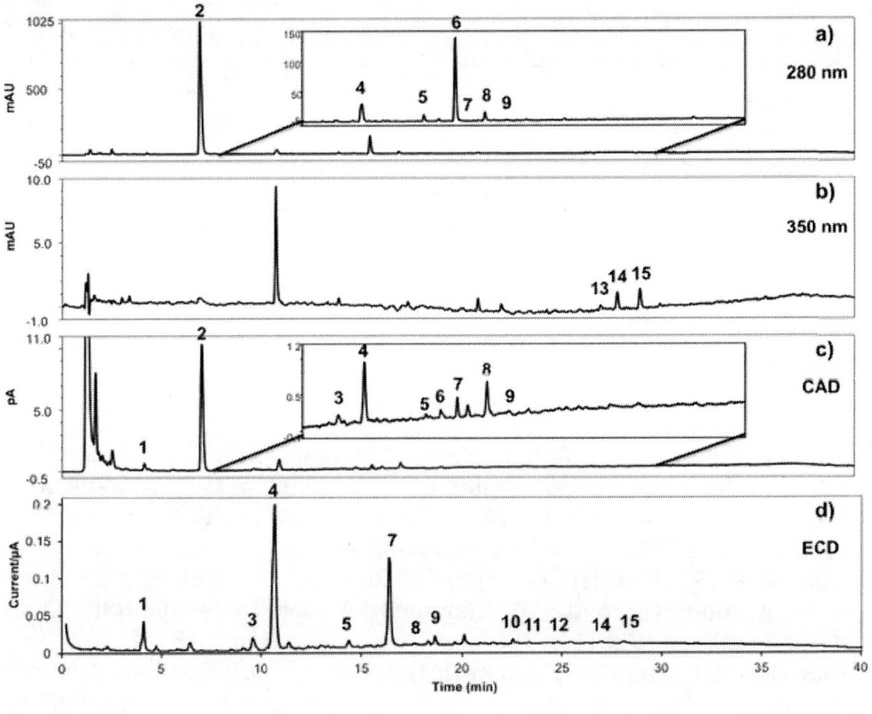

Fig. 4. Chromatograms and amperograms corresponding to the HPLC-DAD-ECD-CAD analysis of cocoa A extract obtained by PHWE at **a** 280 nm, **b** 350 nm, **c** CAD chromatogram, and **d** amperogram. Analytical conditions are given under "Material and Methods." For peak identification, see Table 2.

Table 1 .Quantification (mg/100 g of sample ± standard deviation (sd)) by HPLC-CAD of phenolic compounds and methylxanthines in cocoa and cocoa products

		Cocoa A		Cocoa B		Bread A		Bread B		Nibs	
ID	Name	CSE	PHWE	CSE	PHWE	CSE	PHWE	CSE	PHWE	CSE	PHWE
1	Unknown	27.7 ± 1.3^{f}	57.3 ± 5.5^{g}	8.4 ± 1.0^{ab}	13.4 ± 0.2^{bd}	12.9 ± 1.3^{bc}	17.4 ± 0.9^{cde}	6.2 ± 0.5^{a}	10.1 ± 0.9^{ab}	9.8 ± 0.6^{ab}	26.3 ± 1.9^{f}
2	Theobromine	626.2 ± 18.1^{e}	1039.4 ± 50.0^{g}	314.3 ± 28.6^{d}	813.1 ± 56.0^{f}	184.6 ± 7.3^{ab}	157.3 ± 15.8^{a}	190.1 ± 21.6^{ab}	195.9 ± 1.6^{ab}	152.1 ± 14.4^{ab}	603.3 ± 52.4^{f}
3	Procyanidin B	10.9 ± 0.8^{b}	24.3 ± 1.1^{d}	9.8 ± 0.9^{b}	15.6 ± 0.7^{c}	ND	ND	ND	ND	ND	ND
4	Catechin	32.4 ± 1.8^{f}	88.8 ± 5.5^{i}	18.3 ± 0.5^{bcd}	62.6 ± 5.5^{h}	14.2 ± 0.9^{ac}	11.3 ± 1.1^{ab}	9.1 ± 1.0^{a}	11.9 ± 0.6^{ab}	15.3 ± 1.2^{ac}	27.7 ± 1.6^{ef}
5	Procyanidin B2	12.3 ± 0.1^{ab}	14.2 ± 0.6^{b}	10.0 ± 0.6^{a}	11.7 ± 0.5^{ab}	ND	ND	ND	ND	10.7 ± 0.9^{a}	22.9 ± 1.9^{de}
6	Caffeine	14.9 ± 0.1^{ef}	23.8 ± 1.6^{gh}	8.6 ± 0.1^{abc}	10.5 ± 0.9^{bd}	7.0 ± 0.6^{a}	7.5 ± 0.3^{ab}	7.5 ± 1.6^{ab}	7.6 ± 0.1^{ab}	12.5 ± 0.5^{de}	23.0 ± 1.4^{g}
7	Epicatechin	21.8 ± 0.7^{de}	23.0 ± 2.7^{de}	14.0 ± 1.1^{ab}	20.8 ± 2.3^{cd}	10.6 ± 0.4^{ab}	9.1 ± 0.1^{a}	11.2 ± 1.0^{ab}	10.6 ± 0.2^{ab}	19.7 ± 1.8^{cd}	48.2 ± 2.6^{gh}
8	Procyanidin A glucoside	17.7 ± 1.3^{g}	48.3 ± 3.5^{i}	14.7 ± 0.5^{dg}	40.1 ± 2.6^{h}	9.5 ± 0.5^{ac}	7.6 ± 0.3^{a}	7.5 ± 0.7^{a}	7.8 ± 0.1^{a}	8.8 ± 0.1^{ab}	13.3 ± 1.5^{cdf}
9	Procyanidin C	10.4 ± 0.3^{abc}	10.2 ± 0.2^{abc}	27.7 ± 0.9^{f}	ND	ND	ND	ND	ND	8.0 ± 0.5^{a}	14.6 ± 1.0^{bd}
Total		774.4 ± 24.3	1329.4 ± 56.6	425.7 ± 31.3	987.8 ± 62.2	338.8 ± 8.4	210.3 ± 16.2	231.7 ± 21.7	243.9 ± 0.6	236.9 ± 13.9	779.4 ± 48.0
Total methylxanthines		641.1 ± 18.1^{f}	1063.3 ± 49.2^{h}	322.9 ± 28.6^{ce}	823.6 ± 56.6^{g}	191.6 ± 7.8^{a}	164.9 ± 16.0^{a}	197.7 ± 21.6^{ab}	203.5 ± 1.7^{ab}	164.6 ± 14.6^{a}	626.3 ± 51.2^{f}
Total phenolics		133.2 ± 6.2^{fh}	266.1 ± 7.6^{l}	102.8 ± 2.7^{d}	164.2 ± 6.2^{j}	47.2 ± 0.6^{b}	45.4 ± 0.6^{ab}	34.0 ± 0.1^{a}	40.4 ± 1.1^{ab}	72.3 ± 3.0^{F}	153.1 ± 3.7^{ij}

		Nibs powder		Chocolate B		Chocolate C		Chocolate D	
ID	Name	CSE	PHWE	CSE	PHWE	CSE	PHWE	CSE	PHWE
1	Unknown	18.3 ± 0.3^{de}	28.4 ± 2.6^{f}	11.8 ± 0.3^{f}	19.6 ± 1.6^{e}	13.3 ± 0.3^{bd}	13.3 ± 0.7^{be}	12.0 ± 0.1^{b}	17.6 ± 1.6^{cde}
2	Theobromine	343.8 ± 21.6^{d}	593.5 ± 7.0^{e}	344.7 ± 19.3^{d}	228.5 ± 19.6^{abc}	267.1 ± 5.1^{bd}	315.9 ± 23.7^{cd}	316.9 ± 13.0^{d}	336.4 ± 25.9^{d}
3	Procyanidin B	ND	ND	ND	ND	7.0 ± 0.8^{a}	11.7 ± 0.3^{b}	ND	9.9 ± 0.6^{b}
4	Catechin	14.6 ± 1.5^{ac}	51.4 ± 1.1^{g}	20.1 ± 1.1^{ce}	23.1 ± 3.2^{de}	18.2 ± 0.9^{bcd}	32.0 ± 2.4^{f}	18.2 ± 0.3^{bcd}	27.2 ± 2.5^{ef}
5	Procyanidin B2	25.1 ± 1.1^{de}	43.6 ± 3.7^{f}	10.7 ± 0.3^{a}	18.3 ± 1.9^{c}	10.8 ± 0.9^{a}	22.1 ± 1.6^{d}	25.9 ± 1.5^{e}	18.2 ± 1.4^{c}
6	Caffeine	26.6 ± 2.2^{h}	30.0 ± 0.6^{i}	7.8 ± 0.2^{ab}	11.0 ± 0.7^{cd}	11.3 ± 1.9^{cd}	11.2 ± 0.2^{bd}	17.3 ± 0.6^{f}	12.0 ± 1.0^{de}
7	Epicatechin	49.4 ± 0.8^{gh}	67.4 ± 0.8^{bc}	51.9 ± 2.7^{gh}	28.4 ± 3.2^{cd}	26.5 ± 1.8^{ef}	42.4 ± 2.1^{g}	44.1 ± 1.6^{g}	30.0 ± 0.6^{f}
8	Procyanidin A glucoside	17.4 ± 1.0^{fg}	15.7 ± 1.0^{cd}	13.3 ± 0.8^{cdf}	11.2 ± 1.1^{ad}	14.4 ± 1.8^{de}	10.7 ± 0.4^{ad}	16.1 ± 1.0^{efg}	12.7 ± 1.1^{bcde}
9	Procyanidin C	16.2 ± 1.8^{a}	11.7 ± 1.5^{de}	8.9 ± 0.1^{ab}	7.4 ± 0.6^{a}	12.9 ± 2.0^{cd}	8.1 ± 0.6^{a}	14.0 ± 0.1^{de}	10.7 ± 1.2^{abc}
10	Procyanidin A galactoside	7.9 ± 0.5^{a}	ND	ND	ND	ND	ND	7.2 ± 0.1^{a}	ND
Total		516.5 ± 24.9	841.6 ± 3.5	469.1 ± 22.1	345.0 ± 24.0	381.7 ± 4.8	467.8 ± 31.9	471.7 ± 16.0	474.9 ± 26.4
Total methylxanthines		370.3 ± 23.6^{c}	623.5 ± 6.5^{f}	352.5 ± 19.1^{de}	239.5 ± 20.4^{ac}	278.4 ± 4.^{bcd}	327.0 ± 23.9^{ce}	334.1 ± 19.2^{ce}	348.5 ± 26.8^{de}
Total phenolics		146.2 ± 4.0^{hi}	218.1 ± 3.0^{k}	116.6 ± 2.9^{ef}	105.5 ± 8.3^{de}	103.2 ± 0.7^{d}	140.7 ± 8.0^{fi}	137.6 ± 3.5^{gh}	126.4 ± 1.7^{fg}

All measurements were done in triplicate. Total phenolics: peaks 1, 3, 4, 5, 7, 8, 9, and 10. Total methylxanthines: peaks 2 and 6. *CSE* conventional sonication extraction, *PHWE* pressurized hot water extraction. *ND* not detected. Mean values with different superscript letters in the same row are significantly different ($p \leq 0.05$)

Quantification of the different phenolic compounds and methylxanthines found in the extracts of cocoa and cocoa products was carried out with a CAD detector. The content of phenolics and methylxanthines of each cocoa product using PHWE and CSE is summarized in Table 1. As can be seen, from the 15 identified compounds, just the compounds with higher concentration (around 10 compounds) could be quantified. One disadvantage of CAD is its low sensitivity compared with other detectors such as DAD (Plaza et al. 2014b). Glycosylated procyanidins A and the flavonols found at lower concentrations could not be quantified (see Fig. 4).

Theobromine (peak 2) was the compound with the highest concentration in all extracts (concentrations from 152.1 to 1039.4 mg/100 g of sample), also caffeine (peak 6) was found at considerable concentrations (from 6.6 to 30.0 mg/100 g of sample) (Table 1). The total concentration of methylxanthines obtained in this study agreed with other works found in the literature (Belščak et al. 2009; Langer et al. 2011). These compounds can be of interest regarding their health benefits, as previously described (Akiba et al. 2004). The ratio of theobromine to caffeine found in the cocoa products (if the concentration of theobromine is normalized as 10) was 10:0.1–0.8.

Table 2. Identification (UV-vis and MS characteristics, molecular formula (MF) and mass precision (mDa)) and individual antioxidant capacity (linear coefficient (R^2) and slope of the regression line for a side by side comparison between CAD response (concentration (mg/100 g of sample)) and ECD response (Area (nA*s)/100 g of sample) for each phenolic compound)) of phenolic compounds and methylxanthines in cocoa and cocoa products

ID	Rt (min)	Identification	UV-Vis maxima (nm)	$[M-H]^-$ (m/z), MF, mDa	Main fragments detected (m/z), MF, mDa	Individual antioxidant capacity	
						R^2	Slope
1	4.03	Unknown	276.8			0.761	10.36
2	6.84	Theobromine[a]	205.8, 272.8	181.0731 $[M]^+$, $C_7H_9N_4O_2$, 0.5	138.0666, $C_6H_8N_3O_1$, −0.1	–	–
3	9.66	Procyanidin B	201.8, 279.8	577.1343, $C_{30}H_{25}O_{12}$, −0.3	451.0990, $C_{24}H_{19}O_9$, −3.9 425.0877, $C_{22}H_{17}O_9$, 0.4 407.0768, $C_{22}H_{15}O_8$, 0.1 289.0698, $C_{15}H_{13}O_6$, −1.4 137.0247, $C_7H_5O_3$, 0.8 125.0249, $C_6H_5O_3$, 1.0 109.0306, $C_6H_5O_2$, 1.6	0.835	12.99
4	10.9	Catechin[a]	203.8, 278.8	289.0698, $C_{15}H_{13}O_6$, −1.4	137.0247, $C_7H_5O_3$, 0.8 125.0249, $C_6H_5O_3$, 1.0 109.0285, $C_6H_5O_2$, −0.5	0.846	27.21
5	14.74	Procyanidin B2[a]	205.8, 278.8	577.1343, $C_{30}H_{25}O_{12}$, −0.3	451.0990, $C_{24}H_{19}O_9$, −3.9 425.0877, $C_{22}H_{17}O_9$, 0.4 407.0768, $C_{22}H_{15}O_8$, 0.1 289.0698, $C_{15}H_{13}O_6$, −1.4 137.0247, $C_7H_5O_3$, 0.8 125.0249, $C_6H_5O_3$, 1.0 109.0285, $C_6H_5O_2$, −0.5	0.836	21.76
6	14.91	Caffeine[a]	205.8, 272.8	195.0878 $[M]^+$, $C_8H_{10}N_4O_2$, −0.4	138.0666, $C_6H_8N_3O_1$, −0.1	–	–
7	15.36	Epicatechin[a]	203.8, 278.8	289.0698, $C_{15}H_{13}O_6$, −1.4	137.0247, $C_7H_5O_3$, 0.8 125.0249, $C_6H_5O_3$, 1.0 109.0285, $C_6H_5O_2$, −0.5	0.898	32.04
8	17.47	Procyanidin A glucoside	201.8, 278.8	737.1705, $C_{36}H_{33}O_{17}$, −1.3	611.1370, $C_{30}H_{27}O_{14}$, −3.1 449.0991, $C_{24}H_{17}O_9$, 3.1 289.0698, $C_{15}H_{13}O_6$, −1.4 137.0247, $C_7H_5O_3$, 0.8 123.0226, $C_6H_3O_3$, 1.3 109.0285, $C_6H_5O_2$, −0.5	0.796	11.30
9	18.02	Procyanidin C	209.8, 278.8	865.1942, $C_{45}H_{37}O_{18}$, −3.8	739.1582, $C_{39}H_{31}O_{15}$, −8.1 713.1498, $C_{37}H_{29}O_{15}$, −0.8 695.1346, $C_{37}H_{27}O_{14}$, −5.5 577.1287, $C_{30}H_{25}O_{12}$, −5.3 451.0947, $C_{24}H_{19}O_9$, 0.5 425.0877, $C_{22}H_{17}O_9$, 0.4 407.0768, $C_{22}H_{15}O_8$, 0.1 289.0698, $C_{15}H_{13}O_6$, −1.4 137.0247, $C_7H_5O_3$, 0.8 123.0249, $C_6H_5O_3$, 1.0 109.0285, $C_6H_5O_2$, −0.5	0.595	6.92
10	23.30	Procyanidin A galactoside	278.8	737.1705, $C_{36}H_{33}O_{17}$, −1.3	611.1370, $C_{30}H_{27}O_{14}$, −3.1 449.0861, $C_{24}H_{17}O_9$, −1.2 289.0698, $C_{15}H_{13}O_6$, −1.4 137.0247, $C_7H_5O_3$, 0.8 125.0249, $C_6H_5O_3$, 1.0 109.0285, $C_6H_5O_2$, −0.5	–	–
11	23.90	Procyanidin A arabinoside	278.8	707.1645, $C_{35}H_{31}O_{16}$, −1.7	581.1288, $C_{29}H_{25}O_{13}$, −0.7 449.0861, $C_{24}H_{17}O_9$, −1.2 289.0698, $C_{15}H_{13}O_6$, −1.4 137.0247, $C_7H_5O_3$, 0.8 125.0249, $C_6H_5O_3$, 1.0 109.0285, $C_6H_5O_2$, −0.5	–	–
12	24.65	Procyanidin B	279.8	577.1343, $C_{30}H_{25}O_{12}$, −0.3	451.0990, $C_{24}H_{19}O_9$, −3.9 425.0877, $C_{22}H_{17}O_9$, 0.4 407.0768, $C_{22}H_{15}O_8$, 0.1 289.0698, $C_{15}H_{13}O_6$, −1.4 137.0223, $C_7H_5O_3$, −1.6 125.0249, $C_6H_5O_3$, 1.0 109.0285, $C_6H_5O_2$, −0.5	–	–
13	27.48	Quercetin glucoside	352.8	463.0815, $C_{21}H_{19}O_{12}$, 2.6	301.0337, $C_{15}H_9O_7$, −1.1	–	–
14	28.17	Quercetin galactoside	355.8	463.0858, $C_{21}H_{19}O_{12}$, −1.9	301.0337, $C_{15}H_9O_7$, −1.1	–	–
15	29.26	Quercetin arabinoside	352.8	433.0765, $C_{20}H_{17}O_{11}$, −0.6	301.0337, $C_{15}H_9O_7$, −1.1	–	–

[a] Identified and confirmed using commercial standards

Catechin (peak 4) (9.1–88.8 mg/100 g of sample) and epicatechin (peak 7) (9.1–67.4 mg/100 g of sample) were the phenolic compounds found in higher concentration in almost all extracts of cocoa products. The concentrations detected for catechin and epicatechin agreed with the literature (Gu et al. 2006). Concentrations of procyanidin B2 (peak 5) (8.7–43.6 mg/100 g of sample), procyanidin A glucoside (peak 8) (7.5–48.3 mg/100 g of sample), and procyanidin C (peak 9) (6.7–27.7 mg/100 g of sample) in the extracts of cocoa products were significant. The unknown phenolic compound (peak 1) was one of the main compounds in some cocoa extracts (6.2–57.3 mg/100 g of sample) (Table 1).

Cocoa A was the product with the highest concentration of phenolic compounds followed by nibs powder, cocoa B, nibs, chocolate C, chocolate D, and chocolate B. The lowest concentration of phenolic compounds was found in the cocoa breads (A and B). The total phenolic compounds extracted in cocoas and nibs depended on their manufacture and variety. But the phenolic composition in chocolates and breads depends on the quality of raw cocoa material, variety of the beans, and recipe, for instance, chocolates have higher amount of cocoa powder than bread (15% of cocoa powder base on wet weight).

By comparing both extraction techniques, PHWE seemed to be a more suitable technique than CSE to extract these kinds of phenolic compounds from cocoa and cocoa products (Table 1). The differences in the extraction of the different phenolic compounds between both extraction techniques were significantly different ($p \leq 0.05$) for almost all products in most of the detected peaks, except for breads (A and B) (Table 1). Highly significant differences ($p \leq 0.01$) could be observed in cocoa A, nibs, and nibs powder for peaks 1, 4, and 8, also in chocolate C for peaks 3, 4, 5, 7, and 9. However, non-significant differences ($p \geq 0.05$) were observed for breads A and B, whereas for the different kinds of chocolates (B, C, and D), both significant and non-significant differences were found depending on the phenolic compound detected. Regarding phenolic compounds extraction, PHWE technique revealed to be way more effective than CSE. When compared, PHWE was able to extract twice the amount of phenolic compounds as compared with CSE in cocoa A (266.1 vs. 133.3 mg/100 g of sample) and nibs (153.1 vs. 72.3 mg/100 g of sample). Also, PHWE extracted 1.6, 1.5, and 1.3 times more than CSE in cocoa B (164.2 vs. 102.8 mg/100 g sample), nibs powder (218.1 vs. 146.2 mg/100 g sample), and chocolate C (140.7 vs. 103.2 mg/100 g sample), respectively. Compared with CSE, PHWE was only less efficient (although not significantly ($p \geq 0.05$)) for bread A, chocolate B, and D samples.

The results above are in agreement with the results obtained for total phenolic compounds measured by Folin-Ciocalteu method (section "Total phenolic compounds", Fig. 3). However, the latter method did not measure just phenolic compounds, because it measures any reductive substance as well as possible antagonism and synergism effects between different compounds. For instance, methylxanthines did not show antioxidant capacity (see Table 2) but they could have synergistic interactions with polyphenols. Moreover, comparing total phenolic compounds determined by the Folin-Ciocalteu method (Fig. 3) with the total phenolic compounds obtained by HPLC-CAD analysis (Table 1), there is a decrease using the latter analysis. These differences are likely due to the nonspecificity of the Folin-Ciocalteu assay. Furthermore, some of the phenolic compounds could not be quantified with the employed HPLC-CAD method, and it did not detect all phenolic compounds. Only, the free monomers and dimers that link to small sugars were detected with the HPLC-CAD method.

Moreover, PHWE extracted also higher concentration of methylxanthines than CSE from almost all samples except for bread A and chocolate B (Table 1). Highly significant differences ($p \leq 0.01$) between PHWE and CSE were observed in all products except for breads A and B and chocolates C and D. PHWE was also able to extract 3.9, 2.5, 1.7, and 1.6, times more theobromine than CSE in nibs, cocoa B, nibs powder, and cocoa A, respectively, being less effective only in bread A and chocolate B. Regarding caffeine, again PHWE extracts revealed higher values for almost all samples, except for chocolate C and D, but only significantly ($p \leq 0.05$) for the later (Table 1).

Individual Antioxidant Capacity of Phenolic Compounds

The ECD was used to study the oxidation of phenolic compounds to indicate the antioxidant capacity of each compound. To determine the contribution of the individual phenolic compounds to the total antioxidant capacity of the cocoa extracts, the concentration of each compound as determined by CAD and the peak area obtained by ECD for each compound were correlated in scatter plots (Table 2). The antioxidant capacity was able to detect seven compounds (peaks 1, 3, 4, 5, 7, 8, and 9), because only these compounds were found in high enough concentration possible to carry out the quantification by CAD. The methylxanthines, theobromine (peak 2), and caffeine (peak 6) were found in higher concentration but no antioxidant capacity was associated with these compounds. These results are in line with data reported for these compounds found in tea (Stewart et al. 2005).

As shown in Table 2, the response between both detectors was fairly linear, with the r^2 being ≥0.761. Procyanidin C (peak 9) presented a low r^2 (0.595). Using the slope, it was possible to calculate the individual antioxidant capacity of each phenolic compound in cocoa and cocoa product extracts. The higher the slope, the higher the antioxidant capacity. The results revealed that the compounds with the highest antioxidant capacity were the monomers epicatechin (slope 32.04) and catechin (slope 27.21). The monomers presented higher antioxidant capacity than dimers (procyanidins B2 and B) and trimers (procyanidin C). It is not in agreement with data found in literature that showed that the antioxidant capacity of grape seed procyanidins was related to their degree of polymerization, for instance, the monomers were the compounds with less antioxidant power (Spranger et al. 2008). However, to the best of our knowledge, there is no data that shows this statement in cocoa procyanidins. Other studies suggested that the anthocyanin's monomers were better hydrogen-donating antioxidant compared to oligomeric proanthocyanidins (Kong et al. 2003). Between the dimers, the procyanidins B2 and B had higher antioxidant power that procyanidin A glucoside because the glycosylation of the latter decreased its antioxidant capacity (Plaza et al. 2014a). The unknown compound (peak 1) and procyanidin C (peak 9) possessed the lowest antioxidant capacity (see Table 2).

Determination of Procyanidins by HPLC-FLD

The HPLC-DAD-ECD-CAD-MS method allowed the identification and quantification of monomers, dimmers, and some trimmers of procyanidins. However, this method could not determine larger oligomers of procyanidins. These compounds with different DP belong to the group of flavanols (see Fig. 1). They are the phenolic compounds more abundant in cocoa and have a range of positive physiological effects in humans; most notably, they reduce blood pressure and reduce the risk of cardiovascular disease and stroke (Sudano et al. 2012). Due to their structural diversity and complexity, the analytical methodology employed for the analysis of these compounds is based mainly upon DP instead of separation of individual compounds and specific isomers (Robbins et al. 2009). Therefore, a normal-phase HPLC-FLD method using a HILIC column was carried out to accomplish the analysis of procyanidins with different degrees of polymerization (Robbins et al. 2009; Robbins et al. 2013; Robbins et al. 2012). This method allowed the measurement of the large structural diversity of these compounds by oligomeric size. The procyanidins from cocoa and cocoa products have been quantified up to predefined molecular weight between DP1 and DP10. Procyanidins standards with DP2 and DP3 are commercially available; however, flavanols with DP4–10 are not. Thus, in this work, RRFs were used to achieve the quantification for these compounds due to the lack of commercial standards (see section "Analysis of Procyanidins by HPLC with fluorescence detector"). The RRFs used were specific for this analytical method according to Robbins et al. (2013, 2012). The results and chromatographic profile of cocoa procyanidins are shown in Table 3 and Fig. 5, respectively.

Table 3. Quantification (mg/100 g of sample ± standard deviation (sd)) of procyanidins in extracts obtained by PHWE and CSE from cocoa and cocoa products using HPLC-FLD

Procyanidins	Cocoa A		Cocoa B		Bread A		Bread B		Nibs	
	CSE	PHWE	CSE	PHWE	CSE	PHWE	CSE	PHWE	CSE	PHWE
DP 1	165.0 ± 19.4^{f}	194.9 ± 21.3^{g}	122.3 ± 7.2^{e}	180.2 ± 4.0^{fg}	37.1 ± 0.1^{ab}	39.0 ± 1.4^{ab}	26.9 ± 3.7^{a}	32.8 ± 3.3^{a}	45.0 ± 2.6^{ab}	57.1 ± 3.9^{ab}
DP 2	150.2 ± 19.3^{i}	196.2 ± 14.8^{j}	97.9 ± 7.8^{ef}	132.4 ± 11.8^{hi}	26.0 ± 0.5^{ab}	15.5 ± 2.4^{a}	16.9 ± 0.1^{a}	25.8 ± 1.2^{ab}	47.1 ± 5.4^{ac}	50.2 ± 2.7^{bc}
DP 3	21.5 ± 0.4^{de}	55.6 ± 8.1^{i}	10.2 ± 1.5^{abc}	9.7 ± 1.2^{abc}	3.5 ± 0.2^{a}	4.7 ± 0.7^{a}	3.3 ± 0.5^{a}	3.6 ± 0.5^{a}	8.4 ± 1.1^{ab}	5.7 ± 0.5^{a}
DP 4	10.2 ± 0.2^{b}	24.0 ± 1.2^{d}	1.7 ± 0.2^{a}	1.8 ± 0.3^{a}	ND	ND	ND	ND	0.28 ± 0.02^{a}	0.29 ± 0.03^{a}
DP 5	12.1 ± 1.6^{f}	13.7 ± 2.7^{f}	22.3 ± 0.8^{g}	34.3 ± 3.9^{h}	ND	ND	0.5 ± 0.1^{ab}	0.77 ± 0.04^{ab}	ND	0.023 ± 0.002^{a}
DP 6	122.9 ± 20.4^{f}	99.4 ± 7.6^{e}	46.5 ± 3.6^{cd}	86.2 ± 6.7^{e}	11.4 ± 0.6^{ab}	14.8 ± 1.3^{ab}	2.5 ± 0.1^{a}	2.5 ± 0.3^{a}	16.0 ± 2.3^{ab}	33.4 ± 2.7^{bc}
DP 7	35.5 ± 1.8^{d}	34.5 ± 3.9^{cd}	ND	ND	3.4 ± 0.4^{a}	4.2 ± 0.1^{a}	ND	ND	7.3 ± 0.3^{a}	27.1 ± 2.0^{bd}
≥DP 8	ND	ND	ND	ND	ND	ND	ND	ND	8.5 ± 1.1^{a}	35.8 ± 2.9^{b}
Total	517.4 ± 16.6^{j}	618.4 ± 42.8^{k}	300.9 ± 14.7^{ef}	444.6 ± 13.9^{hi}	81.4 ± 0.2^{ab}	78.2 ± 4.3^{ab}	50.1 ± 3.1^{a}	65.5 ± 2.9^{a}	132.6 ± 4.9^{bc}	209.5 ± 7.9^{de}
Procyanidins	Nibs powder		Chocolate B		Chocolate C		Chocolate D			
	CSE	PHWE	CSE	PHWE	CSE	PHWE	CSE	PHWE		
DP 1	129.1 ± 14.2^{e}	175.7 ± 5.5^{fg}	51.7 ± 5.3^{ab}	62.9 ± 1.7^{bc}	102.8 ± 11.3^{de}	116.9 ± 10.9^{de}	57.8 ± 7.9^{ab}	92.0 ± 3.9^{cd}		
DP 2	153.8 ± 15.5^{i}	207.9 ± 6.8^{j}	62.7 ± 1.6^{cd}	70.7 ± 4.1^{ce}	110.6 ± 9.5^{fgh}	124.9 ± 8.8^{gi}	65.7 ± 8.6^{cd}	81.8 ± 10.0^{def}		
DP 3	45.7 ± 1.7^{h}	62.0 ± 2.6^{i}	18.2 ± 2.5^{cde}	15.4 ± 2.5^{bcd}	36.4 ± 3.0^{g}	35.4 ± 2.0^{g}	26.6 ± 5.1^{ef}	30.2 ± 1.1^{fg}		
DP 4	27.9 ± 2.8^{d}	38.0 ± 2.0^{e}	2.9 ± 0.3^{a}	3.7 ± 0.6^{a}	15.3 ± 2.7^{c}	16.4 ± 0.9^{c}	17.0 ± 1.8^{c}	15.0 ± 2.6^{c}		
DP 5	7.4 ± 0.5^{de}	9.6 ± 1.6^{ef}	1.5 ± 0.1^{abc}	0.8 ± 0.2^{ab}	6.7 ± 0.8^{cde}	4.6 ± 0.6^{bd}	5.8 ± 0.6^{cde}	3.8 ± 0.7^{ad}		
DP 6	12.1 ± 1.4^{a}	92.0 ± 10.2^{e}	34.2 ± 2.0^{bc}	45.9 ± 5.7^{cd}	40.9 ± 0.5^{cd}	61.4 ± 7.5^{d}	4.5 ± 0.8^{a}	6.2 ± 1.2^{a}		
DP 7	65.6 ± 6.1^{e}	80.8 ± 9.3^{f}	23.1 ± 1.5^{bc}	23.4 ± 1.2^{bc}	28.3 ± 1.0^{b}	28.7 ± 0.7^{bd}	26.7 ± 4.6^{bd}	28.2 ± 1.5^{bd}		
≥ DP 8	43.8 ± 4.6^{c}	ND	ND	ND	1.5 ± 0.1^{a}	ND	7.1 ± 0.9^{a}	ND		
Total	485.4 ± 43.4^{g}	666.1 ± 28.3^{k}	194.2 ± 5.3^{cd}	222.8 ± 5.3^{de}	336.5 ± 12.9^{fh}	388.0 ± 29.7^{gh}	211.2 ± 19.6^{de}	257.2 ± 11.2^{de}		

All measurements were done in triplicate. *DP* degrees of polymerization from monomers (DP1) to octamers (DP8). *CSE* conventional sonication extraction, *PHWE* pressurized hot water extraction, *ND* not detected. Mean values with different superscript letters in the same row are significantly different ($p \leq 0.05$)

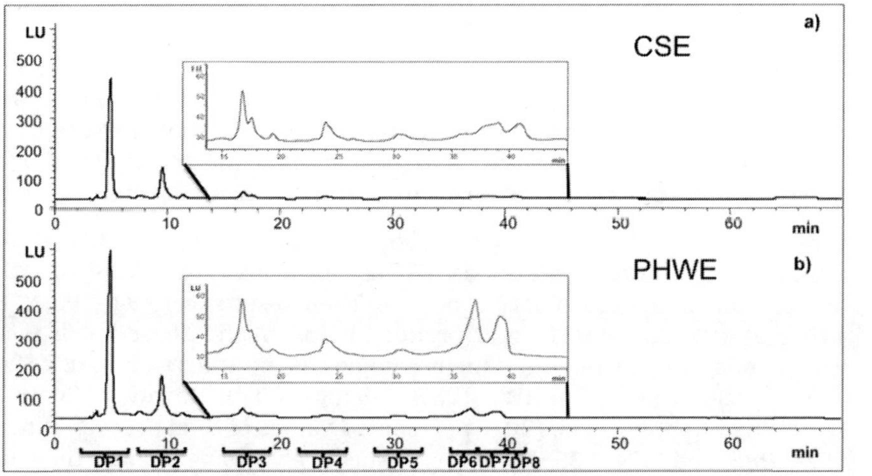

Fig. 5. HPLC-FLD profile of flavanol oligomers from nibs powder extract obtained by **a** CSE and **b** PHWE. FLD detection wavelength was 230 and 321 nm for excitation and emission, respectively. The peaks represent different degrees of polymerization (DP), from monomers (DP1) to octomers (DP8)

As can be seen in Table 3, DP1 and DP2 were the most abundant procyanidins followed by DP6 and DP7. DP7 was in higher concentration than DP6 just in chocolate D. However, ≥DP8, DP4, and DP5 were found in less concentration. Additionally, ≥DP8 was not presented in some samples as cocoas A and B, chocolate B, and the breads A and B, and DP7 was neither in cocoa B nor bread B. Mostly PHWE did not extract ≥DP8, only in nibs and nibs powder was an observed ≥DP8.

In general, PHWE was more efficient and extracted higher amounts of procyanidins from cocoa and cocoa products than CSE. Highly significant differences ($p \leq 0.01$) between both extraction methods in the extraction of total procyanidins were observed in cocoas A and B, and nibs and nibs powder, while non-significant differences ($p \geq 0.05$) were found in chocolates B, C, and D and breads A and B (Table 3). These results matched well with the results achieved for total phenolic compounds and for the phenolic compounds analyzed by reverse-phase HPLC (see Tables 1 and 3 and Fig. 3).

Cocoa A and nibs powder were the samples with more procyanidins. Nibs were a good source of polyphenols. As mentioned previously, the concentration of polyphenols in the different cocoa products depends on the percentage of cocoa, which justifies why the bread samples presented the lowest concentration of phenolic compounds. The percentage of cocoa powder in the bread was of 25% of dry weight and 15% of wet weight. And the total procyanidins on the bread represented an average of 15 ± 2% of the total procyanidins of their respective cocoas and extraction method. Hence, the procyanidin concentration in cocoa and cocoa bread correlate perfectly, also meaning that the procyanidins from the bread were not degraded during its preparation and baking.

Langer et al. (2011), Miller et al. (2006), and Todorovic et al. (2015) used a similar extraction method as CSE to measure the levels of total procyanidin oligomers in dark chocolate. Their results were 94–793, 278–410, and 187–263 mg/100 g of product, respectively. These levels were very similar to the ones found in samples B, C, and D for CSE (194–337 mg/100 g) and for PHWE (223–388 mg/100 g). However, Gu et al. (2006) and Robbins et al. (2013) presented values of total procyanidins of 852–1985 and 654–965 mg/100 g, respectively, being higher than the level detected in our samples.

Usually, higher concentrations of DP1 and DP2 flavanols were found with normal-phase HPLC-FLD method than with reverse-phase HPLC-DAD-ECD-CAD method. These differences could be due to the limited sensitivity of the CAD detector, not able to quantify compounds to low concentration. On the other hand, FLD is a sensitive detector. Also, procyanidins have a tendency to cross-link with proteins, sugars, and others compounds. HPLC-DAD-ECD-CAD method just detected the free monomers and dimers that link to small sugars. Overall, once again, PHWE technique was the most effective in extracting the target compounds.

CONCLUSIONS

The results obtained in this work have supported our hypothesis. PHWE method was more efficient in extracting phenolic compounds and methylxanthines from cocoa and cocoa products as compared to CSE method. The differences between both extraction methods were higher in the procyanidin group for all the tested samples and significantly higher for the cocoas (A and B) and nibs (nibs and nibs powder) samples. In addition, PHWE avoided many of extraction steps (i.e., defatting step) and the use of organic solvents, being a greener extraction method and more controlled. Also a full characterization of the phenolic composition of the cocoa products was possible by using advance analytical tools such as HPLC-DAD-ECD-CAD-MS and HPLC-FLD.

ACKNOWLEDGEMENTS

The present work has been supported by the Antidiabetic Food Centre, a VINNOVA VINN Excellence Centre at Lund University, Sweden. CT acknowledges the Swedish Research Council for funding VR (2010-333) and MP thanks the Spanish Ministry of Economy and Competitiveness (MINECO) for her Juan de la Cierva contract (IJCI-2014-22143).

REFERENCES

1. Adamson GE, Lazarus SA, Mitchell AE, Prior RL, Cao G, Jacobs PH, Kremers BG, Hammerstone JF, Rucker RB, Ritter KA, Schmitz HH (1999) HPLC method for the quantification of procyanidins in cocoa and chocolate samples and correlation to total antioxidant capacity. J Agric Food Chem 47:4184–4188
2. Akiba T, Yaguchi K, Tsutsumi K, Nishioka T, Koyama I, Nomura M, Yokogawa K, Moritani S, Miyamoto K (2004) Inhibitory mechanism of caffeine on insulin-stimulated glucose uptake in adipose cells. Biochem Pharmacol 68:1929–1937
3. Andújar I, Recio MC, Giner RM, Ríos JL (2012) Cocoa polyphenols and their potential benefits for human health. Oxidative Med Cell Longev. doi:10.1155/2012/906252
4. Belščak A, Komes D, Horžić D, Ganić KK, Karlović D (2009) Comparative study of commercially available cocoa products in terms of their bioactive composition. Food Res Int 42:707–716
5. Brcanovic JM, Pavlovic AN, Mitic SS, Stajanocic GS, Manojlovic DD, Kalicanin BM, Veljkovic JN (2013) Cyclic voltammetric determination of antioxidant capacity of cocoa powder, dark chocolate and milk chocolate samples: correlation with spectrophotometric assays and individual phenolic compounds. Food Technol Biotech 51:460–470
6. Chen AY, Chen YC (2013) A review of the dietary flavonoid, kaempferol on human health and cancer chemoprevention. Food Chem 138:2099–2107

7. Cheng CM, Jalil AMM, Ismail A (2009) Phenolic and theobromine contents of commercial dark, milk and white chocolates on the Malaysian market. Molecules 14:200–209
8. de Villiers A, Górecki T, Lynen F, Szucs R, Sandra P (2007) Improving the universal response of evaporative light scattering detection by mobile phase compensation. J Chromatogr A 1161:183–191
9. Gu L, House SE, Wu X, Ou B, Prior RL (2006) Procyanidin and catechin contents and antioxidant capacity of cocoa and chocolate products. J Agric Food Chem 54:4057–4061
10. Hii CL, Law CL, Suzannah S, Misnawi CM (2009) Polyphenols in cocoa (Theobroma cacao L.). As J Food Ag-Ind 2:702–722
11. Iriti M, Varoni EM (2013) Chemopreventive potential of flavonoids in oral squamous cell carcinoma in human studies. Nutrients 5:2564–2576
12. Kong JM, Chia LS, Goh NK, Chia TF, Brouillard R (2003) Analysis and biological activities of anthocyanins. Phytochemistry 64:923–933
13. Koşar M, Dorman HJD, Hiltunen R (2005) Effect of an acid treatment on the phytochemical and antioxidant characteristics of extracts from selected Lamiaceae species. Food Chem 91:525–533
14. Langer S, Marshall LJ, Day AJ, Morgan MRA (2011) Flavanols and methylxanthines in commercially available dark chocolate: a study of the correlation with nonfat cocoa solids. J Agric Food Chem 59:8435–8441
15. Magalhães LM, Segundo MA, Reis S, Lima JLFC (2008) Methodological aspects about in vitro evaluation of antioxidant properties. Anal Chim Acta 613:1–19
16. Martin MA, Goya L, Ramos S (2013) Potential for preventive effects of cocoa and cocoa polyphenols in cancer. Food Chem Toxicol 56:336–351
17. Matissek R (1997) Evaluation of xanthine derivatives in chocolate—nutritional and chemical aspects. Eur Food Res Technol 205:175–184
18. Miller KB, Stuart DA, Smith NL, Lee CY, McHale NL, Flanagan JA, Ou B, Hurst WJ (2006) Antioxidant activity and polyphenol and procyanidin contents of selected commercially available cocoa-containing and chocolate products in the United States. J Agric Food Chem 54:4062–4068
19. Plaza M, Turner C (2015) Pressurized hot water extraction of bioactives. TrAC Trends Anal Chem 77:39–40
20. Plaza M, Abrahamsson V, Turner C (2013) Extraction and neoformation of antioxidant compounds by pressurized hot water extraction from apple byproducts. J. Agric Food Chem 61:5500–5510
21. Plaza M, Pozzo T, Liu J, Gulshan Ara KZ, Turner C, Nordberg Karlsson E (2014a) Substituent effects on in vitro antioxidizing properties, stability, and solubility in flavonoids. J Agric Food Chem 62:3321–3333
22. Plaza M, Kariuki J, Turner C (2014b) Quantification of individual phenolic compounds' contribution to antioxidant capacity in apple: a novel analytical tool based on liquid chromatography with diode array, electrochemical, and charged aerosol detection. J Agric Food Chem 62:409–418
23. Robbins RJ, Leonczak J, Johnson JC, Li J, Kwik-Uribe C, Prior RL, Gu L (2009) Method performance and multi-laboratory assessment of a normal phase high pressure liquid chromatography-fluorescence detection method for the quantitation of flavanols and procyanidins in cocoa and chocolate containing samples. J Chromatogr A 1216:4831–4840
24. Robbins RJ, Leonczak J, Li J, Johnson JC, Collins T, Kwik-Uribe C, Schmitz HH (2012) Determination of flavanol and procyanidin (by degree of polymerization 1–10) content of chocolate, cocoa liquors, powder(s), and cocoa flavanol extracts by normal phase high-performance liquid chromatography: collaborative study. J AOAC Int 95:1153–1160
25. Robbins RJ, Leonczak J, Li J, Christopher Johnson J, Collins T, Kwik-Uribe C, Schmitz HH (2013) Flavanol and procyanidin content (by degree of polymerization 1–10) of

chocolate, cocoa liquors, cocoa powders, and cocoa extracts: first action 2012.24. J AOAC Int 96:705–711
26. Spranger I, Sun B, Mateus AM, Vd F, Ricardo-da-Silva JM (2008) Chemical characterization and antioxidant activities of oligomeric and polymeric procyanidin fractions from grape seeds. Food Chem 108:519–532
27. Stewart AJ, Mullen W, Crozier A (2005) On-line high-performance liquid chromatography analysis of the antioxidant activity of phenolic compounds in green and black tea. Mol Nutr Food Res 49:52–60
28. Sudano I, Flammer AJ, Roas S, Enseleit F, Ruschitzka F, Corti R, Noll G (2012) Cocoa, blood pressure, and vascular function. Curr Hypertens Rep 14:279–284
29. Todorovic V, Redovnikovic IR, Todorovic Z, Jankovic G, Dodevska M, Sobajic S (2015) Polyphenols, methylxanthines, and antioxidant capacity of chocolates produced in Serbia. J Food Compo Anal 41:137–143
30. Wollgast J, Anklam E (2000) Review on polyphenols in Theobroma cacao: changes in composition during the manufacture of chocolate and methodology for identification and quantification. Food Res Int 33:423–447
31. Wollgast J, Pallaroni L, Agazzi ME, Anklam E (2001) Analysis of procyanidins in chocolate by reversed-phase high-performance liquid chromatography with electrospray ionisation mass spectrometric and tandem mass spectrometric detection. J Chromatogr A 926:211–220

CHAPTER 10

Variability in Iron, Zinc and Phytic Acid Content in a Worldwide Collection of Commercial Durum Wheat Cultivars and The Effect of Reduced Irrigation on These Traits

Ana María Magallanes-López[1], Nayeli Hernandez-Espinosa[1], Govindan Velu[1], Gabriel Posadas-Romano[1], Virginia María Guadalupe Ordoñez-Villegas[1], José Crossa[1], Karim Ammar[1], Carlos Guzmán[1]

[1]*Global Wheat Program, International Maize and Wheat Improvement Center (CIMMYT), Apdo. Postal 6-641, Mexico DF., Mexico.*

ABSTRACT

Diets very rich in cereals have been associated with micronutrient malnutrition, and the biofortification of them, has been proposed as one of the best approaches to alleviate the problem. Durum wheat is one of the main sources of calories and protein in many developing countries. In this study, 46 durum varieties grown under full and reduced irrigation, were analyzed for micronutrients and phytate content to determine the potential bioavailability of the micronutrients. The variation was 25.7-40.5 mg/kg for iron and of 24.8-48.8 mg/kg for zinc. For phytate determination (0.462 to 0.952 %), a modified methodology was validated in order to reduce testing costs while speeding up testing time. Variation was detected for phytate:iron and zinc molar ratios (12.1-29.6 and 16.9-23.6, respectively). The results could be useful to generate varieties with appropriate levels of phytate and micronutrients, which can lead to the development of varieties rich in micronutrients to overcome malnutrition.

Keywords: durum wheat; iron; zinc; phytic acid; nutritional quality

1. INTRODUCTION

Malnutrition is a major challenge worldwide and the number of chronically undernourished and malnourished people has been rising (FAO., 2016). Almost 30% of the world population suffers from some form of malnutrition and of these, more than 2 billion people suffer from micronutrient deficiencies, of which 52% are pregnant women and 39% are children under five years of age (FAO, 2016). According to the World Health Organization (WHO) (2002), zinc (Zn) deficiency ranks 11th among the 20 most important risk factors contributing to the burden of disease in the world and 5th among the 10 most important factors in developing countries, while iron (Fe) deficiency ranks 6th. Zinc deficiency is responsible for many severe health complications, including impairments relating to physical growth, the immune system and learning abilities, as well as an increased risk of infections, DNA damage and cancer development (Gibson, 2006). On the other hand, Fe deficiency is the most common cause of anemia globally. Anemia affects around 1.6 billion people worldwide, with pre-school children and pregnant women at the greatest risk (McLean, Cogswell, Egli, Wojdyla, & de Benoist, 2009). Low dietary diversity and diets very rich in cereals have been associated with micronutrient malnutrition (Black, Victoria, Walker, Bhutta, Christian, De Onis, Ezzati, Graham-Mcgregor, Katz, Marterell & Uauy, 2013). These types of diets are commonly observed in populations of low socioeconomic groups in developing countries.

Biofortification of crops, i.e. enhancing micronutrient concentration in the edible part of the crops by plant breeding, has been proposed as one of the most cost effective and environmentally safe approaches to alleviate malnutrition. The Global Wheat Program of the International Maize and Wheat Improvement Center (CIMMYT), works with the HarvestPlus project to develop biofortified wheat varieties with enhanced Zn concentration. So far, the efforts have been focused on bread wheat (*Triticum aestivum* L. ssp. *aestivum*), which nowadays, is the main wheat species cultivated worldwide (around 90% of the total land), particularly in South Asian countries such as Bangladesh, India and Pakistan where Zn deficiency is a major problem.

However, durum wheat (*Triticum turgidum* L. ssp. *durum*) in many other developing countries (particularly in areas with semiarid climates), is one of the main sources of calories and protein. In North Africa, some countries in West Asia, as well as in Ethiopia, durum wheat is used to prepare diverse foods that serve as a staple food for a large part of the population in those countries; additionally it has socio-cultural and religious values. Among the food products are: couscous, which is popular in North Africa and the Middle East and results from the agglomeration of semolina particles; in the case of the Middle East and particularly Turkey, bulgur is a famous food product which results from parboiling, drying and crushing durum wheat grains to make different kinds of leavened breads such as the Algerian Khobz Eddar bread and flat breads such as the Ethiopian kitta or the pancakes named injera; different kinds of porridges as the Ethiopian kinche, prepared with crushed kernels, cooked with milk and water and mixed with spiced butter (Van Damme, 2007); and pasta, a product originally from Italy but that is popular worldwide. Some of these products may be prepared using whole grains or whole-meal flours. Due to this importance as a staple food for different populations, in order to understand the nutritional quality of durum wheat, it is necessary to check the feasibility for developing nutrient-dense durum wheat varieties. The studies carried out so far (Cakmak, Pfeiffer, & McClafferty, 2010; Ficco, Riefolo, Nicastro, De Simone, Di Gesù, Beleggia, Platani, Cattivelli & De Vita, 2009) have shown that the genetic variation for Zn is not high in durum, although more studies, including on diverse germplasm, are necessary.

On the other hand, with biofortification, breeders can achieve the mineral target increment by directly breeding for higher mineral concentration or breeding for increased bioavailability (Cakmak et al., 2010). Phytic acid (myoinositol-1,2,3,4,5,6-hexakisphosphate), as abundant in the aleurone layer as Fe and Zn, affects the bioavailability of minerals because of the possibility of strong chelation between the two (Coudray, Levrat-Verny, Tressol, Feillet-Coudray, Horcajada-Molteni, Demigné, Rayssiguier & Rémésy, 2001). A recent study (Eagling, Wawer, Shewry, Zhao, & Fairweather-tait, 2014) showed that the phytate levels had more influence on Fe bioavailability than total Fe content. Other studies have showed also the relationship between phytate and Zn bioavailability (Frontela, Scarino, Ferruzza, Ros, & Martinez, 2009) Therefore, breeding for low phytate content seems to be a reasonable objective to enhance nutritional quality of any crop.

To achieve new information on the genetic variability for mineral and phytate content in durum wheat, a study was undertaken with the following objectives: 1) describe the variability in grain Fe and Zn and phytic acid concentration in a collection of durum wheat cultivars with worldwide commercial importance; 2) estimate the bioavailability of Zn and Fe in durum whole-meal flours; and 3) examine the effect of reduced irrigation and genotype by environment (GxE) interaction effects on the nutritional quality traits.

2. MATERIALS AND METHODS

2.1. Plant materials

The study was conducted on a collection of 46 durum wheat varieties composed of representative commercial cultivars from main durum wheat growing countries (Electronic Supplementary Table 1). All genotypes were grown in Ciudad Obregon, Sonora (Mexico) during the 2014-2015 cropping season. All the genotypes were sown in November and harvested in May. Plots were arranged in a randomized complete block design with three replicates under two field management conditions: full irrigation (>500 mm) and reduced irrigation (300 mm), the latter to simulate drought stress. Weed, diseases, and insects were well controlled. Nitrogen (N) was applied (pre-planting) at a rate of 50 kg of N/ha and at tillering, 150 additional units of N were applied in the full irrigation management, while in reduced irrigation, only 50 N units were applied. The amount of nitrogen applied was enough to do not consider nitrogen a restricting factor in the study. At maturity, whole plots were harvested and grain yield was calculated.

Table 1. Preparation of enzymatic dephosphorylation reactions for free and total phosphorus determination

	Free Phosphorus (mL)	Total Phosphorous (mL)
Distilled water	0.114	0.094
Solution 1(buffer provided my Megazyme)	0.04	0.04
Neutralized sample extract	0.02	0.02
Suspension 2 (phytase prepared according to Megazyme)		0.02
Subtotal (mL)	*0.174*	*0.174*
Shake the tubes and incubate at 40°C/10min		
Distilled water	0.004	
Solution 3 (buffer provided by Megazyme)	0.04	0.04
Suspension 4 (phosphatase alkaline provided by Megazyme)		0.004
Subtotal (mL)	*0.044*	*0.044*

The meteorology data of the experimental station in Ciudad Obregon was characterized by almost no precipitation during the wheat growing season. Maximum temperatures were between 30 and 35°C in March and April, the grain filling time for both treatments. According to the general growing stages of durum wheat in Ciudad Obregon, drought stress was continuous from stem elongation to grain ripening in the reduced irrigation trial.

2.2. Grain physical parameters

Whole plots were harvested mechanically and grain yield (t/ha) was determined. A sample of 1 Kg of grain was kept for quality analysis. A SeedCount digital imaging system (model SC5000, Next Instruments Pty Ltd, New South Wales, Australia) was used to measure thousand kernel weight (TKW) (g) and test weight (TW) (Kg/hL), as it can rapidly and accurately analyze samples of wheat grains and determine the grain number and its morphological characteristics based on software and flatbed scanner technology.

Grain protein content (GPRO, 12.5 % moisture basis) was obtained by near-infrared spectroscopy (DA 7200 NIR, Perten Instruments, Sweden), validating its calibration with the chemical Kjeldahl method according to the AACC method 46-12 (American Association of Cereal Chemists, 2010).

2.3. Zinc, iron and phytic acid determination

Grain iron (FeC, mg/kg) and zinc (ZnC, mg/kg) concentrations were determined by using a bench-top, non-destructive, energy-dispersive X-ray fluorescence spectrometry (EDXRF) instrument (model X-Supreme 8000, Oxford Instruments plc, Abingdon, UK), which has been standardized for high throughput screening of Zn and Fe in whole grain wheat (Paltridge, Milham, Ortiz-Monasterio, Velu, Yasmin, Palmer, Guild & Stangoulis, 2012).

For phytic acid determination, a Megazyme (Ireland) kit was used by making some modifications to the protocol (Megazyme, 2016) provided. A 5 g grain sample was milled into whole-meal flour using a UDY Cyclone Sample Mill (UDY Corporation, USA) with a 5 mm (1/4 inch) screen. One gram of the resulting whole-meal flour was digested with 20 mL of HCl (0.66M) inside 50 mL Falcon tubes, placed in a mixer with oscillatory agitation (42 oscillations/min) overnight (15h) at room temperature. The day after, 1 mL of the extract was transferred to a 1.5 mL Eppendorf tube and centrifuged at 13,000 r.p.m. for 10 min. Immediately 0.1 mL of the resulting extract supernatant was transferred to a new tube. The solution was neutralized by the addition of 0.3 mL of a 2:1 (NaOH 0.75M: HCl 0.66M) mixture. A control blank sample was included with 0.1mL of HCl 0.66M.

After that, following the official protocol of Megazyme but with smaller amounts, 1.5 mL Eppendorf tubes were prepared as indicated in Table 1 for the enzymatic dephosphorylation reaction to calculate the free and total phosphorus content of the samples. After this, the tubes were incubated in an Eppendorf Thermomixer at 40°C for ten minutes. During the first minute of this incubation time, the tubes were shaken at 1,400 r.p.m. Following this incubation, different solutions, as indicated in Table 1, were added to the tubes for free and total phosphorus determination and incubated at 40°C for 15 minutes with shaking during the first minute at the same speed as mentioned above.

After the solutions were mixed with the help of a vortex, 0.06 mL of trichloroacetic acid were added to all tubes. The tubes were centrifuged at 13,000 r.p.m. for 10 min and 0.25 mL of supernatant were transferred to a new tube. To this, 0.125 mL of solution A+B (prepared according to Megazyme manual) were added. This was mixed by vortex and incubated in a water bath set at 40°C for 1 hour. The preparation of the phosphorus calibration curve was done according to Megazyme protocol but used a final volume of 2 mL. Finally, 0.11 mL of the reaction solutions were transferred to a 96-well plate and the absorbance at 655 nm of each well was read in an Epoch Microplate Spectrophotometer. Finally, the calculation of phosphorus and phytic acid content was carried out following Megazyme instructions.

2.4. Phytic acid:iron and phytic acid:zinc molar ratios

The contents of phytic acid, Fe and Zn, were converted into moles by dividing by their respective molar mass and atomic weight (660.04, 55.85 and 65.4 g mol^{-1}, respectively). The molar ratios of phytic acid:iron (Phy:Fe) and phytic acid:zinc (Phy:Zn) were then calculated.

2.5. Statistical Analysis

Pearson correlation coefficients (r) and statistical significance for each comparison in the entire study were obtained using SAS. Combined analyses of variance (ANOVA) across environments was performed using procedure Proc Anova of the SAS statistical software.

3. RESULTS

3.1. Scaling-down method for phytic acid determination

The protocol of the Megazyme kit for phytic acid determination was scaled-down five times and slightly modified to increase the number of samples that can be analyzed with the same amount of reagents and can handle more samples at the same time. The scaled-down method was validated with 20 wheat samples of the breeding program, which were analyzed with both protocols. The analysis was duplicated, with the average of standard deviation for each duplicate of 0.0122 and of 0.0228 for the official Megazyme method and the scaled-down one, respectively. The results obtained (Fig. 1) showed a highly significant correlation between both methodologies ($r = 0.83$). The range of variation for phytic acid content with the Megazyme official method was 0.546-0.683%, with an average value of 0.625%, while for the scaled-down method, the range was 0.522-0.705 % with an average value of 0.615 %.

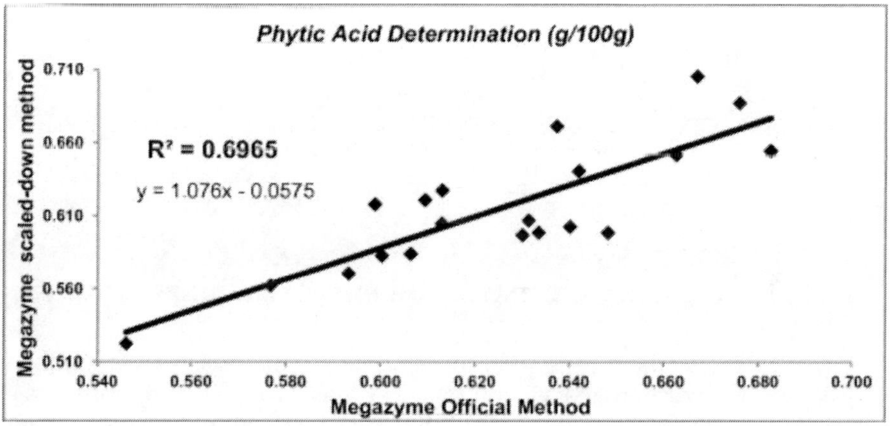

Fig. 1. Correlation between phytic acid content obtained using the Megazyme official and scaled-down method in 20 wheat whole-meal samples.

3.2. Effect of genotype, environment and genotype × environment interaction (G×E) on grain traits

A collection of 46 durum wheat cultivars grown in two different environments was analyzed for diverse grain traits. The combined analysis of variance revealed highly significant effects of the genotype, environment and their interaction (GxE) for all evaluated traits (Table 2). Genotype and environment were the most important factors explaining the variation found followed by GxE. The effect of the environment was particularly high for grain yield (83.2%), phytic acid:Fe (57.8%) and phytic acid (46%). The genotype was the greatest contributor in explaining variation for the rest of traits including FeC and ZnC, except phytic acid:Zn, which was more dependent on the GxE effect (29.8%). GxE was also very important to explain FeC (24.9 %).

Table 2. Effects of genotype, environment and their interaction (GxE expressed as % of the total sum of squares from ANOVA analysis

	DF	Grain yield	TW	TKW	GPRO	FeC	ZnC	Phytic Acid	Phy:Fe	Phy:Zn
Genotype	45	10.0	79.6	87.0	66.8	46.3^{+}	41.6	42.2	31.0	25.5^{+}
Environment	1	83.2	3.3^{++}	0.6▫	22.8	18.3	38.4	46.0^{++}	57.8^{++}	8.4▫
GxE	45	2.8	14.2	9.7	7.9	24.9	8.6	7.9	7.0	29.8^{++}
Error	180	3.4	2.8	2.2	2.5	10.0	9.1	3.9	4.1	29.7

TW, test weight; TKW, thousand kernel weight; GPRO, grain protein content; FeC, iron content; ZnC, zinc content; Phy:Fe, phytic acid:iron molar ratio; Phy:Zn, phytic acid:zinc molar ratio.
All values were highly significant ($p<0.001$), except $^{+}$significant ($p<0.05$); $^{++}$ ($p<0.01$) and '▫' not significant

3.3. Kernel characteristics and micronutrients contents

Table 3 shows the means and ranges of the parameters analyzed in the two different environments where the trial was grown under optimum and reduced irrigation conditions. In the reduced irrigation environment, lower grain yield and a higher GPRO than in the full irrigation trial were observed. TKW was slightly higher in full irrigation than in reduced irrigation and the opposite happened for TW. For micronutrients, the FeC was higher in reduced irrigation with a mean of 33.6 mg/Kg and a range of 29.7-42.6 mg/Kg, while ZnC was higher in the full irrigation environment with a range between 29.4-49.2 mg/Kg and a mean of 37.2 mg/Kg, compared to the range of 23.6-45.7 mg/Kg and average of 30.9 mg/Kg in the reduced irrigation environment. A higher mean value for phytic acid was found in full irrigation, and although maximum levels in the two environments were similar, the reduced irrigation environment had lower minimum values than in the full irrigation.

Table 3. Comparison between traits means and ranges for full irrigation and reduced irrigation environments

	Full Irrigation		Reduced Irrigation	
	Mean	*Range*	*Mean*	*Range*
Grain yield (t/ha)	5.1	2.8 - 6.1	2.6	1.1 - 3.2
TW (kg/hL)	81.6	74.5 - 84.0	82.2	78.9 - 84.4
TKW (g)	45.2	31.6 - 57.7	44.3	34.1 - 53.3
GPRO (%)	13.0	11.3 – 15.8	14.2	12.8 - 18.2
FeC (mg/Kg)	31.3	25.7 - 39.1	33.6	30.2 – 40.5
ZnC (mg/Kg)	37.2	31.8 – 48.8	30.9	24.8 – 44.7
Phytic Acid (%)	0.747	0.654 - 0.945	0.604	0.483 - 0.919
Phy:Fe	20.4	16.3 – 29.6	15.2	12.1 – 20.4
Phy:Zn	20.2	17.4 – 23.6	19.2	16.9 – 22.1

TW, test weight; TKW, thousand kernel weight; GPRO, grain protein content; FeC, iron content; ZnC, zinc content; Phy:Fe, phytic acid:iron molar ratio; Phy:Zn, phytic acid:zinc molar ratio.

Variation for micronutrients and phytic acid was also detected among cultivars (Electronic Supplementary Table 1). In the full irrigation environment cv. *Don Jaime* (39.1 mg/Kg) and cvs. *Iride* and *Rafi C97* (25.7 and 26.3 mg/Kg, respectively) showed the highest and lowest values for FeC, respectively. The largest ZnC was presented in cv. *Normanno* (48.8 mg/Kg) and the lowest content in cv. *Rafi C97* (31.8 mg/Kg). Highest phytic acid contents were obtained in cvs. *Exeldur* and *Normanno* (0.94 and 0.88 %, respectively) and the lowest values were presented in cv. *Altar 84* and cv. *Malavika* (0.66 and 0.65 %, respectively).

With respect to reduced irrigated environment, the highest content of FeC was obtained by cv. *Bellaroi* (40.5 mg/Kg) and the lowest concentrations corresponded with cvs. *Calero* and *Tomouh* (30.3 and 30.2 mg/Kg, respectively). Cvs. *Exeldur* and *Bellaroi* had the highest ZnC (44.7 and 43.0 mg/Kg, respectively), while cvs. *Altar 84* and *Nasr 99* varieties showed the lowest contents (24.8 and 25.7 mg/kg, respectively). For phytic acid, the highest concentration was obtained by the variety *Bellaroi* (0.92 %) and the lowest content was found in cvs. *Calero* and *Don Jaime* (0.49 and 0.48 %, respectively).

Significant variation was also found among cultivars (Electronic Supplementary Table 1) and between environments (Table 3) for phytic acid:micronutrients (Fe and Zn) molar ratios. For Phy:Fe and Zn molar ratios, the reduced irrigation environment showed lower values compared to the full irrigation one, although this difference was much more important for Fe than for Zn. The cvs. *Don Jaime* (16.3) and *Duilio* (17.4) showed the lowest Phy:Fe and Zn molar ratios, respectively, in the full irrigation environment, while the cvs. *Don Jaime* (12.1) and *Nacori C97* (16.) had the lowest Phy:Fe and Zn molar ratios, respectively, in the reduced irrigation environment.

3.4. Correlations between micronutrients content, phytic acid and kernel characteristics

To analyze the relationships among microelement concentrations and phytic acid with kernel characteristics, the Pearson correlation analysis was conducted (Table 4). Several traits showed consistent correlations between them across both environments. For example, the correlations between GPRO and both micronutrients, and between GPRO and phytic acid were highly significant in both environments but stronger in the reduced irrigation environment. Another highly significant correlation in both environments was between ZnC and phytic acid. The correlation between FeC and phytic acid was also highly significant in the reduced irrigation environment but not significant in the full irrigation one. These general associations were found in specific cultivars. For example, cv. *Normanno* had a relatively high FeC (35.0 mg/Kg) and GPRO (15.8 %) and the highest values recorded for ZnC (48.8 mg/kg) and phytic acid (0.88%) in the full irrigation environment.

Grain density (TW) was also negatively correlated with ZnC in both environments but not with FeC. Similarly, TKW was not correlated with FeC in the reduced irrigation environment and with ZnC in both the environments. Grain yield also showed significant negative correlations with several traits including FeC (only significant in the reduced irrigation environment), ZnC and phytic acid. Using the same previous example, cv. *Normanno* had the highest values for ZnC (48.8 mg/Kg) and grain yield slightly below average (4.8 t/ha) in the full irrigation environment. In the case of phytic acid, the Pearson coefficient was -0.55 with grain yield in full irrigation. The cv. *Exeldur* showed the highest phytic acid content (0.94 %) and the lowest value of grain yield (2.8 t/ha). For reduced irrigation environment, lower grain yield was significantly associated with an increase in FeC (r = -0.34), ZnC (r = -0.52) and phytic acid (r = -0.39). An example of this is presented by cvs. *Normanno* and *Bellaroi*, which had high values of FeC (38.5 and 40.5 mg/Kg, respectively), ZnC (37.2 and 43.0 mg/Kg, respectively) and phytic acid (0.79 and 0.92 %, respectively) but with low grain yields (1.8 and 1.5 t/ha, respectively).

Table 4. Pearson correlation coefficients (*r*) among micronutrients grain components and grain yield data obtained in two environments

	Grain yield	TW	TKW	GPRO	FeC	ZnC
	Full irrigation					
Grain yield	1					
TW	0.58**	1				
TKW	0.24*	0.46**	1			
GPRO	-0.39**	-0.47**	0.05	1		
FeC	0.17	0.11	0.47**	0.29**	1	
ZnC	-0.29**	-0.41**	0.10	0.67**	0.35**	1
Phytic Acid	-0.55**	-0.62**	-0.10	0.73**	0.09	0.71**
	Reduced irrigation					
	Grain yield	TW	TKW	GPRO	FeC	ZnC
Grain yield	1					
TW	0.46**	1				
TKW	0.20	0.25*	1			
GPRO	-0.60**	-0.49**	-0.03	1		
FeC	-0.34**	0.14	0.17	0.65**	1	
ZnC	-0.52**	-0.41**	-0.03	0.73**	0.62**	1
Phytic Acid	-0.39**	-0.38**	0.0	0.80**	0.64**	0.77**

TW, test weight; TKW, thousand kernel weight; GPRO, grain protein content; FeC, iron content; ZnC, zinc content.
*,** significant at 5% and 1% probability levels, respectively.

Table 5. Association between Molar Ratio, TW, TKW, GPRO and GY for full irrigation and reduce irrigation traits.

	Full Irrigation		Reduced Irrigation	
	Phy:Fe	Phy:Zn	Phy:Fe	Phy:Zn
Grain yield	-0.57**	-0.23*	-0.31**	0.01
TW	-0.55**	-0.09	-0.42**	-0.06
TKW	-0.44**	-0.22*	-0.08	0.02
GPRO	0.31**	-0.08	0.65**	0.17

TW, test weight; TKW, thousand kernel weight; GPRO, grain protein content; FeC, iron content; ZnC, zinc content; Phy:Fe, phytic acid:iron molar ratio; Phy:Zn, phytic acid:zinc molar ratio.
*,** significant at 5% and 1% probability levels, respectively.

The correlations between phytic acid:micronutrients molar ratios and other grain traits were also analyzed (Table 5). In most cases, there was a negative correlation between grain density (TW) and Phy:Fe or Zn molar ratio. That meant that in most cases the better the grain soundness, the higher the potential bioavailability (less phytic acid for more micronutrients). In full irrigation, the cv. *Exeldur* presented the highest Phy:Fe value (29.6) and the lowest grain density (TW = 74.5 Kg/hL), whereas for reduced irrigation, cv. *Bellaroi* had the lowest value for TW (79.9 Kg/hL) and high molar ratios for both micronutrients (19.7 for Phy:Fe and 21.9 for Phy:Zn).

Significant relationship was also detected between GPRO and Phy:Fe, but not with Phy:Zn. Cvs. *Bellaroi, Exeldur* and *Normanno* showed the highest content of GPRO (15.4, 15.4 and 15.8 %, respectively) and high Phy:Fe (31.7, 28.1 and 35.0, respectively); for reduced irrigation the same varieties presented high Phy:Fe (19.7, 20.4 and 17.2, respectively) and high values of GPRO (18.2, 17.3 and 18.0 %, respectively). In contrast, grain yield showed negative correlations with Phy:Fe in both environments and with Phy:Zn in the full irrigation environment. In full irrigation, cv. *Exceldur* had the lowest grain yield (2.8 t/ha) but corresponded with the highest value of Phy:Fe (29.6); and for reduced irrigation environment cvs. *Exceldur, Normanno* and *Bellaroi* varieties presented the highest values for Phy:Zn (20.4, 17.2 and 19.7, respectively) with the lowest grain yields (1.1, 1.8 and 1.5 t/ha, respectively).

4. DISCUSSION

So far, wheat biofortification breeding efforts for micronutrients (Zn and Fe) have been mainly focused on bread wheat, which have led to the release of several varieties in target countries (India and Pakistan) led by the consortium of the HarvestPlus challenge program. These biofortified varieties have shown competitive grain yields and approximately 30-40% more ZnC compared with the conventional varieties grown in those areas (Velu, Singh, Arun, Mishra, Tiwari, Joshi, & Pfeiffer, 2015). This achievement was possible due to the different wheat genetic resources with high ZnC preserved at CIMMYT's germplasm bank (Guzmán, Medina-Larqué, Velu, González-Santoyo, Singh, Huerta-Espino, & Peña, 2014) and were crossed with modern elite wheat lines, which are not very variable for micronutrients content (Cakmak et al., 2010).

To carry out a similar breeding process with durum wheat, it is necessary to know the current baseline micronutrient levels in commercial cultivars and the magnitude of genetic variability available within the primary genepool. It is also important to have high-throughput methodologies that allow for the fast analysis of hundreds of samples generated by the breeding program at a low cost. The EDXRF (energy-dispersive X-ray fluorescence spectrometry) equipment (Paltridge, Milham, Ortiz-Monasterio, Velu, Yasmin, Palmer, & Stangoulis, 2012) has been an extremely useful tool for analyzing ZnC and FeC and was used in this study for the analysis of grain samples from 46 commercial durum wheat varieties with worldwide economic importance, grown under full and reduced irrigation (simulated drought stress). The range of variation found in this worldwide collection across the whole trial was of 25.7-40.5 mg/kg for FeC and of 24.8-48.8 mg/kg for ZnC. In the case of ZnC, this range is similar to that found by Ficco, Riefolo, Nicastro, De Simone, Di Gesù, Beleggia, and De Vita (2009) (28.5-46.3 mg/kg) in a set of modern durum cvs. from Italy, although their range for FeC (33.6–65.6 mg/kg) was higher than that found in the current study. The study of Ficco et al. (2009) is the only reported study which was carried out under field conditions with a significant number of modern durum cultivars. Other authors have found similar or smaller ranges of variation in studies done with a small number of genotypes and/or under greenhouse conditions (Cakmak *et al.*, 2000 ; Genc and McDonald, 2008; Hakki, et al., 2014; Rachoń, Palys & Szumilo, 2012; Zhao, Su, Dunham, Rakszegi, Bedo, McGrath, & Shewry, 2009). Therefore, it seems clear that the genetic variability available in the modern durum pool is not enough, and it would be necessary to use other wheat genetic resources in the breeding process. In this respect, Cakmak et al. (2000) and Cakmak, Torun, Millet, Feldman, Fahima, Korol, and Ozkan (2004) and Gomez-Becerra, Yazici, Ozturk, Budak, Peleg, Morgounov, and Cakmak (2010) have shown that *T. dicoccoides* could be a good source of high micronutrients concentration along with *T. dicoccum* (Monasterio & Graham, 2000).

To increase micronutrients intake from wheat, it is not only the concentration of the micronutrient that is important, but also the amount that is available for absorption (Frontela et al., 2009). Phytic acid, an abundant component of the wheat grain that serves as phosphorus reservoir, is also considered to be an anti-nutrient because it chelates Fe and Zn during the digestion and avoids their absorption. In fact, phytic acid:micronutrients molar ratios are used to estimate the potential bioavailability of the micronutrients. In general terms, there is higher mineral bioavailability when the molar ratio is low and *vice-versa*. For Phy:Fe, the molar ratio should be <1 or preferably <0.4 to significantly improve Fe absorption (Hurrel & Egli, 2010), while for Phy:Zn molar ratios <5, between 5 and 15 and >15 have been associated with high, moderate and low zinc bioavailability, corresponding to approximately 50%, 30% and 15% of total zinc, respectively (Gibson, 2006). Because of this, the variability for phytic acid was also examined in the current study.

For this purpose, a modified methodology to quantify phytic acid was validated. While different methods to determine the amount of phytic acid in wheat have been described (Dost and Tokul, 2006 ; Haug and Lantzsch, 1983), the costs are high and the methods are designed to evaluate only limited number of samples per day. In breeding programs, the analysis of a large number of samples has to be done in the shortest time possible with the lowest costs. Due to this, we worked in the modification of the simple and accurate commercial kit of Megazyme to determine phytic acid, which was scaled-down in order to reduce testing costs while speeding up testing time. The modifications done were allowed to use smaller disposable components (1.5-2.0 ml Eppendorf tubes) and more efficient equipment for the different steps of the protocol (Thermomixer and Centrifuges for Eppendorf tubes and spectrophotometer for 96 well plates). This implies handling a higher number of samples per day (up to 50 with one technician), significantly reducing the cost of the analysis (5 times), and keeping enough accuracy to make selection in a breeding program (high correlation with the conventional method, $r = 0.83$). The use of a commercial-kit with worldwide distribution (Megazyme) could facilitate the implementation of the described method in other wheat quality labs working for the same (HarvestPlus) or similar projects, making the extrapolation of results found among breeding programs much easier.

The variation found for phytic acid ranged from 0.462 to 0.952 % (2 fold variation) in the whole trial, with an average of 0.675 %. Branković, Dragičević, Dodig, Knežević, Kandić, Šurlan-Momirović, and Sečanski (2015) reported a smaller range of variation for phytic acid but with significantly higher values (1.463-1.678 %) in a set of 15 durum genotypes (nine with CIMMYT origin), which shows the importance of the environmental conditions and methodology used when dealing with this trait. Tabekha and Donelly (1982) also found higher values in six durum cvs. from USA grown in three locations (0.95-1.43 %, average of 1.09%). However, Tavajjoh, Yasrebi, Karimian, and Olama (2011) found more similar values to the ones of the current study in two Iranian durum cvs. (0.879 and 0.740 %), as well as Hussain, Maqsood, and Miller (2012) in 65 bread wheat varieties grown in Pakistan (0.706-1.113 %).

Besides the Phy:Fe and Phy:Zn molar ratios were calculated and an interesting variation was detected (12.1-29.6 and 16.9-23.6, respectively) in the current set of data. This means that there was an almost two fold variation for Phy:Zn and around 1.5 fold for Phy:Fe, although all the varieties fall in the category of low bioavailability for both Fe and Zn according to Gibson (2006) and Hurrel and Egli (2010). The literature about durum wheat grain Phy:Fe is scarce or nonexistent. Salunke, Rawat, Neelam, Tiwari, Randhawa, Dhaliwal, and Roy (2014) showed findings very similar to this study, with a range Phy:Fe of 15.5-31.3 in a set of nine bread wheat varieties. Eagling et al. (2014) reported Phy:Fe around twelve in two bread wheat whole-meal flour samples, while Akhter, Saeed, Irfan, and Malik (2012) gave a range of 1.96-3.86 for the same trait in white flour of twelve bread wheat varieties. For Phy:Zn there is more information available, with Hussain et al. (2012) reporting a range of 23.9-41.4 for durum wheat, Erdal, Yilmaz, Taban, Eker, Torun, and Cakmak (2002) reporting 49-116 in durum and 29-178 in bread wheat, and Tavajjoh et al. (2011) reporting 26.5 and 26.9 in two durum cvs., which were in agreement with our data. Therefore, the concentration of micronutrients and the molar ratios revealed by the current and previous studies are not adequate to meet the daily requirements of humans in countries where durum wheat represents the main source of calories. Durum breeding programs working for target areas where micronutrient deficiency is a problem should be more focused on improving micronutrient concentration and reducing phytic acid to alleviate the malnutrition problems of the region.

This study revealed more information which will be useful for devising an appropriate durum wheat breeding strategy focused on improving nutritional quality. Although the genetic variability found was not very large, the genetic control of most of the traits seems to be high, which probably results in much faster genetic gains through proper selection methods. This would be more difficult to obtain for FeC and Phy:Zn due to the considerable GxE effect on those traits. Ficco et al. (2009) found, in general, a larger environment and, more importantly, GxE effect for those traits, which would significantly slow-down the genetic progress for these traits if confirmed. Another interesting fact found, previously reported by other authors (Kutman *et al.*, 2010; Gomez-Becerra *et al.*, 2010 ; Zhao *et al.*, 2009), is that both micronutrient concentrations are correlated with protein content, which in practice means that increasing the nutritional quality of the durum cultivars would also lead to an indirect increase in the industrial quality of the grain. Neither the protein nor the micronutrients and phytic acid concentration were affected by grain size in most of the cases (only FeC in full irrigation environment was affected by TKW), which removes the presence of a dilution or concentration effect of these components due to grain size. This agrees with Ficco et al. (2009) for FeC and ZnC but not for phytate. However, we could speak of a dilution or concentration effect due to changes in the grain density (significant correlations with TW for most of the components) and in the whole plant grain yield, which was negatively associated with all the grain components in both environments. This has been previously reported by several authors (Ficco *et al.*, 2009; Liu *et al.*, 2014 ; Velu *et al.*, 2016) and would negatively affect the breeding process, where strong selection would need to be applied to break the barrier of the negative association between grain yield and micronutrients. At least this negative association was stronger in most of the cases between grain yield and phytic acid, which means that increasing yield will indirectly contribute to increasing the bioavailability of the micronutrients. This fact was confirmed with the negative correlation found between grain yield and Phy:micronutrients molar ratios, with the exception of Phy:Zn in the reduced irrigation environment.

Lastly, due to drought stress, which is quite frequent in the main durum wheat growing areas (Mediterranean countries and Middle East), it was interesting to observe the effect of water stress on nutritional quality of durum wheat. A greater FeC was found in reduced irrigation, in agreement with Guzman at al. (2016) in a study done in a similar environment but with smaller number of cultivars. However, in the current study, significantly lower ZnC was found in the reduced irrigation environment, which contradicts previous studies in both durum and bread wheats (Guzman et al., 2016; Velu et al., 2016). These results are probably because there was not a remarkable difference in grain size (TKW) across environments, and the zinc uptake was severely reduced by the water stress or lesser grain filling period, which might reduce the loading of more Zn in the grain. The Phy:micronutrients molar ratios were also somehow smaller in reduced irrigation, indicating potentially better bioavailability of Fe and Zn when durum is grown under water stress.

5. CONCLUSIONS

The data generated in the present study has shown differences in micronutrients (Fe and Zn) and phytic acid in a worldwide durum collection, along with the evaluation of the responses to the environments (full and reduced irrigation or drought stress). The results could be useful for breeders to generate varieties with appropriate levels of phytic acid and micronutrients, which can lead to the development of variety-based products rich in the desired minerals to overcome deficiencies in population groups suffering from hidden hunger related issues of micronutrient bioavailability.

6. CONFLICT OF INTEREST STATEMENTS

The authors declare that they do not have any conflict of interest.

ACKNOWLEDGMENTS

We greatly appreciate financial support from the HarvestPlus challenge program and CRP-WHEAT of CGIAR consortium.

REFERENCES

1. Akhter, S., Saeed, A., Irfan, M., & Malik, K. A. (2012). In vitro dephytinization and bioavailability of essential minerals in several wheat varieties. Journal of Cereal Science, 56, 741–746.
2. American Association of Cereal Chemists, 2010. Approved methods of the AACC. St.Paul, MN, USA.
3. Black, R. E., Victora, C. G., Walker, S. P., Bhutta, Z. A., Christian, P., De Onis, M., Ezzati, M., Grantham-Mcgregor, S., Katz, J., Martorell, R., & Uauy, R. (2013). Maternal and child undernutrition and overweight in low-income and middleincome countries. Lancet, 382, 427–451.
4. Branković, G., Dragičević, V., Dodig, D., Knežević, D., Kandić, V., Šurlan-Momirović, G., & Sečanski, M. (2015). Phytic acid, inorganic phosphorus, antioxidants in bread and durum wheat and their associations with agronomic traits. Agricultural and Food Science, 24, 183–194.
5. Cakmak, I., Ozkan, H., Braun, H., Welch, R., Romheld, V. (2000). Zinc and iron concentrations in seeds of wild, primitive, and modern wheats. Food and Nutrition Bulletin, 21, 401–403.
6. Cakmak, I., Torun, A., Millet, E., Feldman, M., Fahima, T., Korol, A., Nevo, E., Braun, H., & Ozkan, H. (2004). Triticum dicoccoides: an important genetic resource for increasing Zinc and Iron concentration in modern cultivated wheat. Japanese Society of Soil Science and Plant Nutritio, 50, 1047–1054.
7. Cakmak, I., Pfeiffer, W., McClafferty, B. (2010). Biofortification of durum wheat with zinc and iron. Cereal Chemistry, 87, 10–20.
8. Coudray, C., Levrat-Verny, M. A., Tressol, J. C., Feillet-Coudray, C., HorcajadaMolteni, N. M., Demigné, C., Rayssiguier, Y., & Rémésy, C. (2001). Mineral supplementation of white wheat flour is necessary to maintain adequate mineral status and bone characteristics in rats. Journal of Trace Elements in Medicine and Biology, 15, 131-137.
9. Dost, K., & Tokul, O. (2006). Determination of phytic acid in wheat and wheat products by reverse phase high performance liquid chromatography. Analytica Chimica Acta, 558, 22–27.
10. Eagling, T., Wawer, A. A., Shewry, P. R., Zhao, F., & Fairweather-tait, S. J. (2014). Iron bioavailability in two commercial cultivars of wheat: comparison between wholegrain and white flour and the effects of nicotianamine and 2 ′ - deoxymugineic acid on iron uptake into Caco - 2 Cells. Journal of Agricultural and Food Chemistry, 62, 10320–10325.
11. Erdal, I., Yilmaz, A., Taban, S., Eker, S., Torun, B., Cakmak, I. (2002). Phytic Acid and Phosphorus Concentrations in Seeds of Wheat Cultivars Grown With and Without Zinc Fertilization. Journal of Plant Nutrition, 25, 113–127.
12. FAO. (2016). El espectro de la malnutricion. www.fao.org/worldfoodsummit /spanish/ fsheets/ malnutrition.pdf. Accessed 18.01.17.
13. Ficco, D. B. M., Riefolo, C., Nicastro, G., De Simone, V., Di Gesù, A. M., Beleggia, R., Platani, C., Cattivelli, L., & De Vita, P. (2009). Phytate and mineral elements concentration in a collection of Italian durum wheat cultivars. Field Crops Research, 111, 235–242.
14. Frontela, C., Scarino, M. L., Ferruzza, S., Ros, G., & Martinez, C. (2009). Effect of dephytinization on bioavailability of iron, calcium and zinc from infant cereals assessed in the Caco-2 cell model. World Journal of Gastroenterology, 15, 1977–1984.
15. Genc, Y., & McDonald, G. K. (2008). Domesticated emmer wheat [T. turgidum L. subsp. dicoccon (Schrank) Thell.] as a source for improvement of zinc efficiency in durum wheat. Plant and Soil, 310, 67–75.

16. Gibson, R. S. (2006). Zinc: the missing link in combating micronutrient malnutrition in developing countries. Proceedings of the Nutrition Society, 65, 51–60.
17. Gomez-Becerra, H. F., Yazici, A., Ozturk, L., Budak, H., Peleg, Z., Morgounov, A., Fahima, T., Saranga, Y., & Cakmak, I. (2010). Genetic variation and environmental stability of grain mineral nutrient concentrations in Triticum dicoccoides under five environments. Euphytica, 171, 39–52.
18. Guzmán, C., Medina-Larqué, A. S., Velu, G., González-Santoyo, H., Singh, R. P., Huerta-Espino, J., Ortiz-Monasterio, I., & Peña, R. J. (2014). Use of wheat genetic resources to develop biofortified wheat with enhanced grain zinc and iron concentrations and desirable processing quality. Journal of Cereal Science, 60, 617–622.
19. Guzmán, C., Autrique, J. E., Mondal, S., Singh, R. P., Govindan, V., Morales-Dorantes, A., Posadas-Romano, G., Crossa, J., Ammar, K., & Peña, R. J. (2016). Response to drought and heat stress on wheat quality, with special emphasis on bread-making quality, in durum wheat. Field Crops Research, 186, 157–165.
20. Hakki, E. E., Dograr, N., Pandey, A., Khan, M. K., Hamurcu, M., Kayis, S. A., Gezgin, S., Ölmez, F., & Akkaya, M. S. (2014). Molecular and elemental characterization of selected Turkish durum wheat varieties. Notulae Botanicae Horti Agrobotanici Cluj-Napoca, 42, 431–439.
21. Haug, W., & Lantzsch, H. J. (1983). Sensitive method for the rapid determination of phytate in cereals and cereals products. Journal of the Science Food and Agriculture, 34, 1423–1426.
22. Hurrell, R., & Egli, I. (2010). Iron bioavailability and dietary reference values. The American Journal of Clinical Nutrition, 91, 1461S–1467S.
23. Hussain, S., Maqsood, M. A., & Miller, L. V. (2012). Bioavailable Zinc in Grains of Bread Wheat Varieties of Pakistan. Cereal Research Communications, 40, 62–73.
24. Kutman, U. B., Yildiz, B., Ozturk, L., & Cakmak, I. (2010). Biofortification of durum wheat with zinc through soil and foliar applications of nitrogen. Cereal Chemistry, 87, 1–9.
25. Liu, H., Wang, Z. H., Li, F., Li, K., Yang, N., Yang, Y., Huang, D., Liang, D., Zhao, H., Mao, H., Liu, J., & Qiu, W. (2014). Grain iron and zinc concentrations of wheat and their relationships to yield in major wheat production areas in China. Field Crops Research, 156, 151–160.
26. McLean, E., Cogswell, M., Egli, I., Wojdyla, D., & de Benoist, B. (2009). Worldwide prevalence of anaemia, WHO Vitamin and Mineral Nutrition Information System, 1993–2005. Public Health Nutrition, 12, 444–454.
27. Megazyme (2016). https://secure.megazyme.com/files/Booklet/K-PHYT_DATA.pdf. Accessed 13.02.17.
28. Monasterio, I., & Graham, R. (2000). Breeding for trace minerals in wheat. Food and Nutrition Bulletin, 21, 392–396.
29. Paltridge, N. G., Milham, P. J., Ortiz-Monasterio, J. I., Velu, G., Yasmin, Z., Palmer, L. J., Guild, G. E., & Stangoulis, J. C. R. (2012). Energy-dispersive X-ray fluorescence spectrometry as a tool for zinc, iron and selenium analysis in whole grain wheat. Plant and Soil, 361, 261–269.
30. Rachoń, L., Pałys, E., & Szumiło, G. (2012). Comparison of the chemical composition of spring durum wheat grain (Triticum durum) and common wheat grain (Triticum aestivum ssp. Vulgare). Journal of Elemntology, 105–114.
31. DOI: 10.5601/jelem.2012.17.1.10 Salunke, R., Rawat, N., Neelam, K., Tiwari, V. K., Randhawa, G. S., Dhaliwal, H. S., & Roy, P. (2014). Effect of grain hardness on bioavailability of iron in wheat as determined using the coupled invitro digestion/Caco-2 model. LWT – Journal of Food Science and Technology, 59, 433–438.
32. Tabekha, M. M., & Donnelly, B. J. (1982). Phytic acid in durum wheat and its milled products. Cereal Chemistry, 59, 105–107.
33. Tavajjoh, M., Yasrebi, J., Karimian, N., & Olama, V. (2011). Phytic acid concentration and phytic acid: Zinc molar ratio in wheat cultivars and bread flours, Fars province, Iran. Journal of Agricultural Science and Technology, 13, 743–755.

34. Van Damme, P. (2007). Plant Resources of Tropical Africa 1. Cereals and Pulses. Economy Botany, 61, 108-118.
35. Velu, G., Guzman, C., Mondal, S., Autrique, J. E., Huerta, J., & Singh, R. P. (2016). Effect of drought and elevated temperature on grain zinc and iron concentrations in CIMMYT spring wheat. Journal of Cereal Science, 69, 182–186.
36. Velu, G., Singh, R., Arun, B., Mishra, V. K., Tiwari, C., Joshi, A., Cherian, B., Virk, P., & Pfeiffer, W. H. (2015). Reaching out to Farmers with High Zinc Wheat Varieties through Public-Private Partnerships – An Experience from Eastern-Gangetic Plains of India. Advances in Food Technology and Nutrition Sciences, 1, (3) 73-75.
37. World Health Organization (WHO). (2002). Ten leading causes of illness and disease in low income countries. http://www.who.int/ mediacentre/ factsheets/fs310_2008.pdf. The World Health Report 2002 Geneva: WHO. Accessed 13.02.17.
38. Zhao, F. J., Su, Y. H., Dunham, S. J., Rakszegi, M., Bedo, Z., McGrath, S. P., & Shewry, P. R. (2009). Variation in mineral micronutrient concentrations in grain of wheat lines of diverse origin. Journal of Cereal Science, 49, 290–295.

CHAPTER 11

Molecular Characterization of an Unauthorized Genetically Modified *Bacillus Subtilis* Production Strain Identified in a Vitamin B_2 Feed Additive

Valentina Paracchini [a], Mauro Petrillo [a], Ralf Reiting [b], Alexandre Angers-Loustau a, Daniela Wahler [c], Andrea Stolz [c], Birgit Schönig [c], Anastasia Matthies [c], Joachim Bendiek [c], Dominik M. Meinel [d], Sven Pecoraro [d], Ulrich Busch [d], Alex Patak [a], Joachim Kreysa a, Lutz Grohmann [c]

[a] *European Commission, Joint Research Centre, Ispra, Italy*

[b] *Hessian State Laboratory Kassel (LHL), Kassel, Germany*

[c] *Federal Office of Consumer Protection and Food Safety (BVL), Genetic Engineering Department, Berlin, Germany*

[d] *Bavarian Health and Food Safety Authority (LGL), Oberschleissheim, Germany*

ABSTRACT

Many food and feed additives result from fermentation of genetically modified (GM) microorganisms. For vitamin B2 (riboflavin), GM *Bacillus subtilis* production strains have been developed and are often used. The presence of neither the GM strain nor its recombinant DNA is allowed for fermentation products placed on the EU market as food or feed additive. A vitamin B_2 product (80% feed grade) imported from China was analysed. Viable *B. subtilis* cells were identified and DNAs of two bacterial isolates (LHL and LGL) were subjected to three whole genome sequencing (WGS) runs with different devices (MiSeq, 454 or HiSeq system). WGS data revealed the integration of a chloramphenicol resistance gene, the deletion of the endogenous riboflavin (*rib*) operon and presence of four putative plasmids harbouring *rib* operons. Event- and construct-specific real-time PCR methods for detection of the GM strain and its putative plasmids in food and feed products have been developed.

Keywords: *Bacillus subtilis*; Genetically modified organism (GMO); Genetically modified microorganisms (GMM); Next generation sequencing (NGS); Polymerase chain reaction (PCR); Detection; Vitamin B_2; Feed additives

1. INTRODUCTION

Riboflavin (7,8-dimethyl-10-[(2*S*,3*S*,4*R*)-2,3,4,5-tetrahydroxypentyl] benzo [*g*] pteridine-2,4-dione, vitamin B_2) is a water-soluble vitamin naturally synthesized by many microorganisms and plants. Since not being produced by higher animals, it is an essential micronutrient in animal and human diets.

The product riboflavin is often used in food as an additive but also finds applications in small amounts as the colouring agent E101 or as a nutritional additive in animal feedstuffs (Abbas & Sibirny, 2011). In the past, riboflavin was mainly chemically synthesised for the production of very pure material. Biotechnological developments have resulted in microbiological processes that can compete with the chemical synthesis and nowadays commercial production of vitamin B_2 is mostly done by fermentation.

In most cases, microbial synthesis of riboflavin involves genetically engineered selected strains of *Escherichia (E.) coli*, *Bacillus (B.) subtilis*, *Ashbya (A.) gossypii*, and *Candida (C.) famata* (Abbas & Sibirny, 2011). Among this plethora of genetically modified micro-organisms (GMMs), GM strains of *E. coli* and *B. subtilis* are probably the best known (see (Burgess, Smid, & van Sinderen, 2009)for a review), as in these bacteria the riboflavin biosynthetic pathway has been studied extensively (reviewed in (Bacher et al., 2001)). *B. subtilis* is known as an aerobic endospore-forming bacterium commonly found in nature and generally not considered to have a pathogenic or toxigenic potential. There is a history of safe use in large-scale fermentation production of speciality chemicals of enzymes used in food production processes, and of several traditional ways of food preparation.

Usually, non-sporulating derivatives of the *B. subtilis* strain 168, which often carry natural mutations inducing riboflavin overproduction, were genetically modified (GM). Introduction of different plasmids harbouring (i) both a (recombinant) *B. subtilis* riboflavin biosynthetic operon (*rib* operon, also known as *ribDEAHT* operon, i.e. including the *ribD, ribE, ribA, ribH, ribT* genes) under the control of a strong promoter and (ii) antibiotic resistance genes as selection markers (e.g. *cat, tet, ermAM*), resulted in GM *B. subtilis* strains with multiple copies of the *rib* operon. These strains are able to amplify the riboflavin expression by a magnitude of 10- to 25-fold (Mander and Liu, 2010; Perkins *et al*., 1999 ; Smolke, 2009).

According to EFSA guidelines for additives produced with GMM, it is necessary to show that, in the final product, neither the production strain nor its recombinant DNA can be detected (EFSA, 2011). In September 2014, it was notified in the European Rapid Alert System for Food and Feed (RASFF) that a German official enforcement laboratory in Hesse detected viable GM *B. subtilis* spores in a consignment of vitamin B_2 feed additive (80% feed grade) imported from China (RASFF, 2014).

In April 2015, a report of a Belgian official control laboratory was published about the genome sequence of a GM *B. subtilis* strain (Barbau-Piednoir, De Keersmaecker, Wuyts, et al., 2015). This strain (isolate 2014-3557) was identified by a French competent authority in a lot of vitamin B_2 (riboflavin, 80% feed grade) imported to France from China. On basis of next generation sequencing (NGS) data of a Belgian institution a TaqMan® qPCR method (named VitB_2-UGM) for specific detection of this EU-unauthorized GM riboflavin-overproducing *B. subtilis* was developed (Barbau-Piednoir, De Keersmaecker, Delvoye, et al., 2015). However, the TaqMan® qPCR method targets a junction between the riboflavin biosynthesis genes and the vector backbone. It remains unsolved whether the targeted sequence is integrated into the bacterial genome or present on a plasmid. For the latter case, the detection might fail if the plasmid is lost and the corresponding target sequence is therefore missing. In addition, the authors have not reported the comprehensive molecular characterization of the GM strains' genome and plasmids.

In the current study, microbiological and molecular analyses of the GM *B. subtilis* strain found in Germany in 2014 are presented. Whole genome sequencing (WGS) was performed with DNA extracted from two independent isolates to characterize in detail the genome of these riboflavin-overproducing GM *B. subtilis* strains, and to reconstruct the putative plasmids present. Subsequently, construct- and event-specific PCR-based methods for its detection in food and feed were developed and applied.

2. MATERIALS AND METHODS

B. subtilis living cells were isolated independently by the Hessian State laboratory (LHL) and the Bavarian Health and Food Safety Authority (LGL) in Germany from a product lot of vitamin B_2 feed additive (80%) powder imported from China and analysed in the framework of the RASFF notification reference number 2014.1249 (RASFF, 2014). Microbiological and molecular methods for the cultivation and identification of the microorganism and procedures for DNA extraction are described as Supplementary Material.

2.1. PCR analyses

Real-time PCR methods were applied to screen for the presence of DNA sequences from recombinant pUC plasmids (Table 1).

Table 1. Primers and probes used in this study.

Target	Name	Oligonucleotide sequence (5′-3′)	Reference
pUC-cloning vectors	401-F_PUC 18-F 401-R_pUC 18-R 401-Tex_Tm-Puc18	TgT CgT gCC AgC TgC ATT A gAg CgA ggA AgC ggA AgA g TexasRed-AAT Cgg CCA ACg CgC gg-BHQ2	(Mäde, Reiting, Strauch, Ketteritzsch, & Wicke, 2008)
	IPC-fw IPC-re IPC-S	TgT gAA ATA CCg CAC AgA Tg AgC Tgg CgT AAT AgC gAA G HEX-gAg AAA ATA CCg CAT CAg gC-TAMRA	(Messelhäusser et al., 2007)
Chloramphenicol acetyl transferase gene (*cat*)	356-F_Cat-Staph-F 356-R_Cat-Staph-R	CAg CTT TTA gAA CTg gTT ACA ATA gCg gCA TgA TAA CCA TCA CAA ACA gAA T	This work
Plasmid pAM-beta1 (*ermAM* adenine methylase gene -)	357-F2 357-R2	CgT CTA TTg AAT TAg ACA gTC ATC TAT TCA Tgg AAC ATC TgT ggT ATg gCg	This work
Integration site of *cat* in *B. subtilis* isolate e871	558-F 558-R 558-Tm	CgA gCT TTT gCg CgT ATA gCC ATT CCA ATA CAA AAC CAC ATA FAM-Cgg ATC TAA CgC ATg CTC CgC A-BBQ	This work
Plasmid pGMBsub01 (junction of pUB110 to pUC19)	690-F 690-R 690-Yak	gAT gAA TTA TAT CAA CAT ATT AAg CCT TTg g gCT Atg ACC ATg ATT Acg CCA Ag Yak-AAg ATC Cgg ggA ATT gCT gCA gg-BBQ	This work
Plasmid pGMBsub01 (junction of *B. amyloliquefaciens rib*-operon to pUC19)	691-F 691-R 691-Tm	CgA TTA AgT Tgg gTA ACg CCA TTC TCT AAA gAA AAC TgC TCg TAC g FAM-ACg gCC AgT gAA TTC gCA AgA Cg-BBQ	This work
Plasmid pGMBsub02 (junction of deleted *B. amyloliquefaciens rib*-operon)	804-F1 804-R1 804-Tm	AgA CCg CgT TTA Cag TCA gCA T CTCg AAT TCT TTT TTC gTT CCA A FAM-ACC ACA AgC TgA CCg AAT ATg Cgg AT-BBQ	This work
Plasmid pGMBsub03 (junction of 3′-*rib*-fragment to pUC19)	693-F 693-R Cy5-693	TCgTgCACAgCTTgAAATCTAgA ggA AAC AgC TAT gAC CAT gAT TAC g Cy5-CCT CTA gAg TCg ACC TgC Agg CAT gC-BBQ	This work
Plasmid pGMBsub04 (junction of *rib*-operon-fragment to pSM19035)	694-F 694-R 694-Tm	CAT TCg ATT gTg CgA gCg Tgg TAT TTT TTg TAT TCA gCg TAA Cag ACA TAA T FAM-Cag gCg AAT TCC AgT TAA ATT CCg TgT Agg--BBQ	This work

Primers for screening and detection of an erythromycin resistance gene (*ermAM*) and of the chloramphenicol acetyl transferase gene (*cat*) were designed using the Primer Express 3.0 software (Life Technologies Inc.) on the basis of the *Streptococcus faecalis* plasmid pAM-beta1 adenine methylase gene (GenBank:Y00116) and the sequence information for plasmid pC194 of *Staphylococcus (S.) aureus* (GenBank:K01998.1), respectively (Table 1).

WGS data were used to develop a GMM event-specific real-time PCR assay targeting the integration site of the chloramphenicol resistance gene (*cat*) in the genome of the *B. subtilis* isolate. Further real-time PCR assays were designed to detect the putative recombinant extra-chromosomal plasmids (see Supplementary Material).

Conventional PCR was done in 25 µL using a 10× PCR buffer (Qiagen Inc.) with 15 mM $MgCl_2$, 0.5 µM of each primer, 0.625 U Taq polymerase (HotStar, Qiagen Inc.) and 5 µL of template DNA corresponding up to 500 ng DNA. For thermal cycling, an initial denaturation step of 15 min at 95 °C was followed by 45 cycles of 30 s at 95 °C, 45 s at 60 °C and 45 s at 72 °C with a final elongation step of 7 min at 72 °C.

Real-time PCR was performed in an ABI PRISM 7500 (Applied Biosystems) and 25 µL PCR buffer (QuantiTect Multiplex PCR Mix, Qiagen Inc.) with 0.4 µM of each primer, 0.1 µM probe and 5 µL of template DNA corresponding up to 500 ng DNA. Thermal cycling conditions were a denaturation step of 15 min at 95 °C followed by 45 cycles of 1 min at 94 °C and 1 min at 60 °C. Template DNAs tested by the different PCR methods were either extracted from the different isolates of LHL and LGL or from isolate 2014-3557 of the French competent authority (kind gift of the Scientific Institute of Public Health, Brussels – WIV-ISP).

2.2. Whole genome sequencing (WGS)

Three WGS experiments using NGS were performed starting from *B. subtilis* DNA isolated by the two German laboratories (LHL and LGL) and by the Joint Research Centre (JRC) of the European Commission (Italy).

The analysis of the DNA sample isolated by the LHL were performed by a NGS service provider (StarSeq Inc., Germany) using a MiSeq apparatus (Illumina Inc.). For library preparation, 1 ng of extracted DNA was used for application in the Nextera XT DNA library preparation kit (Illumina Inc.). The generated genomic library was sequenced using the MiSeq Reagent Nano Kit v2 300 cycles (Illumina Inc.) and the pair-end option of 2 × 150 bp of the MiSeq sequencing system. Sequencing was monitored using the 'Sequencing Analysis Viewer' program (Illumina Inc.).

The NGS analysis of the DNA sample isolated by LGL was also carried out using Illumina Nextera XT library preparation of 1 ng purified DNA. After quality control, the library was sequenced using an Illumina HiSeq 1500 device using the pair-end flowcell v4 and the HiSeq SBS kit v4, 2 × 50 bp chemistry.

The third NGS analysis was performed by the JRC using another DNA sample from LHL. Here, NGS was done using a GS Junior System (GS Junior System, 454 Life Sciences, Roche Applied Sciences). Rapid libraries (medium length 400–600 bp) were prepared using Rapid library preparation kit (Roche) and all the steps were conducted in accordance with the manufacturers' instructions.

2.3. Bioinformatics

2.3.1. Quality of NGS reads

After adapter trimming, the quality of the NGS Illumina reads was analysed with the Fast QC program (version 1.2.10, default setting) (Andrews, 2010).

For the NGS Roche 454 reads, the quality was inspected by using Roche software (gsRunBrowser, version 2.9).

2.3.2. Assembly and mapping of NGS reads

The following assemblers and mapper tools were used to analyse the produced NGS reads:

- Burrows-Wheeler Aligners program (MiSeq Reporter adapted version, default setting) for mapping of Illumina reads. The genome of *B. subtilis* subsp. *subtilis*, strain AG1839 (NCBI No. CP008698) was used as a reference;
- Spades (version 3.6.2, default setting) (Bankevich et al., 2012) and Velvet (version 1.2.10, K-mer length 49, low coverage 0, high coverage 200, maximum contig length 200, expected genome coverage 1) (Zerbino & Birney, 2008) for *de novo* assembly of Illumina reads;
- runAssembly (Newbler, version 2.9, default setting) for *de novo* assembly of 454 reads;
- runMapping (Newbler, version 2.9, default setting) for mapping of 454 reads on *B. subtilis* subsp. *subtilis* strains and for detection of SNPs, indels and inversions.

All cited tools were run with parameters specific for bacterial genomes, as suggested by their developers.

At LGL, data analysis was carried out on a local Galaxy instance and genome assembly was done using Velvet combined with the VelvetOptimiser tool (version 1.0.0; threads = 1, start hash length = 23, end hash length = 45, Kmer optimisation metric = N50, coverage optimisation metric = >1 kb, read type = short paired reads) (Zerbino, 2010). Manual inspections were carried out using the Artemis version 11 (Carver, Harris, Berriman, Parkhill, & McQuillan, 2012) and Mauve version 2.4.0 (Edwards & Holt, 2013) programs with the default settings.

At the JRC, similarity searches and comparisons with other sequences (refseq, nt/nr and pat NCBI databases), were performed by running a local installation of the BLAST (version 2.3.0+) suite (Altschul, Gish, Miller, Myers, & Lipman, 1990). The summary of the similarity searches and the names of the reference strains used are listed in Table 2.

Table 2. Results of similarity searches against known reference strains and their genome sequences.

Species (subsp.)	Strain (substrain)	Number of large deletions	Number of large insertions	Number of large inversions	Number of SNPs	GenBank accession No.
B. subtilis (*subtilis*)	168	4	1	0	541	NC_000964
B. subtilis (*subtilis*)	6051-HGW	4	1	0	638	CP003329
B. subtilis (*subtilis*)	AG1839	5	2	1	887	CP008698
B. subtilis (*subtilis*)	OH 131.1	≥5	≥2	≥1	≥1000	CP007409
B. subtilis (*subtilis*)	JH642 (AG174)					CP007800
B. subtilis (*subtilis*)	BSP1					CP003695
B. subtilis (*subtilis*)	RO-NN-1					CP002906
B. subtilis (*subtilis*)	BAB-1					CP004405
B. subtilis	PY79					CP006881
B. subtilis	BEST7003					AP012496
B. subtilis	BEST7613					AP012495
B. subtilis	QB928					CP003783

2.3.3. 16S and 23S rDNA analyses

The contigs obtained from assembly were identified by using a local and installed copy of RNAmmer (version 1.2, default setting).

3. RESULTS

3.1. Microbe identification

Immediately after the RASFF notification for a lot of vitamin B_2 feed additive (80% feed grade) imported from China, the LHL laboratory and subsequently the LGL laboratory were able to independently isolate living spore-forming bacteria from this batch of vitamin B_2 product. Cultures of isolated cell colonies were producing an intense yellow colour secreted to agar plates or broth suggesting the synthesis of riboflavin. Further microbiological and molecular characterization led to the assignment of the isolated bacteria to the species *B. subtilis* (see Supplementary Material):

- Microbiological analyses of five single bacterial colonies showed mass spectrometry (MALDI-TOF) score values above 2.0 and a 99% confidence value in a biochemically based identification system.
- Sequence analysis of amplified parts of the 16S–rRNA-gene using DNA extracted from the vitamin B_2 product as a template showed a 100% agreement with the corresponding *B. subtilis* sequence.

These results immediately highlighted that the product was not compliant with the European food and feed safety regulatory requirements for feed additives containing GMMs (EU, 2003a ; EU, 2003b).

3.2. Identification of the genetically engineered elements

3.2.1. Screening PCR analyses of the vitamin B_2 product

The first real-time PCR screening tests (with DNA extracted from the *B. subtilis* cultures isolated from the vitamin B_2 product) targeted DNA sequences of recombinant pUC plasmids (Messelhäusser *et al.*, 2007 ; Mäde *et al.*, 2008). The two PCR methods target the left and right DNA junctions of the *lac*-operon from *E. coli* to sequences of the plasmid pBR322, a DNA construct which is present in all pUC plasmids and derivatives thereof. Clear amplification profiles with Cq-values in the range of 12–14 were obtained indicating the presence of these cloning vector sequences (data not shown).

The review of scientific literature on GMM strains in the context of riboflavin production revealed that antibiotic resistance genes are used as selection markers and thus potentially present in production strains. The most likely candidates were the erythromycin resistance gene (*ermAM*) and the chloramphenicol acetyl transferase gene (*cat*). The presence of these antibiotic resistance genes was tested on the *B. subtilis* strain isolated by

LHL using the developed PCR systems (Table 1). The PCR tests resulted in PCR products of 385 bp and of 396 bp corresponding in size to parts of *ermAM* gene and *cat* gene, respectively.

3.2.2. Next generation sequencing (NGS)

At the time the RASFF notification was published (end of 2014), three independent NGS experiments were performed. The DNA of the LHL isolate was analysed in two independent NGS runs (using a MiSeq or a GS Junior System), while DNA of the LGL isolate was sequenced in a single run (using a HiSeq system). The obtained sequence reads were independently assembled by the institutions. The aim was to test the ability of different NGS devices and different throughputs to identify the *B. subtilis* strain and to detect artificially introduced genetic elements. The outcome of these experiments is summarised in Table 3.

Table 3. Summary of the NGS experimental analyses results.

Sample	Institution/NGS platform (specifications)	Read Length	Obtained throughput	BC within Q30	BC error rate	Average assembled contigs depth
LHL	JRC/Roche 454 (WGS modus)	100 bp	83 Mb	99.7%	0.10%	18×
LHL	StarSeq Inc./Illumina MiSeq (paired-end modus)	2 × 150 bp	250 Mb	94.3%	0.18%	58×
LGL	LGL/Illumina HiSeq 1500 (paired-end modus)	2 × 50 bp	12 Gb	94.2%	0.18%	500×

To identify the closest related *B. subtilis* strain, contigs corresponding to the 16S and 23S rRNA genes were used as a query in BLAST analyses (Altschul et al., 1990) on all known complete *B. subtilis* genomic sequences available in the GenBank database at the time of the analyses (end of 2014). The LGL, LHL and JRC complete 16S rDNA sequence assemblies are 100% identical to the sequence of the *B. subtilis* strain 168 and also to 16S rDNA sequence of the French isolate 2014-3557 (Supplementary Material, Fig. S2). Genomes with the best alignment scores (minimum 99% identity over the full length of the locus) were selected and each one was used as a reference genome to map the reads produced (Table 2). It is expected that the better the reads map to a reference genome, the less differences (i.e. indels, inversions and SNPs) are present between the sequenced genome and the reference genome. *B. subtilis* subsp. *subtilis* strains 168, 6051-HGW and AG1839 showed to have genomes with the highest similarity to the *B. subtilis* strain present in the vitamin B_2 lot imported from China. These genomes are also fully covered by the reads, with the exception of the detected indels (Table 2). The genomic sequences of *B. subtilis* strains AG1839, 6051-HGW and 168 show high degree of identity (Kabisch *et al.*, 2013 ; Smith *et al.*, 2014) and the sequence reads generated here presented some of the strain-specific mutations, but also other variations not reported up to now (data not shown). Therefore, the sequenced isolates were considered as a *B.*

subtilis subsp. *subtilis* closely related to strain 168, but with specific differences. It is noted, that strain 168 has a long history of laboratory use because of its competence for DNA uptake and transformation (Barbe et al., 2009).

In 2015, Barbau-Piednoir, De Keersmaecker, Wuyts, et al. (2015) announced the sequencing of a GM *B. subtilis* isolate overproducing riboflavin. These authors did not provide a complete genome sequence but a total of 39 gap-closed scaffolds consisting of 143 contigs with a maximum gap-closed scaffold size of 1,018,461 bp and a minimum size of 370 bp. These contigs were compared to the sequences generated in this study. The comparison (made by BLAST) revealed an almost identical genome with more than 99% sequence identity over the whole length of all the 143 contigs.

In order to detect artificially introduced elements, chromosomal rearrangements were investigated by using the runMapping 454 software and the genome sequence of strain 168 as reference. When compared to this reference, the two sequenced isolates shared the same variations. In particular, four deletions and one insertion were identified:

- Nucleotides 529,421 to 549,933. This deletion involves ICEBs1, an integrative and conjugative element (ICE) integrated in the trnS-leu2 gene in *B. subtilis*, which has been described earlier (Auchtung, Lee, Monson, Lehman, & Grossman, 2005).
- Nucleotides 2,428,481 to 2,436,938. This deletion includes *ribD*, *ribE* and *ribAB* genes, in combination with the *fswa* flavin riboswitch, which is known to be involved in the fine regulation of the *rib* operon (Winkler, Cohen-Chalamish, & Breaker, 2002). As no other *rib*DEAHT operon was found on the main chromosome, this indicates that the strain is unable to produce riboflavin without additional plasmid-encoded *rib* operons. Reads spanning the deletion site are present, while reads that correspond to *ribD* , *ribE* and *ribAB* genes are not included by the software in this region, but are rearranged in a different contig, marked as not present in the reference.
- Nucleotides 3,236,246 to 3,236,442. This deletion involves the 3′ end of the hypothetical *yufK* gene encoding a putative transmembrane protein.
- Nucleotides 3,812,644 to 3,812,747. This deletion occurs in the *rpoE* gene. This gene is known to encode the DNA-directed RNA polymerase subunit delta involved in both the initiation and recycling phases of transcription; in the presence of the delta subunit, RNA polymerase displays an increased specificity of transcription, a decreased affinity for nucleic acids and an increased efficiency of RNA

synthesis because of enhanced recycling (Hiratsu, Amemura, Nashimoto, Shinagawa, & Makino, 1995).

- A unique identified insertion occurs on chromosomal nucleotide position 1,764,920, and it disrupts a gene (*recA*, alternative name *recE*) encoding a multifunctional protein involved in homologous recombination and DNA repair. This insertion is a fragment (1,290 bp) that contains the complete *cat* gene (European Nucleotide Archive (ENA) accession number LT622644). According to the sequence assembly, it has been cloned within a *ClaI* restriction site (AT|CGAT). The *cat* gene encodes a protein that confers chloramphenicol resistance. This finding supports the idea that the chloramphenicol resistance gene present in the chromosome of the analysed *B. subtilis* strain was intentionally introduced by a crossing over recombination event (Fig. 1).

Finally, when compared to the *B. subtilis* strain 168, more than 400 potential SNPs have been identified. Searches for specific mutations conferring resistance to antibiotics or peculiar bacterial features, like auxotrophics for amino acids or related compounds were performed. Some examples are described in the Supplementary Material.

3.3. GMM event-specific detection method

The integration site of the *cat* gene in the genome was identified by bioinformatics analyses on the assembled NGS data. The determined DNA sequence across the junction between *cat* and the *recA* gene was used to develop a real-time PCR-based event-specific method (Fig. 1).

Details for this PCR system (No. 558) and the specificity are given in the Supplementary Material (Table S1). The event-specific method showed equal detection capability for DNA extracts from the LGL and LHL isolates and for DNA of the French isolate 2014-3557 analysed by NGS (Barbau-Piednoir, De Keersmaecker, Delvoye, et al., 2015). The results of the PCR tests are given in the Supplementary Material (Table S2).

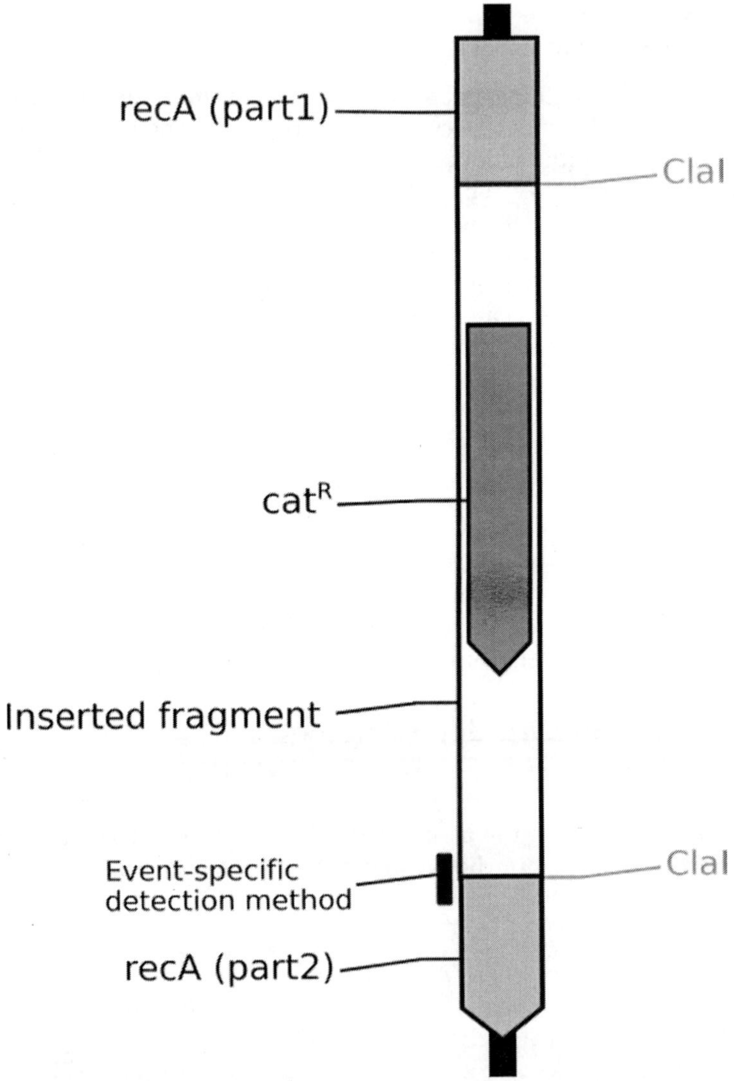

Fig. 1. Graphic representation of the *B. subtilis* genomic region in which a transgenic cassette carrying the *cat* gene has been inserted within *Cla*I site of the genomic *recA* gene, thus disrupted. The location event specific detection method is also represented on the junction between the 3′ prime transgenic cassette and the genome.

3.4. Recombinant extra-chromosomal plasmids

Apart from the described insertion, no other chromosomal integration site could be identified. On the contrary, a fraction of the obtained sequence reads did not map on the reference genome. They could be assembled into four distinct contigs that do not show any read joining them with the genomic sequence of *B. subtilis* strain 168. Moreover, reads overlapping both their ends were identified and thus marked by the assembler tools (such as *newbler*) as potentially circular molecules, suggesting a plasmid-like structure. Similarity searches with publicly available nucleic acid databases revealed correspondences with artificial plasmids. For these reasons, we assume that these contigs correspond to four different recombinant extra-chromosomal plasmids (Table 4):

- *pGMBsub01* (Fig. 2, panel A) (submitted to the European nucleotide Archive – ENA - accession number LT622641). This putative plasmid contains the full *B. amyloliquefaciens rib*DEAHT operon (*ribD, ribE, ribA, ribH, ribT* genes, 11,378 bp), together with a *B. amyloliquefaciens* conserved hypothetical protein (upstream of the *rib* operon in the *B. amyloliquefaciens*) and (downstream) the *B. amyloliquefaciens* genes encoding the segregation and condensation proteins A and B (ScpA and ScpB). The *scpA* open reading frame (ORF) is present as full length, the *scpB* ORF is truncated and directly linked (not in-frame) to the *S. aureus* replication initiation protein B (RepB). The plasmid also carries an ampicillin resistance gene from the pUC19 vector and coding sequences for the kanamycin and bleomycin resistance genes from the vector pUB110 (GenBank M19465.1).
- *pGMBsub02* (Fig. 2, panel A) (ENA accession number LT622641). This putative plasmid (7,581 bp) is a truncated version of pGMBsub01, as mapped reads overlapping the deleted part are present in the NGS data (see Supplementary Material, Fig. S2). In fact, only the *ribD* and *ribE* genes from the *B. amyloliquefaciens rib*DEAHT operon are present. Like pGMBsub01, it carries an ampicillin resistance gene from the pUC19 vector and coding sequences for the kanamycin and bleomycin resistance genes from the pUB110 vector.

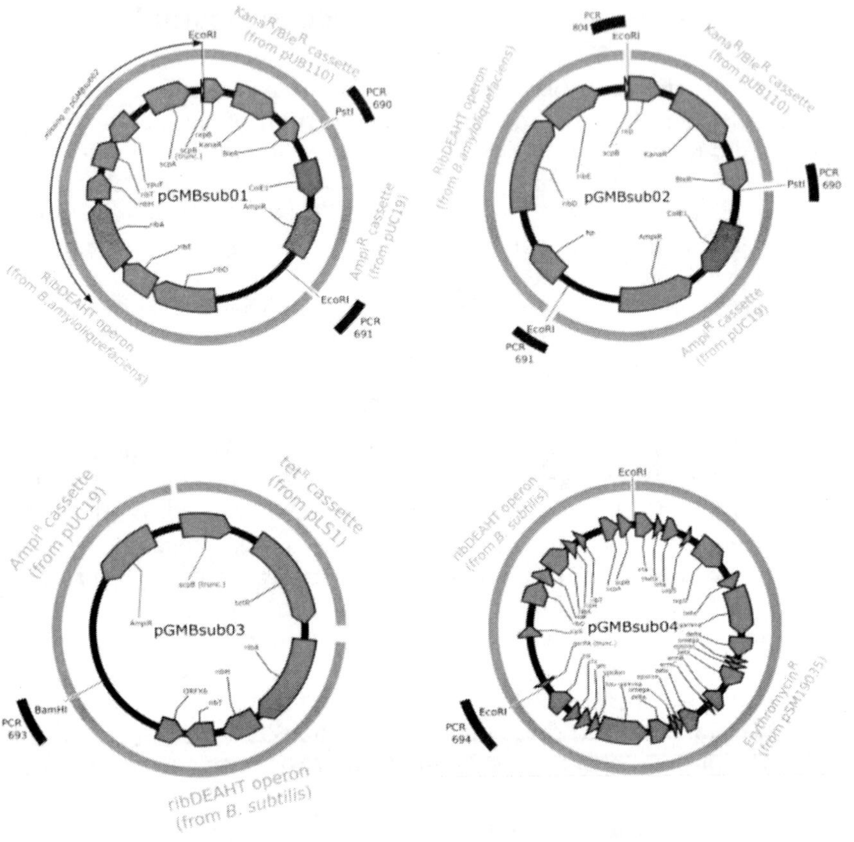

Fig. 2. Schematic representation of the putative recombinant extra-chromosomal plasmids. The different PCR detection methods developed and applied for experimental tests are indicated by the black bars.

- *pGMBsub03* (Fig. 2, panel B) (ENA accession number LT622642). Like pGMBsub01 and pGMBsub02, this putative plasmid (8,544 bp) carries an ampicillin resistance gene from the pUC19 vector (GenBank M77789.2). In addition, a full tetracycline resistance gene (*tet*) cassette is present, together with a (truncated) *scpB* gene upstream, both probably derived from *Streptococcus agalactiae* plasmid pLS1 (Lacks, Lopez, Greenberg and Espinosa, 1986; GenBank M29725.1). Differently from the previous plasmids, it includes part of the *B. subtilis rib*DEAHT operon (only *ribA*, *ribH*, *ribT* genes), together with (downstream) the *B. subtilis* ORFX6 gene.
- *pGMBsub04* (Fig. 2, panel C) (ENA accession number LT622643). This putative plasmid (29,760 bp) includes the full *B. subtilis rib*DEAHT

operon (about 12 Kb) together with genes typical for *Enterococci* and *Streptococci* plasmids and conferring resistance to erythromycin. At the sequence level, the non-*rib* operon sequence is identical to that of plasmid pSM19035 (GenBank AY357120), a low-copy-number theta-replicating plasmid of the pathogenic bacterium *Streptococcus pyogenes*, stably maintained in a broad range of gram-positive bacteria (Lioy, Pratto, de la Hoz, Ayora, & Alonso, 2010). The *B. subtilis rib*DEAHT operon is surrounded by two *Eco*RI restriction sites. pGMBsub04 shares with pSM19035 large inverted repeats which are described in the latter as responsible for the erythromycin resistance (Soberon, Lioy, Pratto, Volante, & Alonso, 2011). Furthermore, the presence of the pSM19035 sequences explains the detection of the erythromycin resistance gene (*ermAM*) at the early stage of the characterization of the isolate.

Table 4. Description of the four extra-chromosomal plasmids identified in this study.

Plasmid name	Size	Genes conferring resistance to	*Rib*DEAHT operon	RibDEAHT origin	Similarity to known vectors	Found in sample	Presence described in Barbau-Piednoir et al. (2015)
pGMBSub01	11,278	ampicillin, kanamycin, bleomycin	Full	*B. amyloliquefaciens*	pUC19, pUB110	LGL, JRC, LHL	Yes (splitted in 5 contigs
pGMBSub02	7,580	ampicillin, kanamycin, bleomycin	Truncated	*B. amyloliquefaciens*	pUC19, pUB110	LGL, JRC, LHL	Yes (splitted in 5 contigs)
pGMBSub03	8,544	ampicillin, tetracycline	Truncated	*B. subtilis*	pUC19, pLS1	LGL	Yes (splitted in 2 contigs)
pGMBSub04	29,760	erythromycin	Full	*B. subtilis*	pSM19035	LGL	Yes (splitted in 5 contigs)

Sequence reads for plasmids pGMBsub01 and pGMBsub02 were present in data obtained from the LHL and the LGL isolate, whereas for pGMBsub03 and pGMBsub04 reads could be found only in the LGL data (Table 4). Presumably these two plasmids were coincidently lost during the enrichments and cultivations at LHL, most probably because the respective antibiotics were not supplemented to the medium. It is noted, that NGS sequence data for all four putative plasmids identified for the LGL isolate are also present in the data of the French isolate 2014-3557.It is thus plausible that there is no incompatibility among the plasmids. In the course of the analyses of the *B. subtilis* isolate, five real-time PCR methods for detection of the extra-chromosomal recombinant plasmids were developed. Details on the specificity tests are included in theSupplementary Material (Table S2). In conformity with the PCR results obtained for the event-specific method, all plasmid specific methods reacted positive for the DNA extracted from isolate 2014-3557 (Barbau-Piednoir, De Keersmaecker, Delvoye, et al., 2015).

4. DISCUSSION

In this study, NGS was performed for the molecular characterization of a GM *B. subtilis* strain overproducing vitamin B_2. The aim was to use NGS coupled with bioinformatics to identify the nucleotide sequence of all genetic modifications inserted into the microorganism's genome and to characterize complementing extra-chromosomal recombinant plasmids. WGS by NGS may represent a promising approach for gaining sequence-based information for detailed molecular characterization of insertions in GM organisms (Pauwels *et al.*, 2015; Wahler *et al.*, 2013 ; Yang *et al.*, 2013). However, as demonstrated by these studies, the approach requires appropriate NGS devices, specific expertise in bioinformatics and the respective analysis tools. Difficulties in the distinction between true variants and sequence alterations and the challenges in making error corrections (for repetitive regions, uncalled bases, ploidy etc.) are currently discussed (Akogwu, Wang, Zhang, & Gong, 2016). On the other hand, NGS may be applied to rapidly generate sequence data in order to identify unknown genomic insertions/deletions and genetic modifications with potential application as basis for developing specific detection methods (Arulandhu *et al.*, 2016 ; Barbau-Piednoir, De Keersmaecker, Delvoye, *et al.*, 2015). The present work is based on the use of two second generation sequencing technologies (Illumina and 454). Both sequencing chemistries deliver unparalleled accuracy, with a vast majority of bases scoring at Q30 and above (see Supplementary Material). This level of accuracy is ideal for a range of sequencing applications, including clinical research. By mapping the generated reads to the *B. subtilis* reference genomes, it was possible to identify in the *recA* gene an insertion site in the genome of the isolated strain where recombinant DNA is integrated. Presumably the *recA* gene was disrupted to avoid homologous recombination of the production strain. The complete nucleotide sequence of this unique insertion could be determined and identified as the complete chloramphenicol resistance gene (*cat*) sequence, which may serve as selection marker. The junction of this artificial and stably integrated cassette as target for the design of an event-specific detection method. The specificity was verified by the experiments with DNA samples derived from the LHL, LGL and the French isolate 2014-3557 (see Supplementary Material).

The genome of the isolates carries a deletion of the endogenous *rib*DEAHT operon, indicating that the strain is unable to produce riboflavin without the recombinant plasmids encoding the *rib* operon. The absence of the endogenous *rib*DEAHT operon is presumably complemented with four recombinant plasmids carrying full or truncated *rib*DEAHT operon from *B. subtilis* and *B. amyloliquefaciens*. Together, these operons carry one or more specific antibiotic resistance genes for selection purposes and stable riboflavin expression during fermentation. In the DNA samples sequenced by JRC and LHL only two of the four putative plasmids could be identified (pGMBsub01 and pGMBsub02). Probably during non-selective cultivation of bacteria without antibiotic supplements, the plasmids pGMBsub03 and pGMBsub04 have been lost. The construction of these recombinant plasmids has not been reported by (Barbau-Piednoir, De Keersmaecker, Delvoye, et al., 2015). This study provides the corresponding sequences as unrelated contigs not distinguished from the chromosomal contigs. Presumably, the local similarities, like the common pUC19 backbone regions, do not allow untangling of ambiguities during the assembling procedure. Only the pGMBsub04 plasmid shows a deletion of 200 bp if compared to the counterpart contigs published previously (Barbau-Piednoir, De Keersmaecker, Wuyts, et al., 2015). This deletion puts the neighbouring additional copies of tau and gamma genes in-frame. The presence of the reads across the point of deletion suggests that pGMBsub04 is presumably present in two forms, i.e. a long one found and reported previously (Barbau-Piednoir, De Keersmaecker, Wuyts, et al., 2015) and a short one sequenced within our study of the isolate, maybe the result of a deletion event naturally occurring.

We further investigated the origin of the GM *B. subtilis* strain. This strain seems not to be described in the literature, apart from the recent study (Barbau-Piednoir, De Keersmaecker, Wuyts, et al., 2015). However, a detailed search in patent databases revealed the presence of an application patent from Russia (Mironov et al., 2004), in which the authors describe a method for producing riboflavin. In this document, the applicants developed and claimed both GM *B. subtilis* strains and plasmids to over-produce riboflavin. No sequence information is provided, but the following detailed descriptions are reported:

- A vector named pEK14 is described which is identical to pGMBsub01, i.e. consisting of the *E. coli* plasmid pUC19, a kanamycin nucleotidyl transferase gene (*kan*) from the pUB110 plasmid, and a desensitized (ribO2 mutation) riboflavin operon from *B. amyloliquefaciens.*
- The development of a GM *B. subtilis* GM51 strain, that is made chloramphenicol resistant by inserting the chloramphenicol acetyl transferase gene (*cat*) from plasmid pBT69 into the *Cla*I site of the *recE* gene.
- The use of a *B. subtilis* GM51 strain, which is auxotrophic for riboflavin.
- The latter two descriptions are in complete agreement with the identified insertion and *rib* operon deletion at the chromosomal level.

Furthermore, the authors describe the transformation of *B. subtilis* GM51 strain with plasmid pMX45 containing the full *rib* operon from *B. subtilis*. pMX45 is referenced as a patented recombinant plasmid described in an older French patent (FR Pat. 2546907) (Stepanov, Kukanova, Glazunov, & Zhdanov, 1977). Here the pMX45 plasmid is described as a 30 Kb DNA molecule carrying the full *B. subtilis rib* operon inserted into the pMX33 plasmid with an erythromycin resistance operon. A bibliographic search revealed that the patent authors published two book chapters (Rabinovich *et al.*, 1985 ; Rabinovich *et al.*, 1984) in which they describe in detail the development of the recombinant plasmid pMX33 by ligation of an *Eco*RI-cut *B. subtilis* DNA fragment containing the full 12 Kb *rib* operon with an *Eco*RI-cut vector pMX30. pMX30 in turn is defined as a deletion mutant of plasmid pSM19035 (Rabinovich et al., 1985) with the large inverted repeats conferring erythromycin resistance, like the sequenced pGMBsub04. For these reasons, we argue that pGMBsub04 corresponds to either the pMX45 plasmid or its parent pMX33.

At the time of the RASFF alert, a request for clarification was sent by the German diplomatic service to the Chinese company that marketed the vitamin B_2 feed additive (80%). A brief description of a strain claimed to be used as the vitamin B_2 production strain was provided, together with the sequence information of a recombinant plasmid that confers resistance to tetracycline. This plasmid sequence was found to be identical to the pGMBsub03 sequence. The company claimed that a plasmid named pMX45 was used for the transformation of the *B. subtilis* production strain, which confirms our hypothesis that plasmid pGMBsub04 correspond to pMX45. A comparison of the information provided by the Chinese company and our findings is shown in Table S2 of Supplementary Material. In particular, we did not find any pUC19 integration at the genomic level, but additional recombinant plasmids neither mentioned by the Chinese company nor expected to be present. The differences in the information provided by the Chinese company and the results of the molecular characterization strongly imply that the production strain must have been contaminated or switched before or during production.

In response to the detection of the presence of GMOs in rice products exported from China, to a case of Bt63 presence in feed additive to the EU and to the case described in our study which lead to a notification in the RASFF system, the European Commission's Food and Veterinary Office (FVO) carried out an audit in China in order to evaluate their control systems for GMO. It is reported that the Chinese competent authorities carried out investigations at all vitamin B_2 producers, but no irregularities were identified (FVO, 2016).

In order to evaluate the possible microbiological risks for human and animal health, the microbiological status of the production process requires consideration with regards to the safety and identity of the producer organism used in the fermentation process. Riboflavin production by fermentation of specific GMMs must therefore involve the approved safety assessment of the genetic modification and the respective strain. The continuous quality control testing of the production process and cultural purity is required to exclude contaminations and/or impurities. Degradation of all DNA needs to be verified by the producer in polymerase chain reaction (PCR) tests (Hermann & Schurter, 1995).

Food and feed additives, produced in closed production systems with GMMs, do not fall under the scope of Regulation (EU) No. 1829/2003 on GM food and feed (EU, 2003a). According to its recital 16, this Regulation covers food and feed produced "from" a GMO but not food and feed "with" a GMO. This implies, and it is laid down in EFSA guidelines, that for additives produced with GMMs it has to be shown that, in the final product, neither the production strain nor its recombinant DNA can be detected (EFSA, 2011). In these cases, an event-specific GMO identification method is not required. The product is not subject to any labelling requirement according to Regulations (EC) 1829/2003 on GM food and feed or No. 1831/2003 (EU, 2003a ; EU, 2003b), assuming that the producing company has verified the purity of the final product. To our knowledge, the finding of a contamination of living GM *B. subtilis* and of recombinant DNA in a marketed product has induced an increase in official controls of additives and enzymes produced with GMMs, at least in Germany. Such measures are based on the precautionary principle and the risk-based approach of enforcement to ensure the high level of food and feed safety in the EU and its Member States (EU, 2002). Since the current case, no other report concerning a GMM has been notified in the RASFF. It is thus assumed that the finding of a GMM in this product group was an exceptional and singular case. However, the lack of a specific detection method for a respective production strain made the analytical work quite challenging.

5. CONCLUSIONS

We have described the value of the use of NGS approaches in the context of assisting food/feed competent enforcement authorities and official control laboratories in their investigations concerning the presence of GMMs in biotechnological products. The approach described in our study allowed to provide a detailed molecular characterization of an unknown GMM and thereby facilitated the risk assessment and as well as the development of specific PCR methods for its detection.

Nevertheless, the study showed that the bioinformatics analysis of the data produced by NGS is still a challenging task and would require the development of adapted bioinformatics tools in order to be implemented for routine data analysis and management in the frame of GMO characterization and detection. The investigation of the GM *B. subtilis* strain showed at the same time, that in-depth literature review is fundamental to interpret the sequence data and to fully understand and characterize the sample, especially if completely unknown.

FUNDINGS

This research did not receive any specific grant from funding agencies in the public, commercial or not-for-profit sectors.

REFERENCES

1. Abbas, C. A., & Sibirny, A. A. (2011). Genetic control of biosynthesis and transport of riboflavin and flavin nucleotides and construction of robust biotechnological producers. Microbiology and Molecular Biology Reviews, 75(2), 321–360.
2. Akogwu, I., Wang, N., Zhang, C., & Gong, P. (2016). A comparative study of kspectrum-based error correction methods for next-generation sequencing data analysis. Human Genomics, 10(Suppl 2), 20.
3. Altschul, S. F., Gish, W., Miller, W., Myers, E. W., & Lipman, D. J. (1990). Basic local alignment search tool. Journal of Molecular Biology, 215(3), 403–410.
4. Andrews, S. (2010). FastQC: a quality control tool for high throughput sequence data Available online at: (Accessed 08.03.2017) http://www.bioinformatics.babraham.ac.uk/projects/fastqc.
5. Arulandhu, A. J., van Dijk, J. P., Dobnik, D., Holst-Jensen, A., Shi, J., Zel, J., & Kok, E. J. (2016). DNA enrichment approaches to identify unauthorized genetically modified organisms (GMOs). Analytical and Bioanalytical Chemistry, 408(17), 4575–4593.
6. Auchtung, J. M., Lee, C. A., Monson, R. E., Lehman, A. P., & Grossman, A. D. (2005). Regulation of a Bacillus subtilis mobile genetic element by intercellular signaling and the global DNA damage response. Proceedings of the National academy of Sciences of the United States of America, 102(35), 12554–12559.
7. Bacher, A., Eberhardt, S., Eisenreich, W., Fischer, M., Herz, S., Illarionov, B., ... Richter, G. (2001). Biosynthesis of riboflavin. Vitamins and Hormones, 61, 1–49.
8. Bankevich, A., Nurk, S., Antipov, D., Gurevich, A. A., Dvorkin, M., Kulikov, A. S., ... Pevzner, P. A. (2012). SPAdes: a new genome assembly algorithm and its applications to single-cell sequencing. Journal of Computational Biology, 19(5), 455–477.
9. Barbau-Piednoir, E., De Keersmaecker, S. C., Delvoye, M., Gau, C., Philipp, P., & Roosens, N. H. (2015). Use of next generation sequencing data to develop a qPCR method for specific detection of EU-unauthorized genetically modified Bacillus subtilis overproducing riboflavin. BMC Biotechnology, 15, 103.
10. Barbau-Piednoir, E., De Keersmaecker, S. C., Wuyts, V., Gau, C., Pirovano, W., Costessi, A., ... Roosens, N. H. (2015). Genome sequence of EU-unauthorized genetically modified Bacillus subtilis strain 2014–3557 overproducing riboflavin, isolated from a vitamin B2 80% feed additive. Genome Announcements, 3(2).

11. Barbe, V., Cruveiller, S., Kunst, F., Lenoble, P., Meurice, G., Sekowska, A., ... Danchin, A. (2009). From a consortium sequence to a unified sequence: the Bacillus subtilis 168 reference genome a decade later. Microbiology, 155(Pt 6), 1758–1775.
12. Burgess, C. M., Smid, E. J., & van Sinderen, D. (2009). Bacterial vitamin B2, B11 and B12 overproduction: An overview. International Journal of Food Microbiology, 133 (1–2), 1–7.
13. Carver, T., Harris, S. R., Berriman, M., Parkhill, J., & McQuillan, J. A. (2012). Artemis: an integrated platform for visualization and analysis of high-throughput sequence-based experimental data. Bioinformatics, 28(4), 464–469.
14. Edwards, D. J., & Holt, K. E. (2013). Beginner's guide to comparative bacterial genome analysis using next-generation sequence data. Microbial Informatics and Experimentation, 3(1), 2.
15. EFSA (2011). Guidance on the risk assessment of genetically modified microorganisms and their products intended for food and feed use. EFSA Journal, 9(6), 2193.
16. EU (2002). Regulation (EC) No 178/2002 of the European Parliament and of the Council of 28 January 2002 laying down the general principles and requirements of food law, establishing the European Food Safety Authority and laying down procedures in matters of food safety. Official Journal L 031, 45, 1–24.
17. EU (2003a). Regulation (EC) No 1829/2003 of the European Parliament and of the Council of 22 September 2003 on genetically modified food and feed. Official Journal L 268, 46, 1–23.
18. EU (2003b). Regulation (EC) No 1831/2003 of the European Parliament and of the Council of 22 September 2003 on additives for use in animal nutrition. Official Journal L 268, 46, 29–43.
19. FVO. (2016). Final report of an audit carried out in China from 18 November 2015 to 26 November 2015 in order to evaluate the controls systems for genetically modified organisms in respect of food and feed intended for export to the European Union. Ref. Ares (2016) 1042407 http://ec.europa.eu/food/fvo/act_getPDF.cfm?PDF_ID=12293 (Accessed 08.03.2017).
20. Hermann, D., & Schurter, W. (1995). PCR-analysis of riboflavin samples from fermentation: Absence of production strain specific DNA. Unpublished report No. B-165'177 from F. Hoffmann La Roche Ltd, dated 9 May 1995. Submitted to WHO by F. Hoffmann La Roche Ltd, Basel, Switzerland.
21. Hiratsu, K., Amemura, M., Nashimoto, H., Shinagawa, H., & Makino, K. (1995). The rpoE gene of Escherichia coli, which encodes sigma E, is essential for bacterial growth at high temperature. Journal of Bacteriology, 177(10), 2918–2922.
22. Kabisch, J., Thurmer, A., Hubel, T., Popper, L., Daniel, R., & Schweder, T. (2013). Characterization and optimization of Bacillus subtilis ATCC 6051 as an expression host. Journal of Biotechnology, 163(2), 97–104.
23. Lacks, S. A., Lopez, P., Greenberg, B., & Espinosa, M. (1986). Identification and analysis of genes for tetracycline resistance and replication functions in the broad-host-range plasmid pLS1. Journal of Molecular Biology, 192(4), 753–765.
24. Lioy, V. S., Pratto, F., de la Hoz, A. B., Ayora, S., & Alonso, J. C. (2010). Plasmid pSM19035, a model to study stable maintenance in Firmicutes. Plasmid, 64(1), 1–17.
25. Mäde, D., Reiting, R., Strauch, E., Ketteritzsch, K., & Wicke, A. (2008). A real-time PCR for detection of pathogenic yersinia enterocolitica in food combined with an universal internal amplification control system. Journal für Verbraucherschutz und Lebensmittelsicherheit, 3(2), 141–151.
26. Mander, L., & Liu, H. W. (2010). Comprehensive Natural Products II: Chemistry and Biology: 10 Volume Set (Vol. 7 chapter 7.04.5.2 "Riboflavin Pathway Engineering"). Elsevier Science & Technology.
27. Messelhäusser, U., Fricker, M., Ehling-Schulz, M., Ziegler, H., Elmer-Englhard, D., Kleih, W., & Busch, U. (2007). Real-time-PCR-System zum Nachweis von Bacillus

cereus (emetischer Typ) in Lebensmitteln. Journal für Verbraucherschutz und Lebensmittelsicherheit, 2(2), 190–193.

28. Mironov, A. S., Korolkova, N. V., Errais, L. L., Semenova, L. E., Perumov, D. A., Kreneva, R. A., Glazunov, A. V., Akishina, R. I., Iomantas, Y. A. V., Abalakina, E. G., Stoynova, N. V., Kozlov, Y. I., & Debabov, V. G. (2004). Method for producing riboflavin. WO 2004/046347 A1. https://www.lens.org/lens/patent/WO_2004_046347_A1 (Accessed 08.03.2017).
29. Pauwels, K., De Keersmaecker, S. C. J., De Schrijver, A., du Jardin, P., Roosens, N. H. C., & Herman, P. (2015). Next-generation sequencing as a tool for the molecular characterisation and risk assessment of genetically modified plants: Added value or not? Trends in Food Science & Technology, 45(2), 319–326.
30. Perkins, B. J., Sloma, A., Hermann, T., Theriault, K., Zachgo, E., Erdenberger, T., ... Pero, J. (1999). Genetic engineering of Bacillus subtilis for the commercial production of riboflavin. Journal of Industrial Microbiology and Biotechnology, 22(1), 8–18.
31. Rabinovich, P. M., Haykinson, M. Y., Arutyunova, L. S., Yomantas, Y. V., & Stepanov, A. I. (1985). The structure and source of plasmid DNA determine the cloning properties of vectors for Bacillus subtilis. In D. R. Helinski (Ed.), Plasmids in bacteria (pp. 635–656). US: Springer.
32. Rabinovich, P. M., Yomantas, Y. V., Haykinson, M. Y., & Stepanov, A. I. (1984). Cloning of genetic material in bacilli. In A. T. Ganesan & J. A. Hoch (Eds.), Genetics and biotechnology of bacilli (pp. 297–308). Academic Press.
33. RASFF (2014). Unauthorised genetically modified (Bacillus subtilis) bacteria in vitamin B2 from China, via Germany (Accessed 08.03.2017) https://webgate.ec.europa.eu/rasff-window/portal/?event=notificationDetail&NOTIF_REFERENCE=2014.1249.
34. Smith, J. L., Goldberg, J. M., & Grossman, A. D. (2014). Complete genome sequences of bacillus subtilis subsp. subtilis laboratory strains JH642 (AG174) and AG1839. Genome Announcements, 2(4).
35. Smolke, C. (2009). Riboflavin pathway engineering. In The metabolic pathway engineering handbook: Fundamentals. CRC Press. Soberon, N. E., Lioy, V. S., Pratto, F., Volante, A., & Alonso, J. C. (2011). Molecular anatomy of the Streptococcus pyogenes pSM19035 partition and segrosome complexes. Nucleic Acids Research, 39(7), 2624–2637.
36. Stepanov, A. I., Kukanova, A., Glazunov, E. A., & Zhdanov, V. G. (1977). Analogs of riboflavin, lumiflavin and alloxazine derivatives. II. Effect of roseoflavin on 6,7-dimethyl-8-ribityllumazine and riboflavin synthetase synthesis and growth of Bacillus subtilis. Genetika, 13(3), 490–495.
37. Wahler, D., Schauser, L., Bendiek, J., & Grohmann, L. (2013). Next-generation sequencing as a tool for detailed molecular characterisation of genomic insertions and flanking regions in genetically modified plants: A pilot study using a rice event unauthorised in the EU. Food Analytical Methods, 6(6), 1718–1727.
38. Winkler, W. C., Cohen-Chalamish, S., & Breaker, R. R. (2002). An mRNA structure that controls gene expression by binding FMN. Proceedings of the National academy of Sciences of the United States of America, 99(25), 15908–15913.
39. Yang, L., Wang, C., Holst-Jensen, A., Morisset, D., Lin, Y., & Zhang, D. (2013). Characterization of GM events by insert knowledge adapted re-sequencing approaches. Scientific Reports, 3, 2839.
40. Zerbino, D. R. (2010). Using the Velvet de novo assembler for short-read sequencing technologies. Current Protocols in Bioinformatics. Chapter 11, Unit 11 15.
41. Zerbino, D. R., & Birney, E. (2008). Velvet: algorithms for de novo short read assembly using de Bruijn graphs. Genome Research, 18(5), 821–829.

CHAPTER 12

Table Olives: A Vehicle for the Delivery of Bioactive Compounds

Dimitrios Boskou

School of Chemistry, Aristotle University of Thessaloniki, Greece

ABSTRACT

Table olives and olive oil are sources of compounds with important biological properties and essential components of the Mediterranean Diet. For a long period the scientific community focused its attention mainly on the beneficial effects of olive oil on health. The importance of the edible processed fruit was rather overlooked. Table olives contain a large amount of high quality fat, essential amino acids, minerals, vitamins, fiber and a group of minor compounds, which include biophenols and pentacyclic triterpenes. In the last decade an effort is seen in the literature to re-evaluate table olives as sources of bioactive compounds. There is an impressive increase of papers dealing with the composition of processed olives, the effect of variety and processing methods on the level of antioxidants and other minor bioactive constituents, and generally the improvement of healthy features. The information, however, is rather scattered and not all the publications are of the same scientific level. There are many repetitions but also discrepancies and blanks due to difficulties deriving from the complex nature of phenols, the biochemical routes related to lactic acid and other fermenting bacteria and the analytical results, considering the lack of standard methods for the quantitation of small or minute quantities of bioactive compounds The present review discusses table olives as a vehicle for the delivery of bioactive compounds. It concentrates mainly on the levels of bioactive compounds in relation to the treatment of olive fruits and the expected healthy features and functionality.

Keywords: Table olives; Diet, Bioactive; Olea europaea L.; Glucoside; Phenolics

INTRODUCTION

Table olives are fermented vegetables very popular and largely consumed in the countries surrounding the Mediterranea basin. They are the fruits of the olive tree (Olea europaea L.) that have undergone the pertinent processes, preserved by natural fermentation, heat treatment or other methods They must comply with the descriptions of the types of olives, trade preparations and styles and with the essential composition and quality criteria laid down in the Trade Standard applying to table olives adopted by the International Olive Council Standard for table olives [1]. Fresh olives are picked when green-ripe, turning color or black-ripe, depending on the mode of processing to be used. They are not comestible because of the presence of the bitter glucoside, oleuropein.

Table oils constitute an essential element of the Mediterranean diet and they are a featured ingredient for hundreds of dishes. They are important from a nutritional point of view for the general population in many Mediterranean countries, especially during the long periods of fasting. They are also of vital importance for the Christian orthodox monks and nuns, who consume large quantities of olives. In other cases organoleptic and other properties are taken into consideration and not the nutritional importance, for example when olives are used as a piece of decoration in a dish, when used in condiments for pasta and pizza or when they are lye-cured and darkened by exposure to air and rapid oxidation.

Processing of raw olives aims at eliminating natural bitterness, mainly due to the biter secoiridoid oleuropein and other phenolics, and improving the taste. It is based on soaking in water, brine, dilute alkali, drying by salting and darkening by oxidation. Large scale processors generally use lye treatments which speed up processing. Small and very small-scale producers favor more "natural" methods. The bitter taste of processed table olives depends on the phenolic compounds present after the processing and it is an important characteristic favored by the consumers. Preparations and styles including procedures based on IOC standards and practices applied for local popular products have been reviewed [2-6].

Many of the health benefits of table olives are related to cardiovascular, respirator, immune, musculoskeletal, nervous and digestive systems. They are associated with fatty acid composition of the lipid fraction and the richness in non-dietary micro-constituents [4,7-11].

For the levels of the latter there are some differences and discrepancies in the literature. This is due to the difference of variety and method of debittering and probably to methods that lead to inaccurate quantifications. This review highlights recent studies and findings related to composition and health benefits expected from the consumption of table olives and the factors influencing the levels of phenolics and other bioactive constituents. Reported values for processed olives were selected mainly from recent, more reliable sources, based on chemical characterization and quantification with modern hyphenated methods that include chemical identification of each peak in the chromatograms.

Nutritional Value of Table Olives

Table olives provide energy, proteins, carbohydrates and vitamins. The fat content ranges from 100 to 300 g/kg. They are a good source of oleic acid, a monounsaturated fatty acid. In 2004 the US Food and Drug Administration (FDA) announced the availability of a qualified health claim for monounsaturated fat from the olive oil stating that eating 2 tablespoons of olive oil a day is beneficial in reducing coronary heart disease (CHD) (http:/www.fda.gov./dms/qholive/htlm).

Table olives provide also vitamin E (13-52 mg/kg), small quantities of the B-complex (thiamin, niacin, pantothenic acid, vitamin B_6), beta-carotene (pro-vitamin A), important minerals (phosphorous, iron, calcium, magnesium, potassium) and microelements (copper, zinc, manganese). Another important characteristic is the high dietary fiber content. Values reported range from 2.5-5%. Most preparations have a content ≥ 3 g/100 g of edible part, so they can be labelled as a source of fiber [4,7].

Bioactive Constituents in Table Olives

Polar phenols

In addition to nutrients table olives are rich sources of compounds that are expected to provide health effects beyond the basic nutritional needs. The most important classes of phytochemicals are phenolics but other compounds such as squalene, terpenoids, phytosterols and betacarotene (in small quantities) are also present. Most of them are compounds with unquestionable biological value.

The major phenolic compounds present in olive fruits are anthocyanins (cyanidin glucosides); flavonols (mainly quercetin-3- rutinoside); flavones (luteolin and apigenin glucosides); phenolic acids (hydroxybenzoic, hydroxycinnamic, others); phenolic alcohols (tyrosol, hydroxytyrosol, 3,4-dihydroxyphenylglycol) (**Figures 1A** and **1B**); secoiridoids (oleuropein, demethyloleuropein, ligstroside, nuzhenide) (**Figure 1C**); verbascoside (caffeoylrhamnosylglucoside of hydroxytyrosol, the main hydroxycinnamic acid derivative found in olive fruits) (**Figure 1D**), lignans and Oleoside-11-methylester [8-17].

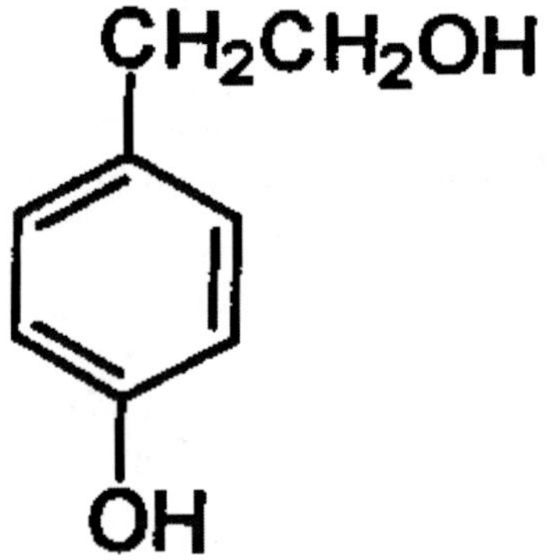

Figure 1A: Structural formula of Tyrosol, an important table olive phenol.

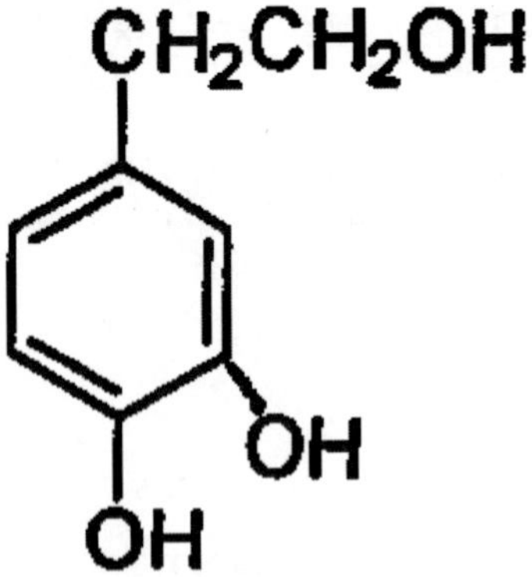

Figure 1B: Structural formula of Hydroxytyrosol.

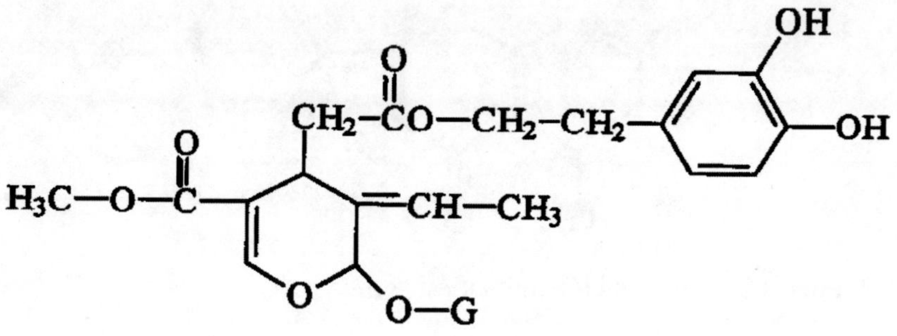

Figure 1C: Structural formula of Oleuropein.

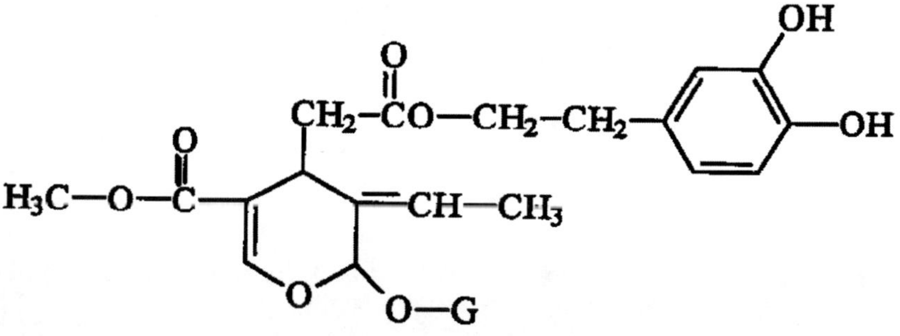

Figure 1D: Structural formula of Verbascoside.

The phenolic composition of table olives is different from that of olive fruits since, depending on the method, phenols diffuse in the lime or decompose. The main polar phenols reported to be found in processed olives are hydroxytyrosol, tyrosol, flavonoids (luteolin, apigenin) (**Figure 1E** and **1F**), anthocyanins (cyanidin-3-O-glucoside and cyanidin-3-O-rutinoside), verbascoside, and phenolic acids [18-22]. Suza and her co-workers [19] and Pereira et al. [20] reported also the presence of luteolin 7-O-glucoside and rutin in different olives from Portugal. Traditionally prepared Throuba Thassos table olives, a Greek table olive variety, have been reported to be rich sources of oleureupein [21,22].

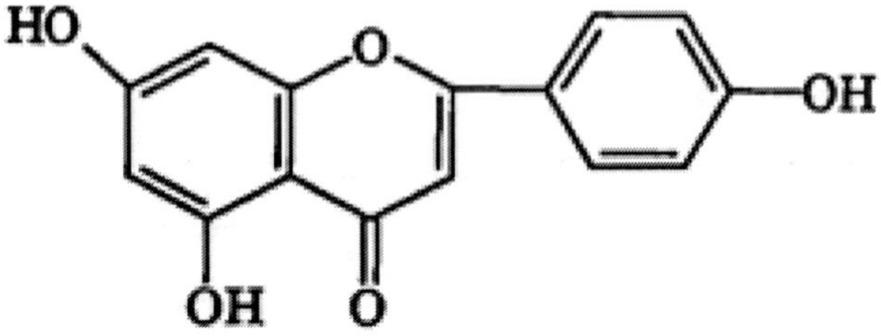

Figure 1E: Structural formula of Apigenin.

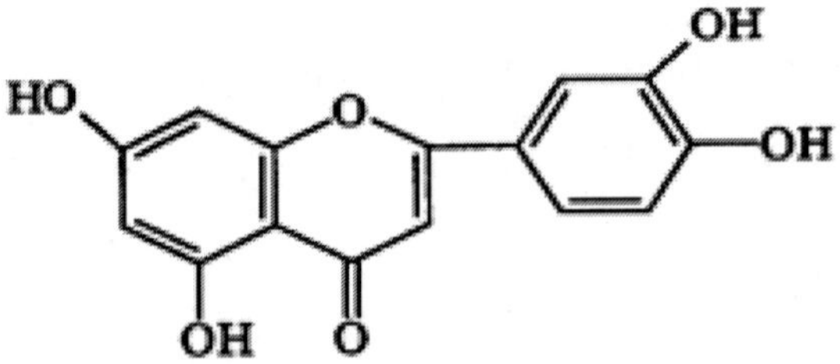

Figure 1F: Structural formula of Luteolin.

Total polar phenols: Blekas et al. [17] and Boskou et al. [18] applied the conventional method Folin-Ciocalteu to samples of commercial table olives from the Hellenic market. Values reported depending on the style and varieties were as high as 1700 expressed in mg caffeic acid per kg. Higher values were found in Greek-style naturally black olives in brine and Kalamata olives in brine. High concentrations of phenolics were also found in Italian table olives (240-1700 mg caffeic acid/kg edible part) [4]. The analyses carried out by Romero et al. [23] on 48 samples of various table olives showed that olives in brine had the highest concentration in total phenols (1200 mg/kg), whereas oxidized olives had the lowest. Melliou et al. [22] quantified phenols using Ultrahigh-Pressure Liquid Chromatography Triple-Quadrupole Tandem Mass Spectrometry. They reported values below 37 mg/kg for California-style black ripe olives.

Hydroxytyrosol, Tyrosol, glycosides: The level of individual phenols in the treated olives differs significantly from those of fresh olives. The glycosides oleuropein, verbascoside, ligstroside are significantly reduced and can reach quantification limits, depending on the method of processing, cultivar, and initial phenol profile. Losses occur due to the dilution to water and hydrolysis. From the hydrolysis of oleuropein and verbascoside an increase in hydroxytyrosol occurs. The latter becomes the major phenol in processed olives. Values have been reported as high as 760 mg/kg by Blekas et al. [17] and 1100 mg/kg by Boskou et al. [18]. Values for tyrosol were found to range from 10-170 mg/kg in various table olives styles. Processing according to Spanish style and California style results in products with lower levels of total phenols, hydroxytyrosol and tyrosol. Zoidou et al. [21] studied carefully throumba olives from Thassos, debittered by a traditional Greek mode based on the use of dry salt. These naturally black wrinkled olives may have a level of oleuropein as high as 550 mg/kg flesh, due to the processing. They are not fermented products but just debittered and this explains why an increased number of phenolic compounds in the flesh remain in their bound forms.

Other minor phenols the presence of which has been confirmed in treated olives are caffeoyl-6'-secologanoside, and comselogoside [24,25]. 3,4-Dihydroxyphenylglycol (DHPG) was also found to be present in the pulp natural table olives from ten different cultivars at levels up to 368 mg/kg dry weight [26]. The presented data for this antioxidant indicate that natural table olives are a rich source of DHPG.

Differences are observed among various reports even for olives of the same style. Romeo et al. [27] determined rather high values for oleuropein (60-95 mg/kg flesh) and Verbascoside (7-14 mg/kg flesh) in green olives naturally fermented. It is difficult to compare various results reported considering the method of fermentation and postfermentation conditions, the different methods of analysis, the lack of standards, and the different mode of expression of the results (as gallic acid, or caffeic acid per flesh weight or dry flesh weight). Additional confusion is due to the characterization of all table biophenols as polyphenols. The major phenols in table olives, hydroxytyrosol, tyrosol, oleuropein are not polyphenols. In spite of the difficulties in evaluating the level of bioactive compounds, however, it can be generally concluded that table olives are good sources of biophenols either in their free or in their bound forms.

Oleuropein, and hydroxytyrosol are among nature's most powerful anti-oxidants and their biological role has been a hot topic of research the last 30 years. Together with vitamin E and carotenoids, they play a vital role fighting against cancer, inflammation, coronary artery disease, degenerative nerve diseases and diabetes. The literature specifically related to their bioaccessibility, absorbability, bioavailability as constituents of table olive phenols is not complete [24] but there is a plethora of publications for the potential health promoting effects of oleuropein, hydroxytyrosol, verbascoside and flavonoids as constituents of olive oil [27-30].

The best-studied antioxidant phytonutrients found in olives are oleuropein and hydroxytyrosol. Their intake is related to LDL cholesterol oxidation, oxidative stress; and protection of nerve cells from oxygen-related damage. The studies are extensive and some health benefits and a protective action against oxidative stress and inflammation are now considered documented.

Bioavailability and mechanisms of action reviews have been presented by [31-36]. Accumulated scientific work focuses on oxidation mechanisms, free radical damage, effect on lipoprotein metabolism, inflammation, endothelial dysfunction, and blood pressure. The potential of certain phenols in oncology, chemoprevention, cell-specific cytotoxic and apoptotic effects has been broadly studied and discussed. The importance of hydroxytyrosol in protection of low-density lipoproteins and consequently its implication in the reduction of cardiovascular disease risk has been highlighted by the European Food Safety Authority, concluding that 5 mg of hydroxytyrosol and its derivatives should be consumed daily to reach this effect at physiological level [36].

Modern studies focus on the clarification of molecular mechanisms accounting for the antioxidant, anti-inflammatory, and anticancer properties through gene transcription profiling and nutrigenomic effects that modulate the expression of decease related genes [31,35,37,38].

The evidence showing the multiple benefits of olive phenolic compounds for the aging brain seems to climb rapidly. The neuroprotective effects of these phenols in in vitro and preclinical studies have been recently highlighted by Syed Haris Omar [38].

Flavonoids: Luteolin levels range from a few mg to 75 mg/kg flesh [18] depending on the variety and the processing method. Higher concentrations were estimated for Greek-style naturally black olives in brine, Kalamata olives in brine and naturally black olives in dry salt. Luteolin 7-O-glycoside was found in air dried Throumba Thassos olives [26] Apigenin-7-glucoside and rutin were reported as constituents of the polar fraction obtained from naturally fermented green olives [20]. Recently Soufi et al. [39] reported high values of luteolin-7-glucoside, quercetin-3-rutinoside (rutin), and cyanidin-3- rutinoside in dry salt processed olives of Algerian varieties.

Lignans: Lignans are phenols found in plants; they are part of the phytoestrogen family, which is known for its health properties. The lignans (+)-1-acetoxypinoresinol, (+)-pinoresinol and syrigaresinol are major components of the phenolic fraction of olive oils [40,41]. Lignans are polyphenols found in plants, and they are part of the phytoestrogen family, which is known for its health properties. López- Biedma et al. [42] reviewed the literature regarding the main lignans found in virgin olive oils and their prominent role in the health effects attributed to virgin olive oils.

Phenolic acids: Small to minute quantities have been reported for processed olives by Boskou and his collaborators [11] who qualified and determined with gas chromatography-mass spectroscopy the following phenolic acids and cinnamic acid in five Greek types of table olives: caffeic acid, ferulic acid, p-hydroxy-benzoic acid, p-hydroxyphenylacetic acid, p-hydroxy-phenyl-propanoic acid, protocatechuic acid, p-coumaric acid, cinnamic acid, 3,4-dihydroxy-phenyl-acetic acid and vanilic acid. According to Romero et al. [23] phenolic acids are detected in small quantities in brine processed olives because they are partly metabolized by the microrganisms during the brining period. Romeo et al. [27] based on gas-chromatograpic analysis provided numerical data for small quantities of 4-hydroxybenzoic, chlorogenic (5-O-caffeoylquinic), syringic, vanillic, caffeic 2,4-dihydroxybenzoic, p-coumaric and ferulic acids in naturally fermented green table olives (cv Nocellara etnea).

Phenolic acids are phytochemicals which exhibit a wide variety of biological activities such as antioxidant, anti-inflammatory and many others.

Lipid soluble phenols

α-Tocopherol values reported for 30 samples of processed olives from the Italian market ranged between 25 and 90 mg/kg [43]. Samples of of "Alcaparras" Stoned Table Olives from Northeast Portugal showed an average of total tocopherols of 12 mg/100 g of fresh weight, being α-tocopherol the most abundant. The other homologues (beta-, gamma-and delta-) were found in ppm or sub-ppm quantities [43].

Non-phenolic compounds

Triterpenic acids and triterpene dialcohols: Maslinic and oleanolic acids are the most abundant triterpenic acids found in olive fruits. Processed olives retain a high level of triterpenic acids although there is diffusion in the brine. During the alkaline treatment oleanolic and maslinic salts are formed which are also diffused. Romero et al. [45] reported levels of triterpenic acids (mainly maslinic and oleanolic) (**Figures 2A** and **2B**) ranging from 460 to 2000 mg/kg in the flesh of many samples of commercial black and green olives from different varieties. Maslinic and oleanolic acids were quantified in naturally processed Conservolea and Kalamon olives [46]. The natural process did not influence their concentration in the final product. Lower concentrations were determined in alkaline-treated Conservolea olives.

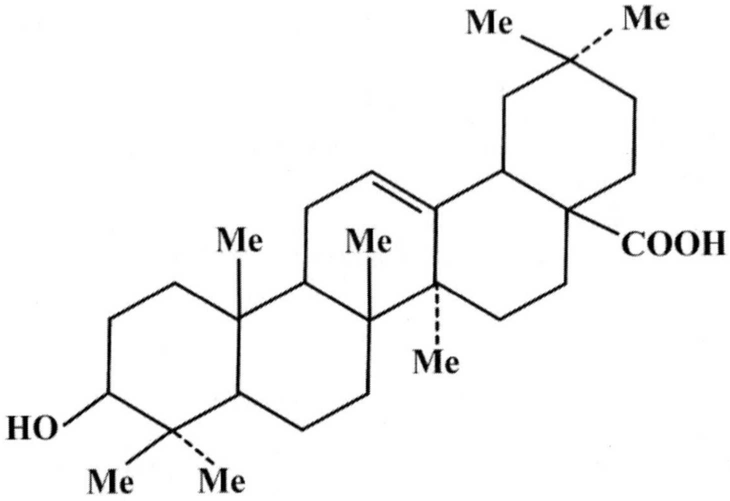

Figure 2A: Structural formula of two triterpenic acids present in Table olives (Oleanolic acid).

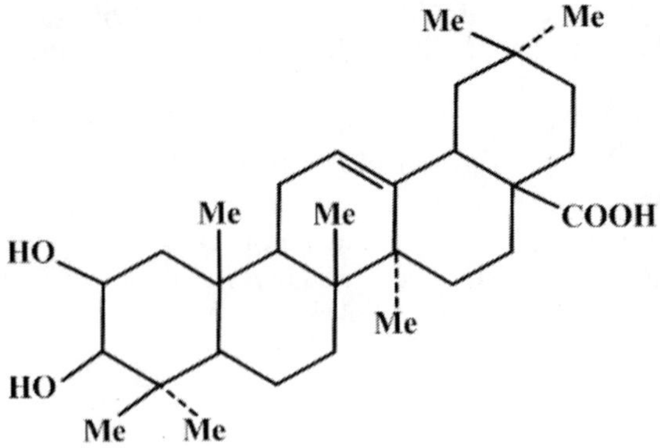

Figure 2B: Structural formula of two triterpenic acids present in Table olives (Maslinic acid).

Triterpenes present in olives and olive oil are described to present different properties, such as an antioxidant protection, and also anti-inflammatory, antitumoral, hepatoprotective, cytotoxic, anti-diabetogenic, cardio protective, antihypertensive antibacterial and anti-HIV activities [47-53].

Two other non-dietary microconstituents, the triterpene dialcohols, erythrodiol and uvaol, have been studied for their biological role but their level in treated olives is very small. Values in table olives are not reported because these compounds are co-eluted in the gas chromatography columns with aliphatic alcohols and reported as a whole.

Squalene: 30 samples from the Italian market were analyzed for squalene contents by Sagratini et al. [43], Values reported ranged between 540 and 1600 mg/kg. Squalene has been reported to possess antioxidant properties. In vitro experimental evidence indicates that squalene is a highly effective oxygen-scavenging agent. It is also found to have chemo-preventive action against colon cancer. Ramirez- Torrest, Gabas-Riverat and Osada [10] have conducted recently a study to explore a proper dose of squalele, a trove of metabolic actions, to be used in a functional food or in the therapy of fatty liver .The authors have discussed their results in relation to the level of squalene in virgin olive oil.

Phytosterols: Fermented olives may contain of 200–520 mg/kg phytosterols [52]. The main phytosterols in the lipid fraction of table olives are β-sitosterol and Δ5-avenasterol followed by campesterol. Epidemiologic and experimental studies suggest that dietary phytosterols, and in particular β-sitosterol, may offer protection from the most common cancers but the concentration of phytosterols in table olives is low.

Beta-carotene: Beta-carotene is a vitamin A provitamin .Sagratini et al. [43] used a validated solid phase extraction (SPE) and high performance-liquid chromatography (HPLC-DAD) method to analyze 30 table olive samples for beta-carotene. Values ranged from 0.4 to 2.6 mg/kg.

Cinnamic acid: Cinnamic acid is not a phenol but it is reported as such by many researchers. Romeo et al. [27] reported a level of approx. 8 mg/kg cinnamic acid in naturally fermented green table olives. Boskou et al. [11] found only ppm quantities in different table olives from the Greek market.

Effect of Processing on the Contents of Bioactive Compounds in Table Olives

The levels of phenolic compounds differ significantly and depend on the cultivar, stage of maturity of the fruit and the mode of processing. These important factors should be taken into consideration in selecting the raw material and the production method in order to obtain a product with increased biological value.

Natural olives (Green olives, olives turning colour or black olives) are placed directly in brine in which they undergo complete or partial fermentation, preserved or not by the addition of acidifying agents. Natural debittering is a complex procedure and not yet fully understood and optimized. Sodium chloride concentration, acetic acid concentration, storage temperature and activity of enzymes are the processing factors that influence the rate the hydrolysis of bitter phenols.

Very recently Ramirez et al. [53] tested a low concentration of sodium chloride and acetic acid and low temperatures for the first two months of fermentation followed by stronger conditions in the rest of the fermentation period. Such advanced techniques may be used to accelerate existing debittering methods and manipulate the level of phenolic compounds in the final product.

Debittering with NaOH solution causes a significant reduction of total polar phenols and especially glycosides, such as such oleuropein, caused by hydrolytic decomposition to hydroxytyrosol and elenolic acid glycositic derivatives and diffusion in the brine. Chemical and biochemical changes due to the effect of lye and the bacteria during the fermentation in the brine have been described by many authors. Poiana and Romeo [54] measured total phenol depletion during the fermentation period. When the Spanish and California methods are applied the losses are higher since the NaOH solution brings about more changes compared to losses due to fermentation. Oxidation by aeration for rapid darkening causes additional losses of hydroxytyrosol. This diminution is related to reactions in the presence of iron salts, used as catalysts for colour fixation. As a result, California type olives (artificial ripening) have the lowest content of phenols. Methods for the preparation of Greek-style naturally black olives in brine and partly the Spanish style green olives in brine methods permit generally a higher retention of phenols. The goal of certain recent investigations is to modify the Spanish and California processes in order to minimize the processing period and improve quality attributes of the obtained olives by using one-lye treatment instead of successive treatments [55]. Such techniques reduce the volume of waste waters and energy consumption, and give rise to a product which is richer in bioactive substances.

Spontaneous fermentation processes used to ferment black or green olives are difficult to monitor or control. New technologies propose microbial starters to improve organoleptic characteristics, to ensure the maintenance and/or improvement of healthy features of the product and to produce table olives that are a carrier of microorganisms with probiotics characters [9]. The use of microbial starters as inoculants should be applied with the currently available technology so that in addition to the control of the fermentation process, nutritional and healthy features of the product are maintained.

The fast debittering with NaOH treatment reduces also the concentrations of triterpenic acids in the flesh because of their solubilisation into alkaline solutions. In contrast, the slow de-bittering during natural fermentation of Greek-style olives had no effect on the content of maslinic or oleanolic acid [8] Tufariello and her co-workers [9] stated that in Greek-style preparations, the fermentation of black table olives driven by selected starter cultures can preserve the triterpenic acid content. The amount of these molecules determined was around 1000-2000 mg/kg olive flesh, much higher than the values observed in extra virgin olive oils These data confirm that certain types of table olives are an appealing food vehicle for the delivery of bioactive compounds.

FUTURE TRENDS

The main preparation methods of table olives are known for a long time but they are not yet sufficiently optimized. Table olive are now reevaluated from a nutritional and functional point of view mainly because of the properties of phenolic compounds, triterpenic acids and other phytochemicals with a biological value which is considered unquestionable. As the concentration of these phytochemicals depends on the olive cultivar and processing methods it is expected that future challenges will be to investigate further bioaccessibility and/or bioavailability of metabolites, considering the changes in the bioactive compounds profile that take place during fermentation. More precise information will be expected from analytical studies with validated methods for better evaluation of the level of bioactive constituents in processed olives and from nutritional trials in humans after table olives ingestion .The latter will indicate the potential health promoting role of olives and their contribution in lowering the incidence of several human diseases. Lowering the level of sodium chloride and the coinoculation of yeasts and Lactic Acid Bacteria as a multifunctional starter and a strategy for carrying probiotics by table olives are also open fields for research in the functional food arena.

REFERENCES

1. International Olive Council (2004) Trade Standard Applying to Table Olives. Resolution No. RES-2/91-IV/04.
2. Kailis S, Harris D (2007) Producing Table Olives, Landlinks Press/CSIRO Publishing: Australia.

3. Sánchez-Gómez AH, García PG, Navarro LR (2006) Trends in table olives production, Grasas Aceites 57: 86-94.
4. Lanza B (2012) Nutritional and Sensory Quality of Table Olives. In Olive Germplasm. The Olive Cultivation, Table Olive and Olive Oil Industry in Italy, Dr. Innocenzo Muzzalupo (Ed.), Intech 343-372.
5. Boskou D, Camposeo S, Clodoveo ML (2015) Table Olives as Sources of Bioactive Compounds, in Olive and Olive Oil Bioactive Constituents, Boskou D (Ed.), AOCS PRESS, 217-261.
6. Arroyo-Lopez FN, Garcıa-Garcıa P, Rodrıguez-Gomez F, Garrido-Fernandez A (2016) Olives: Types and Consumption. In Encyclopedia of Food and Health, Caballero B and Finglas M, (Ed.), Elsevier 4: 167-170.
7. Jimenez A, Rodriguez R, Fernandez-Caro I, Guillen R, Fernandez-Bolanos J, et al. (2000) Dietary fibre content of table olives processed under different European styles: study of physico-chemical characteristics. J Sci Food Agric 80: 1903-1908.
8. Alexandraki V, Georgalaki M, Papadimitriou K, Anastasiou R Zoumpopoulou G, et al. (2014) Determination of triterpenic acids in natural and alkaline-treated Greek table olives throughout the fermentation process. LWT-Food Sci Technol 58: 609-613.
9. Tufariello M, Mita G, Bleve G (2016) Biotechnology can Improve a Traditional Product as Table Olives in Products from olive tree. Boskou D and Clodoveo ML, (Ed.), Intech 235-260.
10. Ramírez-Torres A, Gabás-Rivera C, Jesús Osada J (2016) Squalene: A Trove of Metabolic Action. In "Products from Olive Tree", Boskou D and Clodoveo ML, (Ed.), INTECH 2-17.
11. Boskou D (2015) Olive fruits, olive oil and table olives bioactive constituents, Olive and Olive Oil Bioactive Constituents. AOCS PRESS 1-30.
12. Vinha AF, Ferreres F, Silva BM, Valentao P, Gonçalves A, et al. (2005) Phenolic profiles of Portuguese olive fruits (Olea europea L.): Influence of cultivar and geographical origin. Food Chem 89: 561-568.
13. Silva S, Gomes L, Leitão F, Coelho AV, Vilas Boas L (2006) Phenolic Compounds and Antioxidant Activity of Olea europaea L. Fruits and Leaves. Food Sci Tech Int 12: 385-396.
14. Alagna F, Roberto R, Panara F, Caporali S, Urbani S, et al. (2012) Olive phenolic compounds: metabolic and transcriptional profiling during fruit development. BMC Plant Biology 12: 162-192.
15. Bianco A, Uccella N (2000) Biophenolic components of olives. Food Res Intern 33: 475-485
16. Kanakis P, Termentzi A, Michel T, Gikas E, Halabalaki M, et al. (2013) From Olive Drupes to Olive Oil. An HPLC-Orbitrap-Based Qualitative and Quantitative Exploration of Olive Key Metabolites. Planta Medica 79: 1576-1587.
17. Blekas G, Vassilakis C, Harizanis C, Tsimidou M, Boskou D (2002) Biophenols in Table Olives. J Agr Food Chem 50: 3688-3692.
18. Boskou G, Salta FN, Chrysostomou S, Mylona A, Chiou A, et al. (2006) Antioxidant capacity and phenolic profile of table olives from the Greek market. Food Chem 94: 558-564.
19. Sousa A, Ferreira ICFR, Calhelha R, Andrade PB, Valentao P, et al. (2006) Phenolics and antimicrobial activity of traditional stoned table olives 'alcaparra'. Bioorg Medi Chem 14: 8533-8538.
20. Pereira J, Pereira APG, Fereira IC, Valentao P (2006) Table Olives from Portugal: Phenolic Compounds, Antioxidant Potential, and Antimicrobial Activity. J Agric Food Chem 54: 8425-8431.
21. Zoidou E, Melliou E, Gikas E, Tsarbopoulos A, Magiatis P, et al. (2010) Identification of Throuba Thassos, a Traditional Greek Table Olive Variety, as a nutritional rich source of oleuropein. J Agric Food Chem 58: 46-50.
22. Melliou E, Zweigenbaum JA, Mitchell AE (2015) Ultrahigh-Pressure Liquid Chromatography Triple-Quadrupole Tandem Mass Spectrometry Quantitation of

Polyphenols and Secoiridoids in California-Style Black Ripe Olives and Dry Salt-Cured Olives. J Agric and Food Chem 63: 2400- 2405.
23. Romero C, Brenes M, Yousfi K, García P, García A, et al. (2004) Effect of cultivar and processing method on the contents of polyphenols in table olives. J Agric Food Chem 52: 479-484.
24. D'Antuono I, Garbetta A, Ciasca B, Linsalata V, Minervini F, et al. (2016) Biophenols from Table Olive cv Bella di Cerignola: Chemical Characterization, Bioaccessibility, and Intestinal Absorption . J Agric Food Chem 64: 5671-5678.
25. Mantzouridou F, Tsimidou MZ (2011) Microbiological quality and biophenol content of hot air-dried Thassos cv. table olives upon storage. Europ J Lipid Sci Technol 113: 786-795.
26. Rondriguez G, Lama A, Jaramillo S, Fuentes-Alventosa JM, Guillén R, et al. (2009) 3,4-Dihydroxyphenylglycol (DHPG): An Important Phenolic Compound Present in Natural Table Olives. J Agric Food Chem 57: 6298-6304.
27. Romeo FV, De Luca S, Piscopo A, Poiana M (2009) Effects of post-fermentation processing on the stabilisation of naturally fermented green table olives (cv Nocellara etnea). Food Chem 116: 873-878.
28. Parvaiz M, Hussain K, Shoaib MG, William G, Tufail M, et al. (2013) A Review: Therapeutic Significance of Olive Olea europaea L. (Oleaceae Family). Global J Pharmacol 7: 333-336.
29. Barbaro B, Toietta G, Maggio R, Arciello M, Tarocchi M, et al. (2014) Effects of the olive-derived polyphenol oleuropein on human health. Int J Mol Sci 15: 18508-15524.
30. Servili M, Sordini B, Esposto S, Urbani S, Veneziani G, et al. (2014) Biological Activities of Phenolic Compounds of Extra Virgin Olive Oil. Antioxidants 3: 1-23.
31. Covas MI, Fitó M, de la Torre MR (2015) Minor Bioactive Olive Oil Components and Health: Key Data for Their Role in Providing Health Benefits in Humans. In Olive and Olive Oil bioactive Constituents. Boskou D, (Ed.), AOCS Press 32: 53.
32. Cicerale S, Conlan XA, Sinclair AJ, Keast RS (2008) Chemistry and Health of Olive Oil Phenolics. Crit Rev Food Scie Nutr 49: 218-236.
33. Fernández-Bolaños JG, Lopez O, López-García MA, Marset A (2012) Biological properties of hydroxytyrosol and irs derivatives in Olive Oil. Constituents, Quality, properties and bioconversions. Boskou D (Ed.), Intech 375-396.
34. Vilaplana-Pérez1 C, Auñón D, García-Flores LA, Gil-Izquierdo A (2014) Hydroxytyrosol and potential uses in cardiovascular diseases, cancer, and AIDS. Front Nutr 1: 18.
35. Ray N, Lam N, Luc R, Bonvino N, Karagiannis TC (2015) Cellular and Molecular Effects of Bioactive Phenolic Compounds in Olives and Olive Oil. In Olive and Olive oil Bioactive Constituents. Boskou D, (Ed.), JAOCS- Press, 54-92.
36. EFSA Panel on Dietetic Products, Nutrition and allergies (NDA) (2011) Scientific opinion on the substantiation of health claims related to polyphenols in olive oil and protection of LDL particles from oxidative damage. EFSA Journal 9: 2033.
37. Piroddi M, Albini A, Fabiani R, Giovannelli L, Luceri C, et al. (2016) Nutrigenomics of extra-virgin olive oil: A Review. Biofactors 43: 17-41.
38. Omar SH (2016) Olive: A therapeutic agent against Alzheimer's disease. 5th European Nutrition and Dietetics Conference, Rome, Italy.
39. Soufi O, Romero C, Motilva MJ, Borrás Gaya X, Louaileche H (2016) Effect of dry salting on flavonoid profile and antioxidant capacity of Algerian olive cultivars. Grasas Aceites 67: e132.
40. Owen RW, Mier W, Giacosa A, Hull WE, Spiegelhalder B, et al. (2000) Identification of Lignans as Major Components in the Phenolic Fraction of Olive Oil. Clin Chem 46: 976-988.
41. Dais P, Boskou D (2009) Detection and quantification of phenolic compounds, in olive oil, olives, and biological fluids. In "Olive oil: minor constituents and health. Boskou D, (Ed.), CRC Press Boca Raton 55-105.

42. López-Biedma A, Sánchez-Quesada C, Delgado-Rodríguez M, José J, Gaforio JJ (2016) The biological activities of natural lignans from olives and virgin olive oils: A review. J Functional Foods 26: 36-47.
43. Sagratini G, Allegrini M, Caprioli G, Cristalli G, Dario Giardina D, et al. (2013) Simultaneous Determination of Squalene, α-Tocopherol and β-Carotene in Table Olives by Solid Phase Extraction and High-Performance Liquid Chromatography with Diode Array Detection. Food Analyt Methods 6: 5-60.
44. Sousa A, Casal S, Bento A, Malheiro R, Oliveira M, et al. (2011) Chemical Characterization of “Alcaparras” Stoned Table Olives from Northeast Portugal. Molecules 16: 9025-9040.
45. Romero C, García A, Medina E, Ruíz-Méndez MaV, de Castro A, et al. (2010) Triterpenic acids in table olives. Food Chem 118: 670-674.
46. Herrera MD, Rodriguez-Rodriguez R, Ruiz-Gutierrez V (2006) Functional Properties of Pentacyclic Triterpenes Contained in "Orujo" Olive Oil. Current Nutr Food Sci 2: 45-49.
47. Allouche Y, Beltrán G, Gaforio JJ, Uceda M, Mesa MD (2010) Antioxidant and antiatherogenic activities of pentacyclic triterpenic diols and acids. Food Chem Toxicol 48: 2885-2890.
48. Allouche Y, Warleta F, Campos M, Sánchez-Quesada C, Uceda M, et al. (2011) Antioxidant, antiproliferative, and pro-apoptotic capacities of pentacyclic triterpenes found in the skin of olives on MCF-7 human breast cancer cells and their effects on DNA damage. J Agric Food Chem 59: 121-130.
49. Rodriguez-Rodriguez R, Perona JS, Herrera MD, Ruiz-Gutierrez V (2006) Triterpenic compounds from ‘orujo’ olive oil elicit vasorelaxation in aorta from spontaneously hypertensive rats. J Agric Food Chem 54: 2096-2102.
50. Sánchez-Quesada C, López-Biedma A, Warleta F, Campos M, Beltrán G, et al. (2013) Bioactive Properties of the Main Triterpenes Found in Olives, Virgin Olive Oil, and Leaves of Olea Agric. Food Chem 61: 12173-12182.
51. Lozano-Mena G, Sánchez-González M, Juan EM, Planas JM (2014) Maslinic Acid, a Natural Phytoalexin-Type Triterpene from Olives-A Promising Nutraceutical? Molecules 19: 11538-11559.
52. López-López A, Montaño A, Ruíz-Méndez MV, Garrido-Fernández A (2008) Sterols, Fatty Alcohols, and Triterpenic Alcohols in Commercial Table Olives. J Am Oil Chem Soc 85: 253:262.
53. Ramírez E, Medina E, García P, Brenes M, Romero C (2017) Optimization of the Natural Debittering of Table Olives. LWT-Food Sci Technol 77: 308-313.
54. Poiana M, Romeo FV (2006) Changes in chemical and microbiological parameters of some varieties of Sicily olives during natural fermentation. Grasas Aceites 57: 402-408.
55. Romero C, Garcia-Garcia P, Brenes M (2016) Enrichment of Black Ripe Olives in Bioactive Compounds Using a Single Alkali Treatment. J Food Sci 81: C2686-C2691.

CHAPTER 13

Assessment of Iron Bioavailability from Different Bread Making Processes using an *in Vitro* Intestinal Cell Model

Rodriguez-Ramiro [a], C.A. Brearley [b], S.F.A. Bruggraber [c], A. Perfecto [a], P. Shewry [d], S. Fairweather-Tait [a]

[a]*Norwich Medical School, University of East Anglia, Norwich, UK*

[b] *School of Biological Sciences, University of East Anglia, Norwich, UK*

[c] *MRC Human Nutrition Research, Cambridge, UK*

[d] *Rothamsted Research, Hertfordshire, UK*

ABSTRACT

Myo-inositol hexa*kis*phosphate (IP6), is the main iron chelator in cereals and bread. The aim of this study was to investigate the effect of three commercial baking processes (sourdough, conventional yeast and Chorleywood Bread Making Process (CBP)) on the IP6 content of wholemeal bread, its impact on iron uptake in Caco-2 cells and the predicted bioavailability of iron from these breads with added iron, simulating a mixed-meal. The sourdough process fully degraded IP6 whilst the CBP and conventional processes reduced it by 75% compared with wholemeal flour. The iron released in solution after a simulated digestion was 8-fold higher in sourdough bread than with others but no difference in cellular iron uptake was observed. Additionally, when iron was added to the different breads digestions only sourdough bread elicited a significant ferritin response in Caco-2 cells (4.8-fold compared to the other breads) suggesting that sourdough bread could contribute towards improved iron nutrition.

Keywords: Iron uptake; Myo-inositol hexa*kis*phosphate; Sourdough bread; Caco-2 cells; Simulated gastrointestinal digestion

1. INTRODUCTION

Iron deficiency is one of the most common nutritional deficiencies in humans and is responsible for at least 50% of the 2 billion cases of anaemia worldwide (Lopez *et al.*, 2016 ; Zimmermann and Hurrell, 2007). Iron deficiency anaemia affects maternal and child morbidity and mortality, decreases physical performance and increases the burden of the health care in developing and developed countries (Kassebaum *et al.*, 2014 ; Lopez *et al.*, 2016). Iron deficiency anaemia is therefore a public health concern and nutritional intervention approaches are required to ameliorate its impact.

Interest in vegetarian diets has increased in recent decades because of the proposed health benefits associated with plant-based diets, together with concerns over sustainability and affordability (Gibson, Heath, & Szymlek-Gay, 2014). However, in plant-based diets, iron exists in the form of inorganic non-haem iron, which is less bioavailable than haem iron from animal foods sources, and this results in lower total iron absorption from vegetarian diets (Hurrell & Egli, 2010). Cereals, and in particular wheat and bread, are among the most widely consumed staple foods and are of global importance to human nutrition. However, most of the iron in wholewheat flour is present as phytates, salts of phytic acid (myo-inositol hexa*kis*phosphate (IP6)) which reduce iron bioavailability during the intestinal transit (Davidsson, 2003 ; Hurrell and Egli, 2010). Different strategies have been employed in order to increase bioavailable iron in cereal products, including supplementation, food fortification, biofortification (via plant agronomy, breeding or genetic engineering) and changing dietary habits through consumer education (Abbaspour *et al.*, 2014 ; Zimmermann and Hurrell, 2007). Additionally, in order to counteract the negative effect of phytate, the use of exogenous phytase enzymes and different bread making processes have been explored (Baye *et al.*, 2015; Brune *et al.*, 1992; Lopez *et al.*, 2001 ; Sanz-Penella *et al.*, 2009). In particular, the use of lactic acid bacteria in sourdough bread production has shown a significant degradation of IP6 (Arendt *et al.*, 2007; Haros *et al.*, 2007; Palacios *et al.*, 2008; Sanz-Penella *et al.*, 2009), suggesting that modifications in the processes of fermentation and baking (i.e. temperature, leavening time and microorganism strain) could improve iron bioavailability. It is therefore important to determine how these different technologies affect iron absorption from bread at the intestinal level. However, human intervention trials are lengthy and expensive, which restricts their application when evaluating strategies to reduce iron deficiency. Consequently, the combined *in vitro* digestion/Caco-2 intestinal cell model has been extensively used to predict iron bioavailability from food sources and to investigate the intestinal cellular mechanisms of iron uptake (Fairweather-Tait *et al.*, 2005; Thompson *et al.*, 2010). For example, Eagling, Wawer, Shewry, Zhao, and Fairweather-Tait (2014) compared differences in iron bioavailability between unleavened breads made from wholegrain and white flours from two commercial wheat cultivars and reported an inverse relationship between the cellular iron uptake and phytate content. Similarly, Sanz-Penella, Laparra, Sanz, and Haros (2012) have investigated the use of a *Bifidobacterium* strain, as a starter in the baking process, to increase the available iron through phytate degradation. Their findings suggest that either longer fermentation times or the inclusion of more active phytases are required to completely abolish the phytate content thereby leading to improvement in iron bioavailability from bread. The aims of the present study were therefore i) to compare the effects of three commercial processes of bread making on the reduction of phytate in wholemeal bread, ii) study the resulting impact on iron bioavailability using an intestinal cellular model, and iii) to predict how the different bread matrices might affect cellular uptake of iron from exogenous sources by adding ferrous sulphate to simulate the contribution of non-haem iron from a mixed-meal.

2. MATERIAL AND METHODS

2.1. Reagents

Chemicals, enzymes and hormones were purchased from Sigma-Aldrich, UK unless otherwise stated.

2.2. Grain material and bread making process

Grain (20 kg), from the commercial bread wheat cultivar Hereward, grown in the field of Rothamsted Research, UK, in 2014, was milled using a Buhler–MLU-202 mill at Campden BRI according to TES-CM-01. This resulted in 10 fractions (Table 1), including six white flour fractions: three breaks (B1-B3) and three reductions (R1-R3). These fractions varied in purity, with the first break (B1) and reduction (R1) corresponding to the purest starchy endosperm tissue and the third break (B3) and reduction (R3) to the least pure. The remaining four fractions consisting of bran fractions which were separated into offal flour, bran flour coarse offal and coarse bran. The 10 fractions were recombined and the resultant wholemeal flour was used to make three different breads based on commercial processes: Chorleywood bread process (CBP), conventional, and sourdough bread-making processes. The ingredients for each process are described in Table 2. For the CBP, the dough was mixed and proved for 60 min at 30 °C. The conventional and the sourdough processes followed the two-step sponge and dough method. For the sourdough process, a commercial starter culture (Lesaffre LV1) was mixed with flour, water and salt for 5 min and fermented at 28 °C for 24 h. Additional flour, yeast, water, salt and fat were then added and the sample mixed for 6 min, bulk fermented for 4 h at room temperature, and proved for 5 h at 25 °C. For the conventional system, yeasted dough was mixed for 5 min and fermented 16 h at 14 °C, additional flour, yeast, water, salt and fat were added, mixed again for 10 min, bulk fermented at room temperature for 15 min and proved at 30 °C for one hour. All breads were baked at 230 °C for 30 min. The breads were sliced and stored frozen at −80 °C until used. The crusts of the slices were then removed and the crumb was freeze-dried, finely ground and stored in a desiccator at 4 °C over silica gel.

Table 1. Amount, yield and iron obtained for each mill fractions from 20 kg of the *Hereward* wheat grain.

Mill fraction	Amount (g)	Yield (%)	Cumulative yield (%)	Fe (ppm)
Break 1	2410	12.0	12.0	8.6
Break 2	1567	7.8	19.9	9.0
Break 3	384	1.9	21.8	21.3
Reduction 1	9343	46.7	68.5	7.8
Reduction 2	1302	6.5	75.0	15.3
Reduction 3	309	1.5	76.6	23.7
Bran flour	198	1.0	77.6	63.4
Offal flour	383	1.9	79.4	68.5
Bran	1905	9.5	89.0	115.8
Offal	1729	8.6	97.6	125.3

Table 2. Ingredients (g/kg) employed at the different stages of the three bread making processes.

2.3. Determination of iron in the bread and simulated digestion samples

The concentrations of Fe were determined using an Inductively Coupled Plasma Optical Emission Spectroscopy (ICP-OES, Jobin Yvon Horiba – ULTIMA 2C) equipped with a glass concentric nebulizer (1 ml/min sample flow rate), a 50 ml glass cyclonic spray chamber and a radial torch with a 3 mm i.d. alumina injector. Sample solutions were introduced from an auto-sampler (Jobin Yvon Horiba AS500) using a sample probe with 0.5 mm i.d. sample tubing and 0.76 mm i.d. pump tubing (black/black). Instrument operating conditions are listed in Supplementary Table 1. Peak profiles were used to measure Fe as described in Supplementary Table 2.

The dry ground breads (about 500 mg) were weighed directly into 15 ml TFM® PTFE microwave vials, to which 3 ml of UHP water and 3 ml of 69% HNO_3 were successively added. The vials were placed into a microwave digestion system (UltraWAVE, Analytix, UK) and heated at 230 °C for 15 min. The digested solutions were decanted into pre-weighted 15 ml PP tubes and the microwave vials rinsed twice with 3 ml of UHP water which was then added to the decanted digestates. The tubes were weighed to determine the total mass of digestates. The digestates were then diluted in water to a final acid concentration of 5%. Digestion blank controls and internal quality controls were prepared alongside the food samples and analysed with the digested samples. A series of external calibration standards containing Fe were prepared from commercial standard stock solutions (1000 ± 2 ppm in 2% HNO_3; PerkinElmer Pure Plus), with final concentrations ranging from 0 to 1 ppm in a diluent matched to the digested solution (final concentration of 5% HNO_3). The Fe concentrations were calculated against the linear regression obtained from the calibration standard curve performed with each analysis (Supplementary Fig. 1). The flour fractions used to make the wholemeal bread were analysed by ICP-OES following a similar procedure.

For analysis of the simulated gastrointestinal digestions, the instrument was equipped with a concentric PFA micro-flow nebulizer (0–2 ml/min sample flow rate), a 50 ml glass cyclonic spray chamber and a radial torch with a 3 mm i.d. alumina injector. Sample solutions were introduced from an auto-sampler (Jobin Yvon Horiba AS500) using a sample probe with 0.25 mm i.d. sample tubing and 0.38 mm i.d. pump tubing (orange/green). The instrument running conditions were as described in Supplementary Table 3 and the peak profile measurement parameters were as in Supplementary Table 2.

The simulated gastrointestinal digestions of the bread (400 µl) were placed into (2 ml) cryovials and 800 µl of fresh solution (30%) H_2O_2/(69%) HNO_3 (1:1, v/v) were added and incubated at room temperature overnight and then at 40 °C with the cap unscrewed. The samples were then diluted in water to 5% HNO_3 concentration and the Fe concentrations calculated.

The fractions of bioaccessible and precipitated iron were determined using the following Eqs. (1) ; (2), respectively:

$$[(\%)\text{Bioaccessible Fe}] = [(\text{Fe}_{\text{bioaccessible}})/\text{Total Fe}] \times 100; \quad (1)$$

$$[(\%)\text{Precipitated Fe}] = [(\text{Total Fe} - \text{Fe}_{\text{bioaccessible}})/\text{Total Fe}] \times 100; \quad (2)$$

where $Fe_{bioaccessible}$ represents the fraction of iron released from the bread matrix after the simulated gastrointestinal digestion which remains in solution for the cellular experiments (see below); total Fe is the iron concentration in the final simulated digestion volume and percentages are expressed out of the total iron concentration in the final simulated digestion volume.

2.4. Extraction and analysis of inositol phosphates

Inositol phosphates were extracted from 500 mg of freeze-dried ground bread with 20 ml 0.5 M HCl for 3 h at 20 °C under agitation, as reported previously (Skoglund, Carlsson, & Sandberg, 1997). Then samples were centrifuged at 3000*g* for 10 min and the supernatants were kept at −80 °C until used. Quantification of the inositol phosphates was performed by high performance ion chromatography (HPIC) as described previously (Blaabjerg *et al.*, 2010 ; Phillippy and Bland, 1988). Briefly, prior to the analysis, the samples were centrifuged at 9000*g* for 10 min and the supernatants filtered through 13 mm diameter 0.45 μm pore size polypropylene filters (Kinesis, UK Ltd). Aliquots (50 μl) were injected onto a 3 mm i.d. × 200 mm Carbo Pac PA200 HPLC column (Dionex) fitted with a 3 mm × 50 mm guard column of the same material. The column was eluted at a flow rate of 0.4 ml/min with a gradient of methanesulfonic acid (Acros Organics) delivered from buffers reservoirs containing: A, water; B, 600 mM methanesulfonic acid, according to the following conditions: time (minutes), % B: 0,0; 25,100; 38,100; 39,0; 49,0. The column eluate was mixed using a mixing tee with 0.1% w/v ferric nitrate in 2% w/v perchloric acid delivered at a flow rate of 0.2 ml/min, before passage through a 194 μl volume knitted reaction coil (4 m × 0.25 mm i.d.) obtained from Biotech AB, Sweden. The column, mixing tee and reaction coil were held at 35 °C. Peaks of inositol phosphate were detected at 290 nm in a Jasco UV-2077 Plus UV detector. Chromatographic data were collected from 0 to 49 min and peaks integrated in ChromNav (Jasco) and plotted using GraphPad Prism 5 software. The elution positions of different stereoisomers of different classes of inositol phosphates were determined by the inclusion at regular intervals of a set of standards obtained by extended acid treatment of phytic acid. Inositol phosphates were quantified by reference to the peak area of the standard inositol phosphates.

2.5. Food samples and *in vitro* simulated gastrointestinal digestion

The simulated gastrointestinal digestion was performed as described by Glahn, Lee, Yeung, Goldman, and Miller (1998), with minor modifications. Bread samples were first added to 10 ml of pH 2 buffer saline-solution (140 mmol/l NaCl, 5 mmol/l KCl), followed by the addition of pepsin (0.04 g/ml) and incubated for 90 min on a rolling table at 37 °C to simulate gastric conditions. Once the first phase was completed, the pH of the samples was gradually adjusted to pH 5.5, bile (0.007 g/ml) and pancreatin (0.001 g/ml) digestive enzymes were added to the samples, readjusted to pH 7, and incubated for an additional hour on a rolling table at 37 °C to mimic intestinal conditions. At the end of the simulated gastrointestinal digestion, samples were centrifuged at 3000*g* for 10 min and the supernatants used for subsequent cell culture experiments. A volume of 1.5 ml of the digestion was applied to an upper chamber consisting of a Transwell insert fitted with a 15 KDa molecular weight cut-off dialysis membrane (Spectra/Por 7 dialysis tubing, Spectrum laboratories, Europe) suspended over Caco-2 cell monolayers grown in collagen-coated 6 well plates. The bread simulated digestions were incubated on cells for one hour at 37 °C in a humidified incubator containing 5% CO_2 and 95% air. Inserts were then removed, an additional 1 ml of supplemented MEM was added and cells were incubated for a further 23 h prior to harvesting for ferritin analysis.

In vitro digestions were prepared fresh prior to each cell culture experiment. For the studies of iron bioaccessibility, a fraction of the final digestion used for the cellular experiment was collected and stored at −80 °C until analysis by ICP-OES. In addition, prior to the simulated digestions of the breads, ascorbic acid (AA) was added to the gastric digestion (molar ratio of 1:20, Fe:AA) to ensure that all of the iron released during digestion remained in solution. All experiments were performed using the following controls: (a) a blank digestion without any bread samples or added iron to ensure no iron contamination in the *in vitro* digestion/cellular system and (b) a reference digestion of 100 μM of ferrous sulphate heptahydrate ($FeSO_4$) solubilised in 0.1 M HCl with 2000 μM of AA. In order to determine how the content of IP6 affects the availability of endogenous iron of the breads and the exogenous iron from a potential mixed-meals, two different types of experiments were performed. Experiments to determine the effects of IP6 content on the bioavailability of endogenous iron in freeze-dried bread powder were conducted using equivalent amounts of iron (50 μM) in each digestion. Experiments to determine the effects of IP6 content of the breads on the bioavailability of exogenous iron were conducted with the addition of 100 μM of $FeSO_4$ to the gastric digestion, resulting in a final total Fe concentration of 150 μM.

2.6. Cell culture

Caco-2 cells (HTB-37) were obtained from American Type Culture Collection (Manassas, VA, USA) at passage 20 and stored in liquid nitrogen. Cells were grown in Dulbecco's modified Eagle's medium (DMEM), supplemented with 25 mM HEPES solution, 10% foetal bovine serum, 1% penicillin (5000 u/ml) and 1% 4 mM l-glutamine (ThermoFisher Scientific, UK) and 1% MEM non-essential amino acids solution (Sigma-Aldrich, UK). Cells were maintained at 37 °C in a humidified incubator containing 5% $CO_2$and 95% air. For iron uptake experiments, Caco-2 cells between passages 30–36 were seeded onto collagen-coated 6-well plates (Bio-Greiner, UK) at a density of 4.75×10^5cells/well suspended in 2 ml of supplemented DMEMl which was replaced every 2 days. Cells were used on days 13–15 post-seeding according to the method developed by Glahn et al. (1998). In order to ensure a low basal media iron levels, 24 h prior to the initiation of the *in vitro* digestion experiments, the DMEM medium was replaced by Eagle's minimum essential medium (MEM) without foetal bovine serum supplemented with 10 mmol/l PIPES [piperazine-N,N′ –bis-(2-ethanesulfonic acid)], 26.1 mM $NaHCO_3$, 19.4 mmol/l glucose, 1% antibiotic-antimycotic solution, 11 μmol/l hydrocortisone, 0.87 μmol/l insulin, 0.02 μmol/l sodium selenite (Na_2SeO_3), 0.05 μmol/l triiodothyronine and 20 μg/l epidermal growth factor (Glahn et al., 1998).

2.7. Cell harvest and ferritin analysis

Following 24 h incubation with the bread digestions, the cellular medium was aspirated, the cells were rinsed with 18 Ω MilliQ H_2O and subsequently lysed by scraping in 200 μl of Cellytic M (Sigma-Aldrich, UK). Cell pellets were kept on ice for 15 min and stored at −80 °C. For analysis, samples were thawed and centrifuged at 14,000*g* for 15 min. Cellular debris was discarded and the supernatant containing the proteins was used for ferritin determination using the Spectro Ferritin ELISA assay (RAMCO, USA). The ferritin concentration in the samples was determined using a microplate reader at an excitation wavelength of 500 nm according to the manufacturer's protocol. Ferritin concentrations were normalised to total cell protein using the Pierce Protein BCA protein assay (ThermoFisher Scientific, UK).

2.8. Statistical analysis

Data are presented as mean values with the standard errors of the means (SEM). The data set was tested for homogeneity of variances by the test of Levene. For multiple comparison, one-way ANOVA followed by a Bonferroni test was used when variances were homogenous or by Tamhane test when variances were not homogenous. Statistical significance was set at $P \leqslant 0.05$. The statistical analysis was performed using the SPSS package (version 23; SPSS Inc., Chicago, IL, USA).

3. RESULTS

3.1. Inositol hexakisphosphate was completely degraded by the sourdough bread process

We investigated the effect of three different commercial bread making processes on the IP6 content of wholegrain bread (Fig. 1). Compared to the wholegrain flour used as the control, the CBP and conventional processes significantly reduced (by about 75%) the concentration of IP6. The reduction was greater with the sourdough process, which completely degraded IP6 from the wholegrain flour. IP6 has a strong chelating iron capacity and the molar ratio of IP6 to iron (IP6:Fe) can be used to estimate its effect on iron absorption (Hurrell & Egli, 2010). We therefore determined the IP6:Fe molar ratio of the IP6 in the breads and their endogenous iron contents (Table 3). This showed that the sourdough process was the only baking process which was able to reduce the ratio to below 1.

Table 3. Endogenous iron, *myo*-inositol hexa*kis*phosphate (IP6) content and the molar ratio IP6:Fe of the three bread processes.

Bread type	Endogenous iron mg/100 g (DW)	IP6 g/100 g (DW)	Molar ratio IP6:Fe
Sourdough	3.2 ± 0.06	*n.d.*	⩽1
Conventional	3.0 ± 0.09	0.19 ± 0.06	5.36 ± 0.97
Chorleywood	2.5 ± 0.07	0.17 ± 0.01	6.01 ± 0.35

Data represent means ± SEM (n = 3). DW, dry weight; *n.d.*, non detected (<0.005 g IP6/100 g).

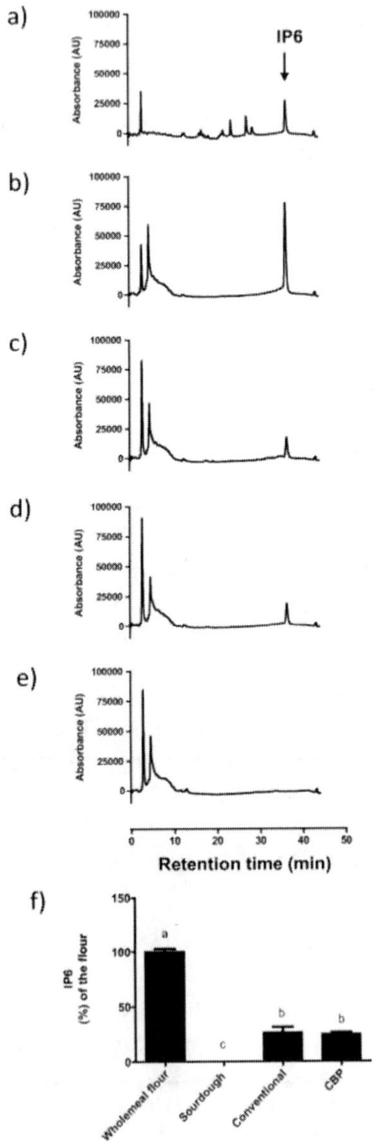

Fig. 1. Representative chromatograms (UV 290 nm detection) of the inositol phosphates (IP) of the flour and the different types of bread: (a) Internal standard for different IPs, the arrow denotes IP6, (b) Hereward wholegrain flour, (c) Conventional, (d) CBP and (e) Sourdough bread. (f) Percentage of IP6 in the different bread types relative to the Hereward flour used for bread making. Data represent means ± SEM ($n = 4$). Means without a common letter differs ($P < 0.05$).

3.2. Sourdough bread releases endogenous iron during the simulated gastrointestinal digestion

Next we investigated whether the greater reduction of IP6 in the sourdough bread was related to an increase in endogenous iron released during the simulated digestion. As shown in Fig. 2a, the sourdough process increased the bioaccesible iron by about 8-fold compared with the other two processes. The percentage of bioaccessible iron increased from the 1.4% and 1.3% of the total iron in the CBP and conventional bread process respectively, to 12% in the sourdough process showing the importance of the reduction of IP6 (Fig. 2b).

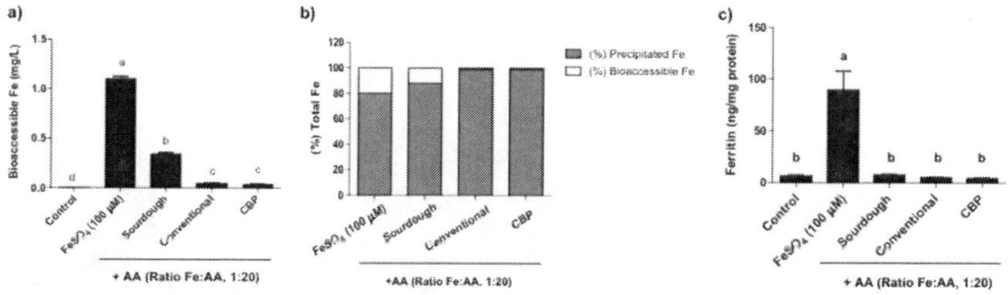

Fig. 2. Bioaccessible and bioavailable endogenous iron in the different type of bread supplemented with ascorbic acid (AA) in a molar ratio 1:20 (Fe:AA). (a) Bioaccessible iron in the final gastrointestinal digestion. (b) Percentage of the precipitated and bioaccessible iron fractions in the different bread digestion. (c) Ferritin response in Caco-2 cells exposed to the different gastrointestinal bread digestion. Data represent means ± SEM (n = 6–8). Means without a common letter differs ($P < 0.05$).

3.3. Endogenous iron released by the sourdough bread does not elicit a ferritin response in Caco-2 cells

In order to determine whether the endogenous iron released by the breads is bioavailable to intestinal cells, we analysed ferritin formation in Caco-2 cells as a measure of cell iron uptake. Caco-2 cells were exposed to different bread digestions, the blank and the positive control (100 μM) (Fig. 2c). The levels of ferritin were 6.6 and 89.4 ng/mg of protein for the blank and $FeSO_4$ control digest, respectively. Nevertheless, the bread digests did not show a significant increase in ferritin formation compared to the blank control, with 7.8, 5.3 and 4.7 ng/mg of protein for the sourdough, conventional and CBP breads, respectively.

3.4. Sourdough bread, but not CBP or conventional breads, results in increased bioaccessibility of added iron in the simulated gastrointestinal digestion

We estimated the impact of reduced IP6 content resulting from the different bread processes on the bioaccessibility of exogenous iron by measuring the effect of adding 100 μM of $FeSO_4$ to the bread digestion. The addition of exogenous iron significantly increased the quantity of bioaccessible iron in the sourdough bread after digestion, reaching similar values to the positive control (100 μM of $FeSO_4$). However, the addition of exogenous iron to the CBP and conventional breads did not increase bioaccessible iron after digestion suggesting that the higher concentration of IP6 in these breads may bind the added iron and limit bioaccessibility (Fig. 3a and b).

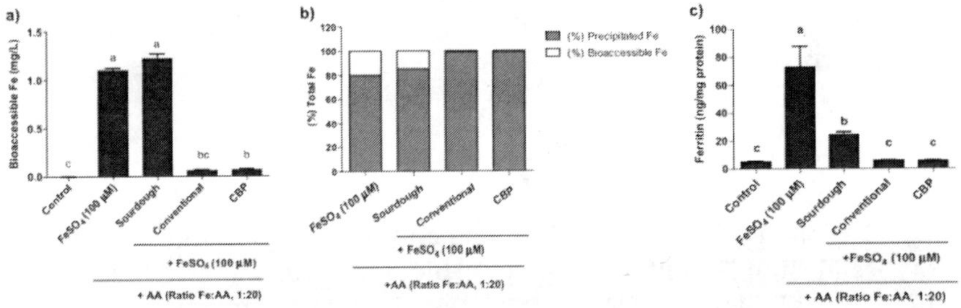

Fig. 3. Bioaccessible and bioavailable iron in a simulated mixed-diet adding 100 μM ferrous sulphate and ascorbic acid (AA) in a molar ratio 1:20 (Fe:AA) to the bread prior the gastrointestinal digestion. (a) Bioaccessible iron in the final gastrointestinal digestion. (b) Percentage of the precipitated and bioaccessible iron fractions in the different bread digestion. (c) Ferritin response in Caco-2 cells exposed to the different gastrointestinal bread digestion. Data represent means ± SEM ($n = 6–8$). Means without a common letter differs ($P < 0.05$).

3.5. The reduction of IP6 in the sourdough bread allows for exogenous iron uptake by the Caco-2 cells

Finally, to confirm that the increased soluble iron resulting from the addition of exogenous iron in the sourdough bread digestion is available for absorption by intestinal cells, we measured ferritin formation in the Caco-2 cell model for these bread digestions. As shown in Fig. 3c, the sourdough bread with added iron resulted in a significant ferritin increase compared with the blank control (24.3 ng/mg *vs* 5.3 ng/mg) whilst the CBP and conventional breads with added iron did not increase ferritin formation.

4. DISCUSSION

This study was performed to compare the effect of three commercial bread making processes on the reduction of phytate in wholemeal flour and to determine iron uptake from these breads after a simulated digestion in the intestinal cells. We further explored the effect of the bread matrix on the absorption of iron from exogenous sources, mimicking a mixed-diet, by the addition of $FeSO_4$ to the simulated gastrointestinal digestion of the breads. Our main findings are that: (1) the sourdough bread process degraded the IP6 present in the wholegrain flour, increasing the bioaccessible endogenous iron in the simulated digestion in comparison with the other bread making processes; (2) none of the three types of bread resulted in increasing ferritin formation, a measure of iron uptake, in the Caco-2 cells; (3) the sourdough bread, but not CBP or conventional, allowed an increase in cellular uptake of iron from exogenous sources.

IP6 is degraded during bread making, but the extent can vary extensively depending on the type of bread (flour) and the fermentation process (Brune *et al.*, 1992 ; Lopez *et al.*, 2001). In the present study, we compared three commercial processes for making wholemeal bread in the UK: the widely used short fermentation CBP and conventional and sourdough long fermentation processes. We observed a 25% greater reduction in the IP6 content of the sourdough bread compared to the conventional and CBP breads. This result is similar to previous studies which reported about 60% reduction in IP6 after the 4 h proving period in sourdough compared with 40% in yeast-fermented dough and in control dough without leavening agent (Leenhardt, Levrat-Verny, Chanliaud, & Remesy, 2005). IP6 has a negative effect on iron absorption, and reducing the molar ratio of IP6:Fe to between 1 and 0.4 has been estimated as the threshold for improving iron bioavailability in cereals (Hurrell & Egli, 2010). However, in general the estimated molar ratio of IP6:Fe for wholegrain bread is higher than 1 (Nielsen, Tetens, & Meyer, 2013), which limits the bioavailability of iron. We found some small differences in the contents of IP6 and iron in comparison with the data reviewed by Nielsen et al. (2013), with a lower IP6 content and higher iron content which probably related to differences in the flour composition and/or bread making processes. However, our phytate contents are similar to those reported in a previous study comparing CBP and a long fermentation (2–3 h) process (Daniels & Fisher, 1981). Similarly, the total iron content in our breads is consistent with Bruggraber et al. (2012), who reported that the iron content of wholemeal bread ranged between 2.4 and 3.4 mg/100 g over the last 70 years. Because of the differences in IP6 and iron contents, we observed a lower molar ratio of IP6:Fe in the CBP and conventional breads than the values reviewed by Nielsen et al. (2013), but in any case, much higher than 1. However, the sourdough process reduced the IP6 content below the limit of detection showing a molar ratio of IP6:Fe below 1, which could explain the 8-fold increase in the bioaccessible iron in the sourdough bread digestion in comparison with the other breads digestion. Despite this increase in the iron released from the sourdough bread digestion, we did not observe an increase in iron uptake by the Caco-2 cells. These findings are in agreement with Sanz-Penella et al. (2012) who reported that a phytase-producing *Bifidobacterium* strain increased bioaccessible iron in bread digestion but had a negligible effect on iron absorption in Caco-2 cells. We therefore suggest that a fraction of the iron released from the bread matrix remains bound and/or encapsulated to other soluble molecules present in the bread digestion, such as dietary fibre, which resist gastrointestinal digestion and reduce its availability for cellular absorption. In fact, it has been recently reported that the mechanical or chemical disruption of the cell walls of the aleurone layer in wheat grain increases ferritin formation in Caco-2 cells after a simulated gastrointestinal digestion (Latunde-Dada et al., 2014). Wholegrain wheat and wholemeal bread also contain ferulic acid, a phenolic compound, which is mainly bound to non-starch polysaccharide (Anson *et al.*, 2009; Li *et al.*, 2008 ; Lu *et al.*, 2014), and could act as an iron chelator (Chandrasekara and Shahidi, 2011 ; Teixeira *et al.*, 2013).

Bread is a staple food which can be consumed in combination with other foods or ingredients. It is therefore necessary to determine the effects of different types of bread on the cellular iron uptake from other food sources of iron consumed in mixed meals. We observed that the addition of ferrous sulphate to the sourdough bread increased the bioaccessible iron in the simulated digestion and the iron uptake by the Caco-2 cells, although no increase was detected with the CBP or conventional breads. This demonstrated the effect of the total IP6 degradation in the sourdough bread and the importance of exceeding the critical molar ratio of IP6:Fe to increase the iron bioavailability. However, it is notable that although the amount of bioaccessible iron in the digestion was similar to that in the ferrous sulphate control (without the bread matrix), the ferritin formation was less than 50%. This supports our suggestion that other soluble chelators in the bread matrix, apart from IP6, also restrict iron uptake by the cells. Consistent with our results, Chaoui, Faid, and Belahsen (2006) have reported a similar improvement in the iron status of mice fed a sourdough bread prepared with fortified flour compared to conventional breads prepared with the same flours.

Hurrell and Egli, 2010; Petry *et al.*, 2010 ; Troesch *et al.*, 2009 have studied the effect of different enhancers and food matrix inhibitors on iron absorption using isotopic Fe labels. They have suggested that the use of a potent phytase enzyme from *Aspergillus niger* in combination with iron fortificants could be a promising approach to improve iron absorption from foods rich in phytic acid and low in iron (Troesch et al., 2009). We have explored this approach *in vitro* using a commercial sourdough bread-making process to reduce the phytate content of the bread, and ferrous sulphate as a fortificant. Our findings are in agreement with those of Hurrell et al. (Troesch et al., 2009) which indicate that reducing the IP6 content in combination with iron fortification can improve iron bioavailability.

In summary, this study demonstrates that the sourdough process leads to the full degradation of IP6 in wholemeal bread and increased the bioavailability of iron in the Caco-2 cell model when the bread was digested in combination with other iron sources, compared with CBP and conventional fermented breads. The wider use of sourdough processes could therefore have significant benefits for iron nutrition.

CONFLICT OF INTEREST

The authors declare that they have no conflict of interest.

ACKNOWLEDGEMENTS

This study was supported by a Biotechnology and Biological Sciences Research Council (BBSRC)-Diet and Health Research Industry Club (DRINC) 2 Grant (BB/L025396/1 and BB/L025515/1). IRR is currently funded by a BBSRC DRINC2 Post-Doctoral fellowship grant. Research at Rothamsted is funded by the BBSRC of the UK. CB is supported by BBSRC Grant BB/M022978/1. The authors would like to thank Campden BRI and in particular Dr Gary Tucker for their support with the milling and baking of the bread. We also would like to thank our DRINC 2 project partners, notably Dr Johathan Powell and Dr Dora Pereira (MRC HNR); Dr Janneke Balk (John Innes Centre); Dr Caterina Dell'Aquila and Dr Andy Neal (Rothamsted); Dr Paul Sharp (Kings College, London) and Dr Katie Moore and Dr Chris Grovenor (University of Oxford) for their contributions to the development of the project.

REFERENCES

1. Abbaspour, N., Hurrell, R., & Kelishadi, R. (2014). Review on iron and its importance for human health. Journal of Research in Medical Science, 19(2), 164–174.
2. Anson, N. M., Selinheimo, E., Havenaar, R., Aura, A. M., Mattila, I., Lehtinen, P., ... Haenen, G. R. (2009). Bioprocessing of wheat bran improves in vitro bioaccessibility and colonic metabolism of phenolic compounds. Journal of Agriculture and Food Chemistry, 57(14), 6148–6155.
3. Arendt, E. K., Ryan, L. A., & Dal Bello, F. (2007). Impact of sourdough on the texture of bread. Food Microbiology, 24(2), 165–174.
4. Baye, K., Guyot, J. P., Icard-Verniere, C., Rochette, I., & Mouquet-Rivier, C. (2015). Enzymatic degradation of phytate, polyphenols and dietary fibers in Ethiopian injera flours: Effect on iron bioaccessibility. Food Chemistry, 174, 60–67.
5. Blaabjerg, K., Hansen-Moller, J., & Poulsen, H. D. (2010). High-performance ion chromatography method for separation and quantification of inositol phosphates in diets and digesta. Journal of Chromatography B: Analytical Technologies in the Biomedical and Life Sciences, 878(3–4), 347–354.
6. Bruggraber, S. F., Chapman, T. P., Thane, C. W., Olson, A., Jugdaohsingh, R., & Powell, J. J. (2012). A re-analysis of the iron content of plant-based foods in the United Kingdom. British Journal of Nutrition, 108(12), 2221–2228.
7. Brune, M., Rossander-Hulten, L., Hallberg, L., Gleerup, A., & Sandberg, A. S. (1992). Iron absorption from bread in humans: Inhibiting effects of cereal fiber, phytate and inositol phosphates with different numbers of phosphate groups. Journal of Nutrition, 122(3), 442–449.
8. Chandrasekara, A., & Shahidi, F. (2011). Inhibitory activities of soluble and bound millet seed phenolics on free radicals and reactive oxygen species. Journal of Agriculture and Food Chemistry, 59(1), 428–436.
9. Chaoui, A., Faid, M., & Belahsen, R. (2006). Making bread with sourdough improves iron bioavailability from reconstituted fortified wheat flour in mice. Journal of Trace Elements in Medicine and Biology, 20(4), 217–220.
10. Daniels, D. G., & Fisher, N. (1981). Hydrolysis of the phytate of wheat flour during breadmaking. British Journal of Nutrition, 46(1), 1–6.
11. Davidsson, L. (2003). Approaches to improve iron bioavailability from complementary foods. Journal of Nutrition, 133(5 Suppl 1), 1560S–1562S.
12. Eagling, T., Wawer, A. A., Shewry, P. R., Zhao, F. J., & Fairweather-Tait, S. J. (2014). Iron bioavailability in two commercial cultivars of wheat: Comparison between wholegrain and white flour and the effects of nicotianamine and 2'- deoxymugineic acid on iron uptake into Caco-2 cells. Journal of Agriculture and Food Chemistry, 62(42), 10320–10325.
13. Fairweather-Tait, S., Lynch, S., Hotz, C., Hurrell, R., Abrahamse, L., Beebe, S., ... Trinidad, T. P. (2005). The usefulness of in vitro models to predict the bioavailability of iron and zinc: A consensus statement from the HarvestPlus expert consultation. International Journal for Vitamin and Nutrition Research, 75 (6), 371–374.
14. Gibson, R. S., Heath, A. L., & Szymlek-Gay, E. A. (2014). Is iron and zinc nutrition a concern for vegetarian infants and young children in industrialized countries? American Journal of Clinical Nutrition, 100(Suppl 1), 459S–468S.
15. Glahn, R. P., Lee, O. A., Yeung, A., Goldman, M. I., & Miller, D. D. (1998). Caco-2 cell ferritin formation predicts nonradiolabeled food iron availability in an in vitro digestion/Caco-2 cell culture model. Journal of Nutrition, 128(9), 1555–1561.
16. Haros, M., Bielecka, M., Honke, J., & Sanz, Y. (2007). Myo-inositol hexakisphosphate degradation by Bifidobacterium infantis ATCC 15697. International Journal of Food Microbiology, 117(1), 76–84.
17. Hurrell, R., & Egli, I. (2010). Iron bioavailability and dietary reference values. American Journal of Clinical Nutrition, 91(5), 1461S–1467S.

18. Kassebaum, N. J., Jasrasaria, R., Naghavi, M., Wulf, S. K., Johns, N., Lozano, R., ... Murray, C. J. (2014). A systematic analysis of global anemia burden from 1990 to 2010. Blood, 123(5), 615–624.
19. Latunde-Dada, G. O., Li, X., Parodi, A., Edwards, C. H., Ellis, P. R., & Sharp, P. A. (2014). Micromilling enhances iron bioaccessibility from wholegrain wheat. Journal of Agriculture and Food Chemistry, 62(46), 11222–11227.
20. Leenhardt, F., Levrat-Verny, M. A., Chanliaud, E., & Remesy, C. (2005). Moderate decrease of pH by sourdough fermentation is sufficient to reduce phytate content of whole wheat flour through endogenous phytase activity. Journal of Agriculture and Food Chemistry, 53(1), 98–102.
21. Li, L., Shewry, P. R., & Ward, J. L. (2008). Phenolic acids in wheat varieties in the HEALTHGRAIN Diversity Screen. Journal of Agriculture and Food Chemistry, 56 (21), 9732–9739.
22. Lopez, A., Cacoub, P., Macdougall, I. C., & Peyrin-Biroulet, L. (2016). Iron deficiency anaemia. Lancet, 387(10021), 907–916.
23. Lopez, H. W., Krespine, V., Guy, C., Messager, A., Demigne, C., & Remesy, C. (2001). Prolonged fermentation of whole wheat sourdough reduces phytate level and increases soluble magnesium. Journal of Agriculture and Food Chemistry, 49(5), 2657–2662.
24. Lu, Y., Luthria, D., Fuerst, E. P., Kiszonas, A. M., Yu, L., & Morris, C. F. (2014). Effect of processing on phenolic composition of dough and bread fractions made from refined and whole wheat flour of three wheat varieties. Journal of Agriculture and Food Chemistry, 62(43), 10431–10436.
25. Nielsen, A. V., Tetens, I., & Meyer, A. S. (2013). Potential of phytase-mediated iron release from cereal-based foods: A quantitative view. Nutrients, 5(8), 3074–3098.
26. Palacios, M. C., Haros, M., Rosell, C. M., & Sanz, Y. (2008). Selection of phytatedegrading human bifidobacteria and application in whole wheat dough fermentation. Food Microbiology, 25(1), 169–176.
27. Petry, N., Egli, I., Zeder, C., Walczyk, T., & Hurrell, R. (2010). Polyphenols and phytic acid contribute to the low iron bioavailability from common beans in young women. Journal of Nutrition, 140(11), 1977–1982.
28. Phillippy, B. Q., & Bland, J. M. (1988). Gradient ion chromatography of inositol phosphates. Analytical Biochemistry, 175(1), 162–166.
29. Sanz-Penella, J. M., Laparra, J. M., Sanz, Y., & Haros, M. (2012). Assessment of iron bioavailability in whole wheat bread by addition of phytase-producing bifidobacteria. Journal of Agriculture and Food Chemistry, 60(12), 3190–3195.
30. Sanz-Penella, J. M., Tamayo-Ramos, J. A., Sanz, Y., & Haros, M. (2009). Phytate reduction in bran-enriched bread by phytase-producing bifidobacteria. Journal of Agriculture and Food Chemistry, 57(21), 10239–10244.
31. Skoglund, E., Carlsson, N. G., & Sandberg, A. S. (1997). Determination of isomers of inositol mono- to hexaphosphates in selected foods and intestinal contents using high-performance ion chromatography. Journal of Agricultural and Food Chemistry, 45(2), 431–436.
32. Teixeira, J., Gaspar, A., Garrido, E. M., Garrido, J., & Borges, F. (2013). Hydroxycinnamic acid antioxidants: An electrochemical overview. BioMed Research International, 2013, 251754.
33. Thompson, B. A., Sharp, P. A., Elliott, R., & Fairweather-Tait, S. J. (2010). Inhibitory effect of calcium on non-heme iron absorption may be related to translocation of DMT-1 at the apical membrane of enterocytes. Journal of Agriculture and Food Chemistry, 58(14), 8414–8417.
34. Troesch, B., Egli, I., Zeder, C., Hurrell, R. F., de Pee, S., & Zimmermann, M. B. (2009). Optimization of a phytase-containing micronutrient powder with low amounts of highly bioavailable iron for in-home fortification of complementary foods. American Journal of Clinical Nutrition, 89(2), 539–544.
35. Zimmermann, M. B., & Hurrell, R. F. (2007). Nutritional iron deficiency. Lancet, 370 (9586), 511–520.

CHAPTER 14

Protein Profile Characterization of Donkey Milk

Paolo Polidori[1] and Silvia Vincenzetti[1]

[1] *Università di Camerino, Italy*

1. INTRODUCTION

Milk is a biological fluid designed to contain all nutritional requirements of a specific mammalian newborn; therefore, the composition of milk differs by the needs of the neonate of different species. Although much research has been devoted to milk composition in the domestic horse, donkey's milk has recently aroused scientific interest, above all among paediatric allergologists and nutritionists. Clinical studies have demonstrated that donkey milk may be considered a good replacer for dairy cow's milk in feeding children with severe Ig-E mediated cow's milk protein allergy, when human milk can not be given [1]. For these patients, donkey milk is not only useful [2], but also safer compared with milk obtained by other mammalian species [3], due to the high similarity with human milk, especially considering protein fractions content [4, 5].

2. MILK PROTEINS

In 1838, a Swedish scientist called Jacob Berzelius suggested the term "protein" after the greek word "proteios", which means "primary" or "of the first rank"; since then, many scientific discoveries have been made about these large molecules indispensable for the good functioning of the body's cells, tissues and organs. In the past, it was believed that the function of proteins was restricted to tissue-formation, while carbohydrates and lipids provided energy to the body. Another out-of-date concept was the belief that dietary proteins were completely hydrolyzed in the gastro-intestinal tract and only free amino acids could be adsorbed from the gut. The new concept is that macro- and microelements (such as vitamins and minerals) may interact to perform different functions in the body [6]; amino acids and peptides formed in the digestion of natural proteins are adsorbed and incorporated (anabolism) into various tissues and organs as body proteins.

Milk protein is a very heterogeneous group of molecules and, for ease description, could be classified into five main categories: caseins, whey proteins, milk fat globule proteins, enzymes and other miscellaneous minor proteins [7]. Milk proteins appear to be an exciting link between nutrition, dietetics and therapy; today, consumers can expect more than just nutrition from intake of milk proteins. In fact, milk contains a variety of bioactive compounds with special properties associated with the development, growth and survival of infants beyond those provided by nutrition alone [8]. The major antimicrobial proteins in milk are immunoglobulins, lactoferrin, lactoperoxidase and lysozyme [9]. Immunoglobulins (IgG, IgM and secretory IgA) act by a specific mode of action involving antigen-antibody reactions. The other three proteins are non-specific protective factors, and their antimicrobial mechanisms of action differ from each other.

Lactoferrin, also called lactotransferrin, is an iron binding protein present in milk, saliva, tears and mucus secretions, its bacteriostatic effects are usually attributed to lactoferrin's ability to bind environmental iron ions. Lactoferrin inhibits the growth of many kinds of Gram-negative and Gram-positive bacteria as well as some species of fungi and yeast [9]. Lactoperoxidase, a heme-containing protein, catalyzes an oxidation reaction involving hydrogen peroxide (H_2O_2) and functions as a component of host defense system. The lactoperoxidase system is known to be effective to preserve raw milk without refrigeration and it is used in the production of many dairy products [10].

Lysozyme exerts its antimicrobial activity by the hydrolysis of glycosidic bonds of mucopolisaccarides in bacterial cell walls [11]. Lysozyme, together with other peptides including immunoglobulins, lactoferrin and lactoperoxidase, is active in the infant's digestive tract in order to reduce the incidence of gastro-intestinal infections [12].

3. COW MILK ALLERGY

Food allergy is the clinical syndrome resulting from sensitization of an individual to dietary proteins or other food allergens present in the intestinal lumen [13]. Food allergy is much more common among children than adults, and is more common among younger children than older children [14]. Cow Milk Allergy (CMA) is a frequent disease in infants, but its etiologic mechanisms are not clear [15]. Clinical symptomology for patients allergic to bovine milk proteins include: rhinitis, diarrhea, vomiting, asthma, anaphylaxis, urticaria, eczema, chronic catarrh, migraine, colitis and epigastric distress. Cow milk allergy is clinically an abnormal immunological reaction to cow milk proteins, which may be due to the interaction between one or more milk proteins and one or more immune mechanisms, and resulting in immediate IgE-mediated reactions [16]. On the other side, reactions not involving the immune system are defined as cow milk protein intolerance.

Cow milk is one of the most common food allergies in children, occurring in between 0.3 and 7.5% of the infant population [17]. The clinical diagnosis of milk allergy differs widely due to the multiplicity of symptoms. Cow milk contains more than 20 proteins (allergens) that can cause allergic reactions [18]. The main proteins are casein and whey protein; casein is fractionated into α-, β- and k-casein, whey proteins include α-lactalbumin, β-lactoglobulin, bovine serum albumin (BSA) and immunoglobulin (Igs). In addition to those, several minor proteins are also present in cow milk. Most studies revealed that casein and β-lactoglobulin are the main allergens in cow milk. The type of immune response after intrusion of foreign proteins is extremely variable, depending on the animal species, the age of the host, the quality and the quantity of antigens absorbed, the location of the absorption, the pathophysiological state and the genetic background [19].

For human beings cow's milk represents the most common feeding during the infant weaning, but also the first allergen in life. The European Academy of Allergy and Clinical Immunology distinguishes allergy from intolerance [20]. Allergy is an adverse reaction to food with an involvement of the immune system; intolerance is an adverse reaction to food that does not involve the immune system, does not reply to a precise and single fault and shows different symptoms. In many countries cow's milk is the most important food allergen in babies and children [21]. Adverse reactions to cow's milk were found in 2% of babies during the first year of life: 30% of cases at the first month, 60% before the third and 96% within the twelfth

[22, 23]. Symptoms can even appear during the breast-feeding because newborn reacts against a small amount of cow milk proteins present in maternal milk [24]. Children followed for the first 3 years of life, 56% of cases had recovered from cow's milk allergy at 1-year age, 77% at 2 years and 87% at 3 years age [25] (Host and Halken, 1990). However allergy can persist for all life.

Considering the possible use of alternative milk sources for human in cases of cow's milk allergy, the use of goat's milk should be avoided because of the high risk of cross-reactivity, while mare's and donkey's milks, used in popular practice for allergic children, are valid alternative protein sources when appropriately evaluated from the hygienic point of view [26].

4. DONKEY MILK

The donkey (*Equus asinus*) is a member of the horse family and its progenitor was the small gray donkey of northern Africa (*Equus africanus*) domesticated around 4000 BC on the shores of the Mediterranean Sea. It worked together with humans for centuries; the most common role was for transport. It still remains an important work animal in the poorer Regions.

Compared with ruminant's milk, donkey milk has been studied less in the past, but in the last years research interest and capital investment in donkey milk have increased because its composition is similar to that of human milk (see Table 1).

The protein composition is significantly different from cow's milk: the total content is lower (13-28 mg/ml) and quite similar to that of human and mare milk: this condition avoids an excessive renal load of solute [28]. The main difference is the proportion of whey proteins: they are 35-50% of the nitrogen fraction while they represent only 20% in cow's milk [29]. Comparing donkey's and mares milk, the casein to whey protein ratio in mares milk is 0.2:1 immediately post-partum, and changes to 1.2:1 during the first week of lactation [30].

The donkey's three major whey proteins are α-lactalbumin, β-lactoglobulin and lysozyme. Donkey's milk α-lactalbumin has two isoforms with different isoeletric point [31]. Recently, it has been shown that α-lactalbumin presents antiviral, antitumor, and anti-stress properties. In particular in human breast milk it was shown that the α-lactalbumin forms a complex with oleic acid called HAMLET (Human Alpha-lactalbumin Made Lethal to Tumor cells) that proved to be able to induce tumour-selective apoptosis. This complex may be considered as a potential therapeutic agent against various tumour cells [32]. Furthermore it was shown that α-lactalbumin possesses anti-inflammatory activity exerted by the inhibition of cyclooxygenase-2 (COX-2) and phospholipase A2 [33].

TABLE 1.able 1.Comparison of chemical composition and physical properties of donkey and human milk [27] (copyright permission obtained).

	Donkey	Human
pH	7.0-7.2	7.0-7.5
Protein (g/100g)	1.5-1.8	0.9-1.7
Fat (g/100 g)	0.3-1.8	3.5-4.0
Lactose (g/100 g)	5.8-7.4	6.3-7.0
Ash (g/100 g)	0.3-0.5	0.2-0.3
Total Solids (g/100 g)	8.8-11.7	11.7-12.9
Caseins (g/100 g)	0.64-1.03	0.32-0.42
Whey Proteins (g/100 g)	0.49-0.80	0.68-0.83

One of the main allergens in children is β-lactoglobulin that is the major whey protein in cow milk [17] while it is absent in human milk [34]. In donkey milk the content of β-lactoglobulin is approximately 40% of the whey proteins equal to the level in mare milk and lower than that in cow milk [34]. This condition may be related to the hypoallergenic characteristic of donkey milk [1, 12, 35]. The mechanism for tolerance may be related to the specific levels of the major allergenic components in the milk. Donkey's milk has three genetic variants for β-lactoglobulin: one presents three amino acid substitutions while the others have two amino acid exchanges [29]. Donkey milk β-lactoglobulin is a monomer whereas this protein is a dimer in ruminant's milk. β-lactoglobulin is a protein of the lipocalin family and has high affinity for a wide range of compounds opening the way to various suppositions about its function. In fact it has been shown that this protein is involved in hydrophobic ligand transport and uptake, enzyme regulation, and the neonatal acquisition of passive immunity, other authors demonstrated that β-lactoglobulin forms complexes with folic acid suggesting that these complexes could be used as an effective carrier of folic acid in functional foods [36].

The high content of lysozyme may be responsible for the low bacterial concentration in donkey milk [4, 5]. Donkey milk lysozyme presents two isoforms that differ in three amino acid substitutions at position 48, 52 and 61 [29]. The concentration of lysozyme in human milk increases strongly after the second month of lactation, suggesting that this enzyme plays an important role in fighting infections in breast-fed infants during the late lactation [30].

The lactose content (7%) of donkey milk is similar to that of human milk and is much higher if compared with cows milk. The high content is responsible for the good palatability and facilitates the intestinal absorption of calcium that is essential for infant's bone mineralization. Donkey milk shows a lower fat content compared to human milk, presenting for this reason a reduced energetic value [37]. The large number of fatty acids present in the lipid fraction of milk makes it one of the most complex naturally occurring fats. Saturated fatty acids are the most represented class in donkey's milk compared to monounsaturated and polyunsaturated fatty acids, even though a wide variability can be observed in the data available in literature, most likely related to dietary and/or body condition differences [4]. Within a well balanced and integrated diet, donkey's milk is a good source of essential fatty acids; this category of fatty acids are very important in the diet of patients with Cow Milk Allergy (CMA), especially if affected by multiple food allergy. These subjects are in fact at risk of developing a deficiency in essential fatty acids and particularly in PUFA n – 3, which are absolutely necessary for adequate growth, neurological development and cardiovascular health [38]. Donkey's milk shows an high content of both linoleic (C18:2) and linolenic (C18:3) acids, respectively 9.0 g/100 g and 5.1 g/100 g of total fatty acids, when compared with ruminant species milks, in which the contents of the above mentioned Polyunsaturated Fatty Acids are always lower. To increase the total fat content in donkey milk, clinical studies [17, 35] suggested to enrich donkey milk with medium-chain triglycerides, in order to obtain a final fat content similar to human milk.

The mineral composition is very close to that of human milk except for the highest level of calcium and phosphorus but the Ca-P ratio is similar. The milk produced in the first month of lactation, when it is the only nutritional source for the foal, contained the highest levels of mineral elements that may be related to the fast growth stage of the foal. Afterwards, the mineral supply in milk decreases considerably.

Basically, donkey milk has nutritional properties that make it more similar to human milk than another mammalian one. Therefore it could be used not only as a breast milk substitute for allergic children but also as a new dietetic food for human consumption. This chapter would be a further contribution to increase the characterization of donkey milk, evaluating the nutritional qualities of donkey milk using different proteomic approaches.

5. PROTEOMIC APPROACHES

The term proteome was introduced to describe "all proteins expressed by a genome or tissue [39]. Another definition of proteome, but similar, is: "a set of all expressed proteins in a cell, tissue or organism at a certain point in time" [40]. After genomics and transcriptomics, proteomics is considered the next and more articulate step in the study of biological systems. In fact, while the genome is more or less constant, any proteins may exist in multiple forms that vary from cell to cell and from time to time. In fact the proteome analysis reveals translational, post-translational modification, regulatory and degradation processes that affect protein structure, localization, function, and turnover. Proteomics is the study of multi-protein systems and their roles as part of a larger system or network. Therefore the context of proteomics is system biology, rather than structural biology, since it is a tool used to characterize the behaviour of the system rather than the behaviour of any single component.

Milk proteins have been studied in depth for well over 50 years and lot of studies were performed in order to analyse the various milk protein components, in various milk from different mammals. However, many questions concerning milk protein expression, structure and protein modifications remain still not completely covered such as some details of protein modifications due to disease and processing. The milk proteome is extremely complex because of the presence of post-translational modifications, alternative splicing and different genetic variants. The post-translational modifications are: glycosylation, phosphorylation, disulphide bonds formation and proteolysis, they create a large number of different protein variants from a single gene product. In milk the molecular composition of proteins is very important since it influences the functional properties of milk proteins such as solubility, clotting aptitude, thermal denaturation and the nutritional properties of the milk. Usually, the analytical procedures used for structural analysis of milk proteins are based on chromatographic techniques (ionic exchange, reversed phase chromatography, size-exclusion chromatography) followed by one-dimensional electrophoretic techniques (PAGE, Urea-PAGE and SDS-PAGE), or more efficiently on bi-dimensional electrophoresis technique (2-DE).

In our work we approached the study of donkey milk protein profile by different techniques for protein separation: initially they were based on chromatographic techniques followed by sodium dodecyl sulphate polyacrylamide gel electrophoresis SDS-PAGE [5]. Successively, the milk was analyzed through two-dimensional electrophoresis (2-DE) followed by N-terminal sequencing, in order to give a more detailed panoramic view of the proteins that are present in donkey milk [41].

5.1. Chromatographic Approaches Followed By Sds-Page

Donkey's milk casein fraction was characterized by different chromatographic approaches using an Äkta Purifier HPLC system: ion-exchange chromatography, and reversed-phase. After chromatography, each protein was subjected to SDS-PAGE. The purified caseins were identified by N-terminal sequencing [5]. By cation-exchange chromatography (Mono S HR 5/5 column, GE Healthcare,1.0 ml bed volume), performed at pH 5 and 7, followed by 15% SDS-PAGE (Mini Protean III apparatus, Bio-Rad) it was possible to separate 9 peaks that were identified as β-caseins (sequence: REKEELNVSS) and α_{S1}-caseins (sequence: RPKLPHRQPE), having different molecular weight, as shown in figure 1 and table 2.

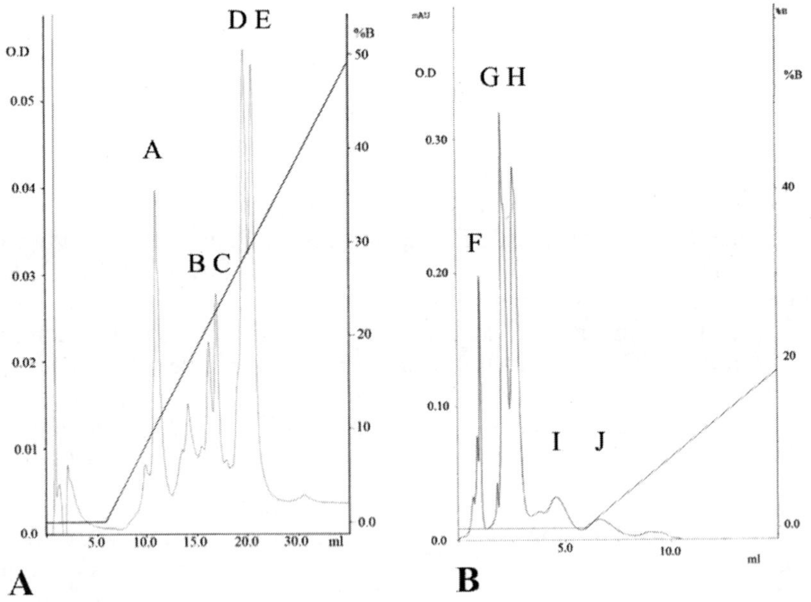

FIGURE 1. Cationic-exchange chromatography on HPLC (Mono S HR 5/5) analysis on whole casein performed at pH 5.5 (A) and at pH 7.0 (B) [5] (copyright permission obtained).

TABLE 2. (copyright permission obtained).Table 2.Donkey's milk caseins identified by cationic exchange chromatography [5].

Peak/Chromatography	Protein	kDa	N-terminal sequence
A: cationic exchange, pH 5.5	B-casein	35.40	REKEELNVS
C, D, E: cationic exchange, pH 5.5	αs_1-casein	33.00	RPKLPHRQPE
	αs_1-casein	30.70	RPKLPHRQPE
F: cationic exchange, pH 7.0	B-casein	33.30	REKEELNVS
G, H, I, J: cationic exchange, pH 7.0	αs_1-casein	31.30	RPKLPHRQPE
	αs_1-casein	29.40	RPKLPHRQPE

(copyright permission obtained).

Reversed-phase chromatography on HPLC (RP-HPLC) followed by 15% SDS-PAGE and N-terminal analysis was performed on the skimmed donkey's milk giving as a result three main peaks (K, L, M) identified as lysozyme (sequence, KVFSKXELA), α-lactalbumin, (sequence, KQFTKXELSQVLXSM), and β-lactoglobulin (sequence TNIPQTMQ), respectively (Figure 2A andtable 3). RP-HPLC was also performed on the donkey's milk casein fraction after their precipitation from skimmed milk at pH 4.6. Five peaks were recovered (N-R) each of them submitted to 13% SDS-PAGE and N-terminal analysis and the results, showed in figure 2B and in table 3, indicated mainly the presence of αS1-caseins and β-caseins. Furthermore, the β-casein sequence of peak R (REKEALNV) showed an E→A substitution in the fifth aminoacid [5].

This study revealed the presence of β-caseins (sequence: REKEELNVSS) and αs_1-caseins (sequence: RPKLPHRQPE), which presented marked homology with αs_1- and β-caseins from mare's milk [42], while the presence of other types of caseins, such as αs_2-, γ- and k- were not determined in donkey milk. This result show another high similarity between donkey and human milk: in fact, the presence of αs_2-caseins in human milk has not been demonstrated [34].

TABLE 3. able 3.Donkey's milk protein fraction identified by reversed phase chromatography in HPLC (see also figure 2 A and B). [5] (copyright permission obtained).

Peak/RP-HPLC	Protein	kDa	N-terminal sequence
K	Lysozyme	14.60	KVFSKXELA
L	α-lactalbumin	14.12	KQFTKXELSQVLXSM
M	β-lactoglobulin	22.40	TNIPQTMQ
P	αs_1-casein	33.30	RPKLPHQPE
Q	β-casein	37.50	REKEELNVS
R	β-casein	37.50	REKEALNVS

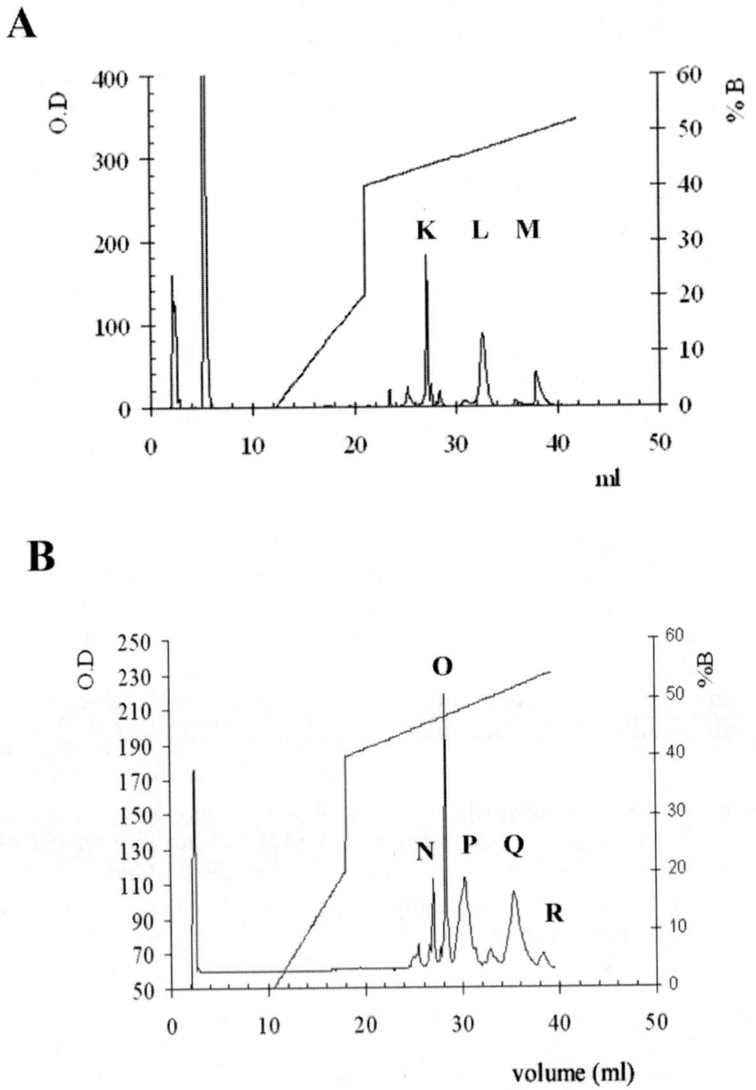

FIGURE 2. A) Reversed-phase HPLC of: (A) skimmed donkey's milk and (B) casein fraction [5] (copyright permission obtained).

5.2. Quantitative Determination of Lysozyme, Β-Lactoglobulin, Α-Lactalbumin And Lactoferrin.

Thanks to RP-HPLC analysis, it was possible also to calculate the lysozyme, β-lactoglubulin and α-lactalbumin concentrations (in mg/ml) at different stages of lactation (60, 90, 120, 160 and 190 days after parturition), the results are shown in Table 4.

TABLE 4. able 4.Quantitative determination of lysozyme, β-lactoglobulin, α-lactalbumin in different stages of lactation [5] (copyright permission obtained).

Days after parturition	Lysozyme (mg/ml)	β-lactoglobulin	α-lactalbumin
60	1.34	Not determined	0.81
90	0.94	4.13	1.97
120	1.03	3.60	1.87
160	0.82	3.69	1.74
190	0.76	3.60	1.63

The amount of lysozyme in donkey's milk varied considerably during the different stages of lactation, with a mean value of 1.0 mg/ml, and proved to be higher with respect tothat in bovine (traces), human (0.12 mg/ml) and goat'smilk (traces), whereas, it was very close to mare's milk (0.79 mg/ml) [43]. The mean β-lactoglobulin content in donkey's milk (3.75 mg/ml) was very close to that of bovine milk (3.3 mg/ml) and pony mare's milk (3.0 mg/ml), whereas in human milk the β-lactoglobulin is absent [34]. The α-lactalbumin content increased in the three months after parturition till the value of 1.8 mg/ml, close to the α-lactalbumin content in human milk (1.6 mg/ml) but lowest compared to the pony mare's α-lactalbumin content (3.3 mg/ml) [5].

Lactoferrin was purified by a cationic exchange chromatography (Mono S HR5/5 column) and its identity was confirmed by N-terminal sequencing and by western blot analysis using anti-lactoferrin antibodies (see figure 3A) [41]. The quantitative determination of donkey's milk lactoferrin (Figure 3B) gave a result of 0.080 ±0.0035 g/L, similar to that found in mare (0.1 g/L), cow (0.02-0.2 g/L), goat (0.06-0.40 g/L), and sheep milk (0.135 g/L), but lower when compared with the lactoferrin content in human milk, in which values are usually in the range 1.0-6.0 g/L [44, 45].

Lactoferrin is an iron-binding protein that displays many biological functions: regulation of iron homeostasis, cellular growth, anti-microbial and anti-viral functions, and protection against cancer development and metastasis [46]. Lactoferrin exerts its antibacterial effect by two different mechanisms involving two separate domains of the protein. In the first one, the antimicrobial effect is due to the high iron binding affinity of the protein that deprives some iron-requiring bacteria of iron and consequently inhibits their growth [46, 47]. The second antimicrobial property is due to the cationic domain at N-terminus directly responsible for the bactericidal effect [48].

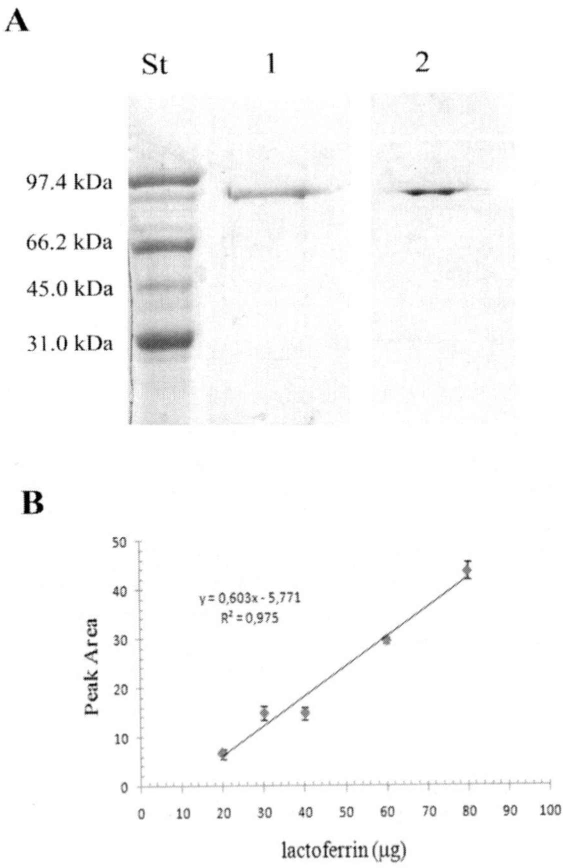

[41] (copyright permission obtained).

FIGURE 3.A) 12% SDS-PAGE (lane 1) and immunoblotting using anti-lactoferrin antibodies (lane 2) of lactoferrin purified by Mono S column. St: Bio-Rad low-molecular-weight standard (phosphorylase b, 97.4 kDa; bovine serum albumin, 66.2 kDa; ovalbumin 45.0 kDa; carbonic anhydrase, 31 kDa; soybean trypsin inhibitor, 21.5 kDa; lysozyme, 14.4 kDa. B) Calibration line of standard solutions of lactoferrin (20-100 final µg).

Lactoperoxidase is a glycoprotein consisting of a single peptide chain with a molecular weight of 78.0 kDa. This enzyme exerts its antimicrobial action through the oxidation of thiocyanate ions (SCN-) by hydrogen peroxide, both present in biological fluids and also in milk. Lactoperoxidase activity in skimmed donkey milk was evaluated by a continuous spectrophotometric rate determination using as substrate 2,2'-Azinobis (3-Ethylbenzthiazoline-6-Sulfonic Acid) [49]. In donkey milk the activity of lactoperoxidase is very low, 4.83±0.35 mU/mL. The enzyme quantification was achieved by a calibration line obtained by plotting the nanograms of peroxidase standard solutions against the enzymatic activity The mean (± SD) concentration of donkey milk lactoperoxidase was calculated to be 0.11±0.027 mg/L, close to the value obtained with human milk (0.77±0.38 mg/L) [50]. In table 5 the concentration of three proteins with antimicrobial effect are compared from donkey, human and cow milk. From these data is evinced that human and donkey milk contain considerable amounts of lysozyme and lactoferrin but lactoperoxidase is present only in small amounts.

TABLE 5.able 5.Content of lactoperoxidase, lactoferrin and lysozyme from bovine, donkey and human milk [5, 50, 51] (copyright permission obtained).

Milk	Lactoperoxidase (mg/L)	Lysozyme (g/L)	Lactoferrin (g/L)
Human	0.77	0.12	0.3-4.2
Donkey	0.11	1.0	0.080
Bovine	30-100	trace	0.10

5.3. Two-Dimensional Electrophoresis (2-De) Analysis

Whole casein was obtained from skimmed milk by adjusting the pH to 4.6 with 10% (v/v) acetic acid and centrifuging at 3000xg for 10 min in order to obtain a supernatant of whey proteins and the isoelectrically precipitated caseins. In the 2-DE analysis [41], the first dimension was an isoelectric focusing (IEF) performed using a pre-cast immobilized pH gradient gel strip Immobiline DryStrip, (IPG-strip, length 18 cm) in the pH range of 4-7 for the casein fraction and in the pH range of 3-11 for the whey proteins. The second dimension, performed using a Protean II apparatus (Bio-Rad, 180 x 200 x 1.5 mm), consisted of a 13% SDS-PAGE for the casein fraction and a 15% or a 7.5% SDS-PAGE for the whey proteins. After staining, the gel were analyzed using PD-Quest software (Version 7.1.1; Bio-Rad) in order to define spot-intensity calibration, spot detection, background abstraction, calibration, and calculation of molecular mass and pI. The spots considered interesting and quantitatively significant were subjected to N-terminal analysis for their identification. The results of donkey milk 2-DE analysis is shown in figure 4.

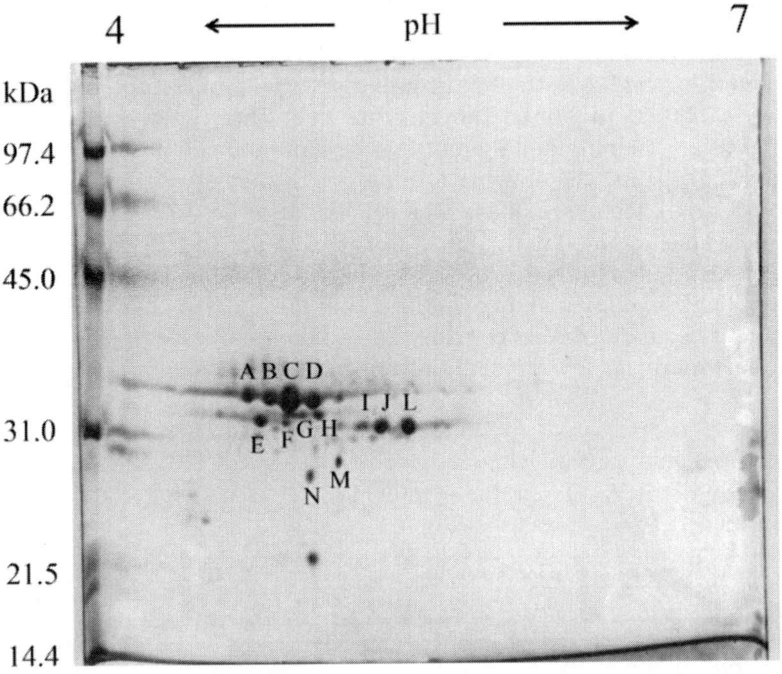

FIGURE 4. Two-dimensional electrophoresis analysis of donkey milk casein fraction. The first dimension was performed in the pH range of 4-7, the second dimension consisted of a 13%SDS-PAGE. St: Bio-Rad low molecular weight standard (phosphorylase b, 97.4 kDa; bovine serum albumin, 66.2 kDa; ovalbumin 45.0 kDa; carbonic anhydrase, 31 kDa; soybean trypsin inhibitor, 21.5 kDa; lysozyme, 14.4 kDa). [41] (copyright permission obtained).

The casein fraction (pH range 4-7, figure 4) showed the presence of 13 major protein spots: eight of them were identified as β-caseins (N-terminal sequence: RKEELNVSS) with a pI values ranging from 4.63 to 4.95. Furthermore four β-caseins spots (from spot A to D) showed molecular weights ranging from 33.10 to 33.74 kDa whereas the other four β-caseins spots (from spot E to H) displayed molecular weights ranging from 31.15 to 32.15 kDa, with a difference of about 1000 aminoacids. This results are in good agreement with [52] who demonstrated the presence of a full-length β-casein variant carrying 7, 6, 5 phosphate groups, with a pI of 4.74, 4.82,

4.91 respectively and a spliced variant (-923 aminoacids), carrying 7, 6, 5 phosphate groups with a pI of 4.64, 4.72, 4.80 respectively. On the basis of these observations, and looking at figure 4, it may be evinced that the spots B, C and D may correspond to the full-length forms of β-caseins (pI: 4.72, 4.82, 4.92 respectively) whereas the spots E, F, and G may correspond to the spliced variants of β-caseins (pI: 4.68, 4.80, 4.88). The other remaining five spots (from I to N) were identified as αs_1-caseins (N-terminal sequence: RPKLPHRPE) with a pI values ranging from 4.92 to 5.36 (see figure 4). In donkey milk an heterogeneity for the αs1-casein was found [52], assigned to either discrete phosphorylation (5, 6 and 7 phosphate/mole) or non-allelic spliced forms. In our work we found in donkey milk five αs1-caseins: three of them showed a high molecular weight (about 31. 3 kDa) and probably correspond to the full-length phosphorylated forms, whereas two αs1- caseins showed a lowest molecular weight (about 28 kDa) therefore they may correspond to the spliced variants. In our study the presence of αs2-casein and κ-casein were not demonstrated probably because of their low amount in donkey milk. Another group of authors [53] identified in donkey milk the presence of a weak spot identified as αs2-casein and three very weak spots identified as κ-casein. Therefore, the heterogeneity shown in the whole casein analysis by 2-DE may be due to a variable degree of phosphorylation and to spliced forms of αs1- and β-caseins [54-56].

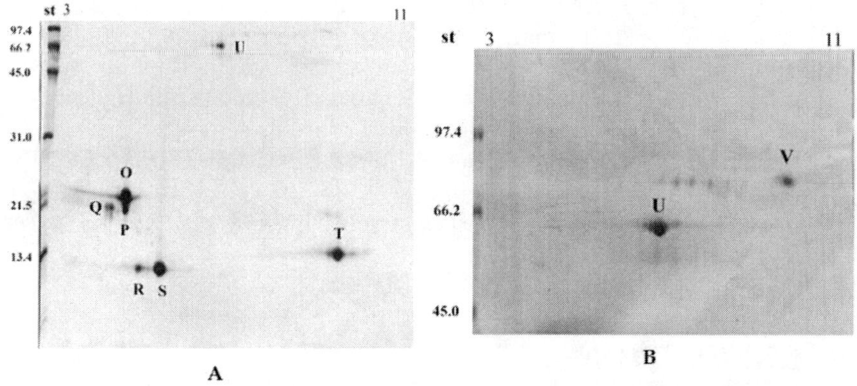

FIGURE 5. Two-dimensional electrophoresis analysis of donkey milk whey protein fraction. The first dimension was performed in the pH range of 3-11. The second dimension was carried out by: A) 15% SDS-PAGE for the identification of the low molecular weight whey proteins, B) 7.5 % SDS-Page for the identification of high molecular weight whey proteins. The standard (st) was the same as figure 4. [41] (copyright permission obtained).

The whey fraction was analyzed by 2-DE in a pH range of 3-11 for the first dimension but with two different polyacrylamide gel percentages in the second dimension in order to have a better differentiation and identification of the low- and high-molecular weight whey proteins as shown in figure 5 A and B.

The separation of low-molecular-weight whey proteins achieved by 2-DE (first dimension: IPG-strip, pH 3-11, second dimension: 15% SDSPAGE) revealed the presence of two isoforms of α-lactalbumin (Figure 5A) corresponding to the spots R and S. This result is in agreement with [57] who observed oxidized methionine forms for α-lactoalbumin (Met 90), due to in vivo oxidative stress that give rise to two α-lactalbumin isoforms. Furthermore from 2-DE, three isoforms of donkey milk β-lactoglobulin (Figure 5A), corresponding to the spots O, P, and Q, were observed. In this case from literature it is known that in donkey milk, this protein exists under two different forms, named β-lactoglobulin I, that constitutes the major form (80%), and β-lactoglobulin II [58, 59] that constitutes the minor form (20%). Successively, a genetic variant for β-lactoglobulin I (named β-lactoglobulin I B) with three amino acid substitutions explained by the degeneracy of the genetic code was found [29], and two genetic variants for β-lactoglobulin II (named β-lactoglobulin II B and C). Successively another β-lactoglobulin II variant (named D) was detected as minor component in the whey fraction of donkey milk [57]. Finally, after 2-DE experiments, only one spot corresponding to donkey milk lysozyme was detected (Figure 5A, spot T) even if from literature the presence of two donkey milk lysozyme genetic variants that arise from an oxidized methionine residue at position 79 [29, 57]. Finally, Figure 5B shows the electropherogram for the donkey milk high molecular-weight whey proteins separated by 2-DE (first dimension: IPG-strip pH 3-11, second dimension: 7.5% SDS-PAGE). By N-terminal sequencing it was possible to assign the spot U to serum albumin (kDa/ pI: 62.7/7.1) and the spot V to lactoferrin (kDa/pI: 77.0/9.8), already discussed in the section 5.2 [41].

6. CONCLUSION

Recent clinical evidence has renewed the interest in donkey milk because of high tolerability in infants with cows' milk protein allergy. To be successful as a substitute for human milk in infant nutrition, donkey milk must be capable of performing many biological functions associated with human milk. The specific protein fraction in donkey milk can be a good indicators of its potential role. In this study, donkey milk whey proteins were analyzed by 2-DE and were also quantified. From the proteomic map was revealed the presence of two isoforms of α-lactalbumin, three isoforms of β-lactoglobulin, lysozyme, albumin and lactoferrin.

The high lysozyme and α-lactalbumin content found in donkey milk may be responsible for the low bacterial count reported in literature. Lysozyme, lactoperoxidase and lactoferrin have been recognized as antimicrobial and bacteriostatic agents and could be useful to prevent intestine infections in infants. Their action may extend the conservation of fresh donkey milk and the relative potential commercial supply.

On the basis of results obtained donkey milk may be considered suitable for feeding young children affected by severe cow's milk allergy. In the past it has been widely used to replace human milk because its chemical composition and particularly protein content are close to that of human. Great attention must be obviously given to the hygienic characteristics of donkey milk production, in order to consider this milk a valid substitute of hydrolysed proteins or soy-bean derived formulae in the treatment of infants with cow's milk protein allergy.

Food is called functional if it contains one or more components that can provide a benefit to human health, beyond their traditional nutritional role. Donkey milk may be configured as functional food in early childhood and not only.

REFERENCE

1. A. Carroccio, F. Cavataio, G. Montalto, D. D'Amico, L. Alabrese, G. Iacono, 2000Intolerance to hydrolysed cow's milk proteins in infants: clinical characteristics and dietary treatment. Clin. Exp. Allergy 3015971603
2. G. Monti, E. Bertino, M. C. Muratore, A. Coscia, F. Cresi, L. Silvestro, C. Fabris, D. Fortunato, M. G. Giuffrida, A. Conti, 2007Efficacy of don key's milk in treating highly problematic cow's milk allergic children: an in vivo and in vitro study. Pediatr. Allergy Immun. 18258264
3. P. Polidori, D. Beghelli, P. Mariani, S. Vincenzetti, 2009Donkey milk production: state of the art. Ital. J. Anim. Sci. 8(Suppl. 2): 677-683.
4. E. Salimei, F. Fantuz, R. Coppola, B. Chiofalo, P. Polidori, G. Varisco, 2004Composition and characteristics of ass's milk. Anim. Res. 536778
5. S. Vincenzetti, P. Polidori, P. Mariani, N. Cammertoni, F. Fantuz, A. Vita, 2008Donkey's milk protein fractions characterization. Food Chem. 106640649
6. Pacheco MTB, Costa Antunes AE, Sgarbieri VC2008New Technological and physiological functional properties of milk proteins. In: Boscoe AB, Listow CR, editors, Protein Research Progress. 117-168, New York: Nova Science Publishers Inc. 117168
7. Ng-Kwai-Hang KF2003Milk proteins- Heterogeneity, Fractionation and Isolation. In: Roginski H, Fuquay JW, Fox PF, editors, Encyclopedia of Dairy Sciences. London: Academic Press. 18811894
8. Schanbacher FL, Talhouk RS, Murray FA, Gherman LI, Willett LB1998Milk-borne bioactive peptides. Int. Dairy J. 8393403

9. T. Tanaka, 2007Antimicrobial activity of lactoferrin and lactoperoxidase in milk. In: Ling JR, editor, Dietary Proteins Research Trends. New York: Nova Science Publishers Inc. 101115
10. L. Bjorck, 1978Antibacterial effect of the lactoperoxidase system on psychotrophic bacteria in milk. J. Dairy Res. 45109118
11. S. Vincenzetti, P. Polidori, A. Vita, 2007Nutritional characteristics of donkey's milk protein fraction. In: Ling JR, editor, Dietary Proteins Research Trends. New York: Nova Science Publishers Inc. 207225
12. L. Businco, P. G. Gianpietro, P. Lucenti, F. Lucaroni, C. Pini, G. Di Felice, P. Iacovacci, C. Curadi, M. Orlandi, 2000Allergenicity of mare's milk in children with cow's milk allergy. J. Allergy Clin. Immunol. 10510311034
13. McClenathan DT, Walker WA1982Food allergy. Cow milk and other common culprits. Postgraduate Medicine 72233239
14. Park YW, Haenlein GFW2006Therapeutic and hypoallergenic values of goat milk and implication of food allergy. In: Park YW, Haenlein GFW, editors. Handbook of milk of non-bovine mammals. Oxford: Blackwell Publishing. 121135
15. Park YW1994Hypo-allergenic and therapeutic significance of goat milk. Small Rumin. Res. 14151159
16. El-Agamy EI2007The challenge of cow milk protein allergy. Small Rumin. Res. 686472
17. A. Carroccio, F. Cavataio, G. Iacono, 1999Cross-reactivity between milk proteins of different animals. Clin. Exp. Allergy 2910141016
18. G. H. Docena, R. Ferandez, F. G. Chirdo, Fossati, 1996Identification of casein as the major allergenic and antigenic protein of cow's milk. Allergy 51412416
19. M. Heyman, J. F. Desjeux, 1992Significance of intestinal food protein transport. J. Pediatr. Gastroent. Nutr. 154857
20. Sampson HA2004Update on food allergy. J. Allergy Clin. Immunol. 113805819
21. Hill DJ, Hosking CS1996Cow milk allergy in infancy and early childhood. Clin. Exp. Allergy 26254261
22. G. Stintzing, R. Zetterstrom, 1979Cow's milk allergy, incidence and pathogenetic role of early exposure to cow's milk formula. Acta Paediatr. Scandin. 68383387
23. Bock SA1987Prospective appraisal of complaints of adverse reactions to foods in children during the first 3 years of life. Paediatrics 79683688
24. A. Host, S. Husby, O. Osterballe, 1988A prospective study of cow's milk allergy in exclusively breast-fed infants. Incidence, pathogenic role of early exposure to cow's milk formula, and characterization of bovine milk protein in human milk. Acta Paediatr. Scandin. 77663670
25. A. Host, S. Halken, 1990A prospective study of cow's milk allergy in Danish infants during the first 3 years of life. Clinical course in relation to clinical and immunological type of hypersensitivity reaction. Allergy 45587596
26. P. Restani, B. Beretta, A. Fiocchi, C. Ballabio, C. L. Galli, 2002Cross-reactivity between mammalian proteins. Ann. Allergy Asthma Immunol. 89(Suppl.): 1115
27. H. Y. Guo, K. Pang, X. Y. Zhang, L. Zhao, S. W. Chen, M. L. Dong, F. Z. Ren, 2007Composition, Physiochemical properties, nitrogen fraction distribution, and amino acid profile of donkey milk. J. Dairy Sci. 9016351643
28. M. Malacarne, F. Martuzzi, A. Summer, P. Mariani, 2002Protein and fat composition of mare's milk: Some nutritional remarks with reference to human and cow's milk. Intern. Dairy J. 12869877
29. M. Herrouin, D. Mollé, J. Fauquant, F. Ballestra, J. L. Maubois, J. Léonil, 2000New genetic variants identified in donkey's milk whey proteins. J. Protein Chem. 19105115
30. T. Uniacke-Lowe, T. Huppertz, P. F. Fox, 2010Equine milk proteins: chemistry, structure and nutritional significance. Intern. Dairy J. 20609629
31. M. G. Giuffrida, A. Cantisani, L. Napoletano, A. Conti, J. Godovac-Zimmerman, 1992The amino-acid sequence of two isoforms of α-lactalbumin from donkey (Equus asinus) milk is identical. Biol. Chem. Hoppe-Seyler 373931935

32. M. Zhang, F. Yang, F. Jr Yang, J. Chen, C. Y. Zheng, Y. Liang, 2009Cytotoxic aggregates of alpha-lactalbumin induced by unsaturated fatty acid induce apoptosis in tumor cells. Chem. Biol. Interactions 180131142
33. M. Yamaguchi, K. Yoshida, M. Uchida, 2009Novel functions of bovine milk derived alpha-lactalbumin: anti-nociceptive and anti-inflammatory activity caused by inhibiting cyclooxygenase-2 and phospholipase A2. Biol. Pharm. Bull. 32366371
34. G. Miranda, M. F. Mahé, C. Leroux, P. Martin, 2004Proteomic tools to characterize the protein fractions of Equidae milk. Proteomics 424962509
35. G. Iacono, A. Carroccio, F. Cavataio, G. Montalto, M. Soresi, V. Balsamo, 1992Use of ass's milk in multiple food allergy. J. Pediatric Gastroent. Nutr. 14177181
36. L. Liang, H. A. Tajmir-Riahi, M. Subirade, 2008Interaction of beta-lactoglobulin with resveratrol and its biological implications. Biomacromolecules 95056
37. P. Dugo, T. Kumm, Presti. M. Lo, B. Chiofalo, E. Salimei, A. Fazio, A. Cotroneo, L. Mondello, 2005Determination of tryacylglycerols in donkey milk by using high performance liquid chromatography coupled with atmospheric pressure chimica ionization mass spectrometry. J. Sep. Sci. 2810231030
38. L. Aldamiz-Echevarria, A. Bilbao, F. Andrade, J. Elorz, J. A. Prieto, J. Rodriguez-Soriano, 2008Fatty acid deficiency profile in children with food allergy managed with elimination diets. Acta Paediatr. 9715721576
39. Wilkins MR, Appel RD, Williams KL, Hochstrasser DF2007Proteome research: concepts, technology and application. Berlin: Springer-Verlag, 240 p.
40. Pennington SR, Wilkins MR, Hochstrasser DF, Dunn MJ1997Proteome analysis: from protein characterization to biological function. Trends Cell Biol. 7168173
41. S. Vincenzetti, A. Amici, S. Pucciarelli, A. Vita, D. Micozzi, F. M. Carpi, V. Polzonetti, P. Natalini, P. Polidori, 2012A Proteomic Study on Donkey Milk. Biochem. Anal. Biochem. Available: http://dx.doi.org/10.4172/21611009
42. Miclo. L. Egito, C. Lopez, A. Adam, J. M. Girardet, J. L. Gaillard, 2002Separation and characterization of mare's milk αs1-, β-, k-caseins, γ-casein-like and proteose peptone component 5-like peptides. J. Dairy Sci. 85697706
43. K. Stelwagen, 2003Milk protein. In: Roginski H, Fuquay JW, Fox PF, editors. Encyclopedia of Dairy Sciences. London: Academic Press. 18351842
44. Hennart PF, Brasseu DJ, Delogne-Desnoeck JB, Dramaix MM, Robyn CE1991Lysozyme, lactoferrin, and secretory immunoglobulin A content in breast milk: Influence of duration of lactation, nutrition status, prolactin status, and parity of mother. Amer. J. Clin. Nutrit. 533239
45. Kanyshkova TG, Buneva VN, Nevinsky GA2001Lactoferrin and its biological functions. Biochem. (Moscow) 6617
46. P. P. Ward, E. Paz, O. M. Conneely, 2005Multifunctional roles of lactoferrin: a critical overview. Cell. Molecular Life Sci. 6225402548
47. P. Valenti, G. Antonini, 2005Lactoferrin: an important host defense against microbial and viral attack. Cell. Molecular Life Sci. 6525762587
48. W. Bellamy, M. Takase, K. Yamauchi, H. Wakabayashi, K. Kawase, M. Tomita, 1992Identification of the bactericidal domain of lactoferrin. Biochem. Biophys. Acta 1121130136
49. Pruitt KM, Kamau DN.1993Indigenous Antimicrobial Agents of Milk. In: International Dairy Federation Editions. Bruxelles: IDF. 7387
50. K. Shin, H. Hayasawa, B. Lönnerdal, 2001Purification and quantification of lactoperoxidase in human milk with use of immunoadsorbent with antibodies against recombinant human lactoperoxidase. Amer. J. Clin. Nutrit. 73984989
51. Masson PL, Heremans JF1971Lactoferrin in milk from different species. Comp. Biochem. Physiol. 39119129
52. L. Chianese, M. G. Calabrese, P. Ferranti, R. Mauriello, G. Garro, C. De Simone, M. Quarto, F. Addeo, G. Cosenza, L. Ramunno, 2010Proteomic characterization of donkey milk "caseome". J. Chrom. A 121748344840

53. E. Bertino, D. Gastaldi, G. Monti, C. Baro, D. Fortunato, Garoffo. L. Perono, A. Coscia, C. Fabris, M. Mussap, A. Conti, 2010Detailed proteomic analysis on DM: insight into its hypoallergenicity. Frontieres in Biosciences E2526536
54. S. Visser, R. Jenness, R. J. Mullin, 1982Isolation and characterization of β- and γ-caseins from horse milk. Biochem. J. 203131139
55. B. Ochirkhuyag, J. M. Chobert, M. Dalgarrondo, T. Haertlè, 2000Characterization of mare caseins. Identification of αs1- and αs2-caseins. Lait 80223235
56. A. Criscione, V. Cunsolo, S. Bordonaro, A. M. Guastella, R. Saletti, A. Zuccaro, G. D'Urso, D. Marletta, 2009Donkey milk protein fraction investigated by electrophoretic methods and mass spectrometry analysis. Intern. Dairy J. 19190197
57. V. Cunsolo, R. Saletti, V. Muccilli, S. Foti, 2007Characterization of the protein profile of donkey's milk whey fraction. J. Mass Spectrom. 4211621174
58. J. Godovac-Zimmermann, A. Conti, L. James, L. Napolitano, 1988Microanalysis of the amino-acid sequence of monomeric beta-lactoglobulin I from donkey (Equus asinus) milk. The primary structure and its homology with a superfamily of hydrophobic molecule transporters. Biol. Chem. Hoppe-Seyler 369171179
59. J. Godovac-Zimmermann, A. Conti, M. Sheil, L. Napolitano, 1990Covalent structure of the minor monomeric beta-lactoglobulin II component from donkey milk. Biol. Chem. Hoppe-Seyler 371871879

CHAPTER 15

Changes in the Arabinoxylan Fraction of Wheat Grain During Alcohol Production

Ondrej Kosik [a], Stephen J. Powers [b], Afroditi Chatzifragkou [c], Parvathy Chandran Prabhakumari [c], Dimitris Charalampopoulos [c], Linde Hess [d], James Brosnan [d], Peter R. Shewry [a], Alison Lovegrove [a]

[a]*Department of Plant Biology and Crop Science, Rothamsted Research, Harpenden AL5 2JQ, Hertfordshire, UK*

[b]*Department of Computational and System Biology, Rothamsted Research, Harpenden AL5 2JQ, Hertfordshire, UK*

[c]*Department of Food and Nutritional Sciences, University of Reading, Whiteknights, PO Box 226, Reading RG6 6AP, UK*

[d] *The Scotch Whisky Research Institute, Research Avenue North, Riccarton, Edinburgh EH14 4AP, UK*

ABSTRACT

Laboratory produced DDGS samples were compared with commercial samples from a distillery and a biofuel plant. Changes in structure, solubility and content of arabinoxylan (AX) was determined. The distillation process results in a relative increase of AX content compared to the starting material. The heating and drying processes involved in the production of DDGS lead to an increased solubility and viscosity of water-extractable AX. Production of DDGS results in structural changes to the AX. There is a decrease in 2- and 3-linked arabinose oligosaccharides, that contributes to around a 50% reduction in arabinosylation in DDGS compared with the starting grains. The current study shows that laboratory-scale DDGS provide an accurate representation of the commercial scale and that the AX composition of DDGS is consistently uniform irrespective of starting material. The uniformity of DDGS and thin stillage makes them a good potential source of AX for production of prebiotics or other novel products.

Keywords: DDGS, distillers dried grains with solubles; Dietary fibre; AX, arabinoxylan; Viscosity; Solubility

1. INTRODUCTION

Distillers' Dried Grains with Solubles (DDGS) and Wet Distillers Grains (WDG) are the main co-product of distilling processes during which starch from grains (wheat, barley, maize or rye) is converted into glucose and then fermented into ethanol. Bioethanol is produced for two main purposes; traditionally as beverages or potable ethanol and more recently as a biofuel for automotive transportation (Chatzifragkou et al., 2015). The whole process is optimised to gain the highest possible yield of ethanol and the composition of DDGS varies depending on differences in the grains used, their composition, and variation in processing and production (Liu, 2011). For example, up to 85% of the thin stillage (the liquid remaining after removal of solids) is concentrated, mixed with wet solids and dried down to produce DDGS (Chatzifragkou et al., 2016). It has been reported for maize that the concentrations of components which are not fermented (such as oil, proteins, minerals and non-starch polysaccharides) are increased approximately three fold when compared to the starting feedstock and are structurally relatively unchanged (Han and Liu, 2010; Kim *et al.*, 2008 ; Liu, 2011). Similar trends have been reported for wheat and mixed cereal DDGS (Nuez Ortín & Yu, 2009).

Arabinoxylan (AX) is the major fibre present in DDGS of maize and wheat, the total content being similar despite maize DDGS containing more insoluble AX than wheat (Pedersen et al., 2015). The AX content of maize DDGS, calculated as the sum of its constituent arabinose and xylose, is around four times higher than in the starting grains; in the case of soluble AX this increase is even more significant, six times higher when compared to the starting material. There is also a significant decrease in arabinose:xylose ratio when compared to the grain AX (Pedersen, Dalsgaard, Knudsen, Yu, & Lærke, 2014), indicating structural changes in the AX. Whereas the water-extractable AX comprises between 5 and 10% of total AX in maize DDGS, in wheat DDGS the portion of water-extractable AX is about 28–36% of the total (Pedersen *et al.*, 2014 ; Widyaratne and Zijlstra, 2007).

These differences in AX solubility may relate to differences in their molecular structure and in their cross-linking with other components in the cell wall. The basic structure of AX in wheat and maize grains is similar, with 1,4-linked β-D-xylopyranoses making up the backbone and α-L-arabinofuranose decorations at the O-2 or/and O-3 positions of xylose units (Rose, Patterson, & Hamaker, 2010). However, more complex forms of AX occur in both species. In wheat, the AX present in the outer layers of the grain (pericarp and testa) is more highly substituted, including with glucuronic acid and *p*-coumaric acid, and is highly cross-linked (see below) resulting in low solubility. In maize bran, arabinoxylan is more complicated as the backbone can be additionally decorated with α-glucopyranosyl uronic acids and with branched side chains comprising arabinose, xylose, galactose residues and ferulic acid (Saulnier & Thibault, 1999). In both species, the O-3 linked arabinose units present as single substitutions on the xylose backbone can be esterified at the O-5 position with ferulate which can oxidatively cross-link to ferulate on an adjacent AX chain or other components to form diferulates (Huisman *et al.*, 2000 ; Saulnier *et al.*, 2007). Feruolylation is more abundant in the AX present in the outer layers (aleurone, pericarp, testa) of the grain than in the starchy endosperm. Diferulates (DFA) may play a significant role in cross-linking various cell wall components which generates insoluble complexes. It has been estimated that insoluble AX contains 8–39 times more DFA compared to soluble AX and the ratio of DFA to xylose is about five times higher in maize than in wheat insoluble fibre (Bunzel, Ralph, Marita, Hatfield, & Steinhart, 2001).

In addition to varying between grain tissues, the amount and structure of AX also varies between different wheat cultivars (Saulnier et al., 2007), and the proportion of water-extractable AX is also affected by environmental conditions during grain development (Shewry *et al.*, 2010 ; Shewry *et al.*, 2013).

The second major fibre component in cereal grains is mixed-linkage glucan (MLG) which comprises stretches of β-1,4-linked glycopyranoses interspersed with β-1,3-linkages (Buckeridge, Rayon, Urbanowicz, Tiné, & Carpita, 2004). In wheat starchy endosperm, MLG accounts for 20% of the total non-starch polysaccharides (Nemeth et al., 2010) whereas in maize the MLG content is four times lower. When the composition of the aleurone layer is taken into account, the content of MLG in wheat endosperm is ten times higher than in maize (Knudsen, 2014).

The European Food Safety Authority (EFSA) defines dietary fibre (DF) as non-digestible carbohydrates plus lignin (European Food Safety Authority, 2010). The main types of dietary fibre include resistant starch, cellulose, hemicelluloses (including xylans, pectins and β-glucans) and oligosaccharides (notably fructans). In general, these are carbohydrate polymers with 10 or more monomeric units that are not hydrolysed or absorbed in the small intestine but fermented in the colon to give short chain fatty acids (SCFAs) (Jones, 2014 ; Nantel, 1999). Dietary fibre has a number of health benefits (Scientific Advisory Committee on Nutrition, 2015) which are recognised by regulatory authorities in the form of accepted health claims. Furthermore, in many countries consumer intakes of fibre fall far short of recommended daily intakes, for example, approximately 18 g per day in the UK compared with a recommendation of 25–30 g (Scientific Advisory Committee on Nutrition, 2015). One strategy to exploit the health benefits of fibre in the colon is the development of prebiotics. These are types of fibre that are selectively fermented and stimulate the growth of beneficial bacteria producing SCFAs (Gibson, Probert, Van Loo, Rastall, & Roberfroid, 2004); mainly acetate, propionate and butyrate (Hopkins et al., 2003), that are considered to be responsible for a range of health benefits (Broekaert *et al.*, 2011 ; Verspreet *et al.*, 2016). We are therefore exploring the potential to use in-production DDGS fractions as a novel source of dietary fibre for human health, including the development of prebiotics, and report here the changes in AX structure which occur during DDGS production.

2. MATERIALS AND METHODS

2.1. Raw material

Four wheat cultivars, Claire, Istabraq, Viscount and Warrior grown at Rothamsted Research in two consecutive years (2012 and 2013) were used for laboratory scale production of Distillers' Dried Grains with Solubles (DDGS) by The Scotch Whisky Research Institute Edinburgh, potential spirit yield was determined (Agu, Bringhurst, & Brosnan, 2006) and wet distillers grains (WDG) and thin stillage (TS) samples provided for analysis. Commercial samples, including wheat and barley grains, WDG, TS and DDGS, were kindly supplied by a UK-based potable alcohol producer (blending 95% wheat and 5% barley) and DDGS from a leading UK biofuel producer (unknown blending). Samples were freeze-dried to determine dry matter content and dried WDG and DDGS samples were ball-milled prior to analyses.

2.2. Alcohol yield analysis

The method used (Agu et al., 2006) simulates conditions of the production process in a 'typical' Scotch whisky grain distillery in respect of temperature, mashing and fermentation. Thirty g of cereal flour obtained by milling the grains in a Bühler Miag disc mill (0.2 mm setting) was transferred into a stainless steel mashing beaker and slurried with 81 ml of water and 25 µL of Termamyl 120L (industrial bacterial α-amylase; Novozymes) added. The contents of the beaker were heated to 85 °C, (2 °C/min temperature rise) in a water bath before being pressure cooked at 142 °C in an autoclave for 15 min. The cooked slurry was cooled down to 85 °C and given a second Termamyl treatment for 30 min to prevent starch retrogradation. The mash was then transferred to a 65 °C water bath and mashed for 1 h with a calculated amount of high enzyme grain distilling malt grist (Miag setting 0.2 mm), an equivalent to a malt inclusion ratio of 20% (dry weight basis) to 80% wheat. After cooling to room temperature, the mash was transferred to a fermentation vessel, pitched with distiller's yeast ('M' type; Kerry Bioscience Ltd.) at a pitching rate of 0.4% (w/w) pressed yeast and adjusted to 250 g with water. Finally, the mash was fermented at 30 °C for 68 h and distilled to collect alcohol (Bringhurst & Brosnan, 2014). Alcohol yield or potential spirit yield (PSY) was determined from alcohol strength of the distillate that was measured using a Paar 500 density meter and is quoted as litres of alcohol per tonne (LA $tonne^{-1}$) on a dry weight basis.

2.3. Determination of arabinoxylan content and solubility

A colorimetric method based on Finnie, Bettge, and Morris (2006) measuring pentose sugars content was used to determine total AX (TOT-AX) and water-extractable AX (WE-AX). Briefly, 125 mg of sample were suspended in 25 mL of water, vortexed and 1 ml aliquots, in duplicate, were transferred to new Pyrex tubes (fraction 1, total AX). The remainder of the suspended sample was mixed on a Spiramix tube roller for 30 min and then centrifuged at 2500×*g* for 10 min. One mL aliquots of the supernatant, in duplicate, were transferred to new Pyrex tubes (fraction 2, water-extractable AX). All fractions were diluted to 2 ml and 10 ml of freshly prepared extraction solution (93.2% (v/v) acetic acid (Sigma), 1.69% (v/v) hydrochloric acid (Fisher Scientific), 0.85% (w/v) phloroglucinol (Sigma) and 0.017% (w/v) glucose (Sigma) was added to each sample, according to the method described by Douglas (1981). Samples were boiled for 25 min and vortexed frequently. After rapid cooling the absorbance of the resulting solution was measured. The pentosan content was determined by comparing the differences in absorbance measured at 552 and 510 nm (Jenway, Bibby Scientific). Values were estimated based upon a calibration curve generated using known amounts of xylose (Sigma) standard. Water un-extractable AX (WU-AX) was calculated as the difference between TOT-AX and WE-AX and AX solubility was calculated as the ratio of WE-AX to TOT-AX.

2.4. Enzymatic fingerprinting of arabinoxylan and β-glucan

The protocol to analyse the AX and β-glucan structure was adapted from (Saulnier et al., 2009) using highly specific recombinant glycosylhydrolases. 100 mg of sample was suspended in 1 mL of 80% (v/v) ethanol and heated at 95 °C for 10 min to inactivate endogenous enzymes. After centrifugation at 13,400×*g* for 2 min the step was repeated with 80% (v/v) and 95% (v/v) ethanol, discarding the supernatant after each wash. The washed pellet was dried *in vacuo*. Two μL of xylanase 11 (≈11 μg, NpXyn11A, Prozomix UK) and 1 μL of lichenase (≈0.35 U, CtGH26, Prozomix UK) were added to the dried sample and made up to 1 mL with water. Samples were incubated at 40 °C in a Thermomixer (Eppendorf) with continuous shaking (750 rpm) for 16 h. Samples were then centrifuged at 13,400×*g* for 5 min, the supernatants boiled for 30 min to inactivate hydrolases and filtered through 0.45 μm PVDF filter (Whatman). Finally samples were diluted 1:20 with water prior to separation by HPAEC-PAD (Dionex ICS-3000, Thermo Scientific) equipped with a CarboPac PA1 analytical column (3 × 150 mm, Thermo Scientific) and guard (3 × 30 mm) column using the method of (Ordaz-Ortiz, Devaux, & Saulnier, 2005) modified by (Nemeth et al., 2010). Chromeleon 7 software (Thermo Scientific) was used to interpret the chromatograms. The percentage of each oligosaccharide originating from the fingerprinted AX (for nomenclature used see Fauré et al., 2009), the percentages of xylose, 3- and 2-linked arabinose, as well as total arabinosylation and the arabinose:xylose ratio were calculated. The% arabinosylation was calculated using the full fingerprint data (Table S1). The peak areas of those oligosaccharides with arabinose additions was divided by the total peak area for all of AXOS peaks generated by the enzyme digests and quoted as percent. The G3:G4 ratio (Toole et al., 2010) reflecting the changes in β-1,3- and β-1,4-linkages in mixed-linkage glucan, was also calculated.

2.5. Dynamic viscosity

The dynamic viscosity of aqueous extracts prepared according to the protocol of (Saulnier, Peneau, & Thibault, 1995) were measured using a Micro-Ostwald capillary viscometer (AVS370, SI Analytics Germany) at 30 °C (Freeman et al., 2016). Briefly, 1 g of sample was suspended in 4 ml of water, vortexed and mixed on a Spiramix tube roller for 15 min. After centrifugation at 5000×*g* for 10 min the supernatant was collected and boiled for 10 min. Precipitated proteins were removed by centrifugation at 5000×*g* for 10 min. The supernatant was centrifuged again (10,000*g* for 10 min) filtered through a 0.45 μm PVDF filter and the dynamic viscosity measured.

2.6. Statistical analyses

The method of principal coordinate analysis (PCoA) was used to give an overall low-dimensional description of the data. This method analyses the matrix of similarities between *n* samples calculated using a simple Euclidean distance measure in multivariate space between the samples. A number of principal coordinates (PCos) are retained that account for the majority of the variance in the distances, thus allowing visualisation of the differences between samples in 2-d plots (Krzanowski, 2000). Afterwards, relating the retained principal coordinates (PCos) to the original variables via correlation analysis enabled us to see the variables most correlated with the separation in each dimension. This was done by obtaining the F-statistics for each correlation.

Correlations between response variables were also calculated, and tested using the F-test. All analyses were done using the GenStat (17th edition, VSN International Ltd, Hemel Hempstead, UK) statistical package. SIMCA 14 (MKS Data Analytics Solutions, Sweden) was used to generate the plots.

3. RESULTS AND DISCUSSION

3.1. Potential spirit yield and variation of TOT-AX in grains

Four wheat cultivars grown at Rothamsted Research in 2012 and 2013 and were used to determine changes in AX amount and composition during alcohol production. Three of these cultivars are or were commonly used for distilling (Claire, Istabraq and Viscount) and were expected to give high yields of alcohol. The fourth cultivar, Warrior, is known to give low alcohol yields and is therefore not usually used for distilling. Commercial samples, including wheat and barley grains, WDG, TS and DDGS from a UK-based potable alcohol producer and DDGS from a leading UK biofuel producer, all from 2012, were also analysed. Samples of the grains from the commercial biofuel plant were not available for comparison.

The results are summarised in Table 1, including total (TOT-AX), water-extractable (WE-) AX and AX solubility, calculated as the percentage of water-extractable AX. The dynamic viscosities of aqueous extracts were also measured and the potential spirit yields were determined. The mean total AX content of these samples was 1.5-fold higher than in the grain used in the commercial distillery (which comprised 95% wheat and 5% barley grains). As expected, Warrior gave the lowest spirit yield in 2012, but this was slightly higher than for Claire in 2013 (both cultivars being substantially below Istabraq and Viscount)

Table 1. Analysis of grain fractions from laboratory scale alcohol production (wheat cultivars Claire, Istabraq, Viscount and Warrior grown in 2012 and 2013) and a commercial distillery and biofuel plant.

Fraction	Cultivar	Year	Total-AX [mg g^{-1} DW]	Water-extractable AX [mg g^{-1} DW]	AX solubility [%]	Dynamic viscosity [mPa s^{-1}]	Potential spirit yield [LA tonne^{-1}]
Wholemeal (WM)	Claire	2012	32.27	6.02	18.7	1.256	444
		2013	32.45	6.44	19.8	1.420	417
	Istabraq	2012	36.87	6.33	17.2	1.291	437
		2013	25.66	5.66	22.1	1.378	436
	Viscount	2012	30.09	6.42	21.4	1.341	449
		2013	25.15	5.89	23.4	1.306	449
	Warrior	2012	28.71	6.35	22.1	1.598	419
		2013	26.62	6.96	26.1	1.535	420
	Distillery	2012	19.19	4.37	22.8	1.401	–
Thin stillage (TS)	Claire	2012	64.10	64.10	100	1.290	–
		2013	63.15	63.15	100	1.380	–
	Istabraq	2012	62.85	62.85	100	1.310	–
		2013	57.13	57.13	100	1.330	–
	Viscount	2012	67.24	67.24	100	1.290	–
		2013	48.96	48.96	100	1.340	–
	Warrior	2012	72.38	72.38	100	1.330	–
		2013	42.31	42.31	100	1.420	–
	Distillery	2012	58.61	58.61	100	1.323	–
Wet distillers grains (WDG)	Claire	2012	55.16	15.25	27.6	0.853	–
		2013	100.38	15.35	15.3	0.986	–
	Istabraq	2012	61.64	13.26	21.5	0.854	–
		2013	98.44	9.36	9.5	0.894	–
	Viscount	2012	68.82	22.35	32.5	0.856	–
		2013	93.45	20.45	21.9	1.037	–
	Warrior	2012	75.77	24.24	32.0	0.857	–
		2013	110.14	12.18	11.1	0.868	–
	Distillery	2012	75.88	7.82	10.3	0.861	–
Distillers' dried grains with solubles (DDGS)	Claire	2012	91.43	20.57	22.5	1.151	–
		2013	68.67	21.10	30.7	1.195	–
	Istabraq	2012	66.02	20.35	30.8	1.200	–
		2013	71.52	19.40	27.1	1.134	–
	Viscount	2012	72.82	24.11	33.1	1.343	–
		2013	64.73	21.13	32.6	1.207	–
	Warrior	2012	62.67	22.80	36.4	1.289	–
		2013	75.24	21.80	29.0	1.205	–
	Distillery	2012	85.78	48.07	56.0	2.316	–
	Biofuel	2012	90.22	34.50	38.2	2.304	–

The wholemeal samples are the raw material which for the distillery comprised 95% wheat and 5% barley (cultivars not known). Only DDGS was available from the biofuel plant. Arabinoxylan (AX) solubility is expressed as a percentage of WE-AX. The potential alcohol yield from the laboratory scale samples is also given. It is notable that the alcohol yield for Warrior was lower than for the other cultivars grown 2012, but similar to that of Claire (both being low) in the samples grown in 2013.

The wholemeal samples are the raw material which for the distillery comprised 95% wheat and 5% barley (cultivars not known). Only DDGS was available from the biofuel plant. Arabinoxylan (AX) solubility is expressed as a percentage of WE-AX. The potential alcohol yield from the laboratory scale samples is also given. It is notable that the alcohol yield for Warrior was lower than for the other cultivars grown 2012, but similar to that of Claire (both being low) in the samples grown in 2013.

3.2. Compositional analysis of fractions from alcohol production

The production of alcohol from wheat grain (Fig. 1) results in two "waste streams", corresponding to soluble components (thin stillage, TS) and the insoluble grain residue (wet distillers' grains, WDG). In commercial distilleries and biofuel plants a proportion of the TS is sprayed onto the WDG and dried to give distillers' dried grains with solubles (DDGS). For the laboratory scale preparations individual fractions were retained and the WDG alone dried to give DDGS, as it was not known what proportion of TS is sprayed onto WDG in the commercial distillery and biofuel plant. All fractions from the laboratory scale and commercial preparations were analysed for TOT-AX and WE-AX and the dynamic viscosity of aqueous extracts measured and AX solubility calculated. The results are included in Table 1.

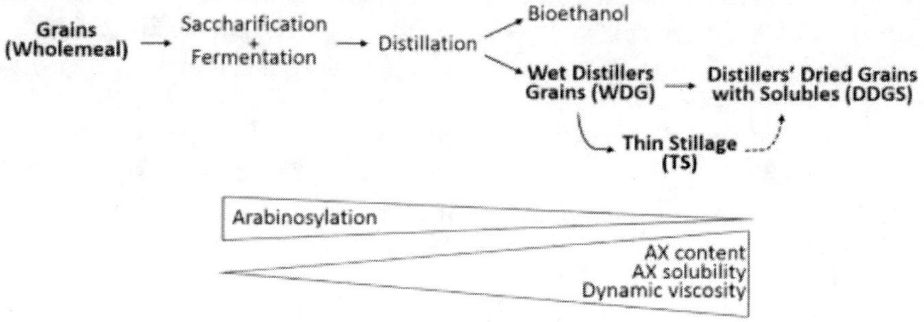

Fig. 1. Flowchart showing the production of bioethanol and DDGS and the effects on arabinoxylan (AX) content and structure.

3.3. Increase in AX content and AX solubility

All three processed fractions (TS, WDG, DDGS) are richer in TOT-AX than the starting material which is to be expected as starch accounts for about 70% of the mature dry grain and would be removed by hydrolysis and fermentation. The increase in total AX in the DDGS from the commercial distillery was 4.5-fold, (Table 1) compared with the raw material (95% wheat, 5% barley). A similar increase was previously reported for commercial maize DDGS (3.9-fold; Pedersen et al., 2014). The increase in TOT-AX was less for the four laboratory samples, about 1.8 to 2.8-fold.

The DDGS fractions also had somewhat higher contents of WE-AX than the starting material. There was an 11-fold increase in water-extractable AX (WE-AX) for the commercial distilling samples and the WE-AX content of the DDGS from the biofuel production was also high (although in this case a comparison with the starting grains could not be made). The DDGS produced in the laboratory had between 3.2 and 3.7-fold more WE-AX than the starting grain. A 2.7-fold increase in WE-AX content was previously reported for wheat DDGS (Widyaratne & Zijlstra, 2007) from a Canadian ethanol plant.

The differences in the degree of enrichment of TOT-AX and WE-AX between the commercial and laboratory scale DDGS samples probably results from the fact that the TS was sprayed back onto the DDGS in the commercial samples but not those prepared in the laboratory (Chatzifragkou et al., 2015).

When expressed as a percentage of TOT-AX, the proportion of WE-AX increased from 22.8% in flour to 56% in the distillery DDGS and was 38.2% in the biofuel sample (where no starting material was available). For the laboratory scale samples, WE-AX increased on average (over cultivars) from 19.9% for wholemeal flour to 30.7% for DDGS in 2012 and from 22.8% to 29.9% in 2013.

3.4. Increase in dynamic viscosity

The values for dynamic viscosity of extracts of the laboratory scale and distillery samples of WM, TS and WDG were very similar, as were those of the DDGS samples from the distillery and biofuel plants (summarised in Table 1). The differences between the dynamic viscosities of extracts of the commercial and laboratory scale samples of DDGS probably relate to the absence of the TS from the latter.

3.5. Changes in AX and glucan structure

The structures of AX and MLG in the samples were determined by enzymatic fingerprinting and HPAEC-PAD. The oligosaccharides released by enzymatic digest had previously been identified (Anders *et al.*, 2012 ; Ordaz-Ortiz *et al.*, 2005) allowing the proportions of xylose and 3- and 2-linked arabinose, the total arabinosylation, the arabinose:xylose ratio, the total AX peak area and the proportions of the individual fragments to be compared. The total MLG peak area (the sum of G3 and G4 peak areas), the G3:G4 ratio and the total AX and glucan peak areas were also calculated. The data discussed below are summarised in Table 2 while complete datasets are provided in Supplementary Table S1.

Arabinosylation decreased during processing, from means of 12.1% (2012) and 13.7% (2013) over cultivars for the raw material to 6.4% (2012) and 6.8 (2013) for the WDG and to 6.2% (2012) and 6.8% (2013) for DDGS. Similar values of percentage arabinosylation were obtained for the commercial distillery, biofuel samples and laboratory-scale samples, indicating that the laboratory-scale samples replicate the commercial process. The TS samples had similar% arabinosylation to the starting material. Although this may indicate that they have similar structures, all of the AX in the TS was water-extractable, whereas only between 17% and 26% of the total AX is water-extractable in the raw material (Table 1). This demonstrates that the solubility of AX in wheat is not determined by the

Table 2. Analysis of the structures of arabinoxylan (AX) and mixed-linkage glucan (MLG) by enzyme fingerprinting. For nomenclature see (Fauré et al., 2009). Arabinosylation data is also included, along with AX + MLG and MLG peak area from HPAEC-PAD. Only data which are discussed in the text are included, with full datasets being provided as Supplementary data (Table S1).

Fraction	Cultivar	Year/grain	Xyl + Xyl_2 [%]	XA^3XX [%]	$XA^{2+3}XX$ [%]	XA^3XA^3XX [%]	$XA^3A^{2+3}XX$ [%]	Arabinosylation [%]	AX + MLGPeak area	MLGPeak area
Wholemeal (WM)	Claire	2012	49.5	21.2	13.7	1.6	2.9	13.6	169.96	37.28
		2013	48.6	16.6	15.9	2.3	3.8	14.5	111.08	27.13
	Istabraq	2012	63.0	17.1	9.5	1.2	2.0	9.8	222.25	52.07
		2013	53.1	17.9	13.1	1.9	2.9	12.9	108.09	33.12
	Viscount	2012	59.3	16.7	11.7	1.5	2.5	11.0	164.17	34.95
		2013	54.2	15.5	14.2	1.6	3.6	12.9	131.46	34.88
	Warrior	2012	48.1	22.3	12.3	1.8	2.7	13.8	119.62	28.54
		2013	48.5	15.9	14.9	2.1	3.8	14.6	120.92	35.47
	Distillery	Wheat	56.2	18.1	12.5	1.4	2.5	11.9	137.55	22.64
		Barley	63.4	18.3	8.6	2.6	2.6	9.5	111.16	21.70
Thin stillage (TS)	Claire	2012	54.9	16.3	14.3	2.6	2.6	12.4	419.36	45.27
		2013	38.5	18.2	21.5	2.7	5.7	17.7	429.90	116.15
	Istabraq	2012	59.9	15.5	9.7	2.6	2.9	10.9	359.20	41.97
		2013	45.2	19.9	17.4	2.7	4.0	15.3	513.75	188.57
	Viscount	2012	61.3	13.6	11.1	2.1	3.0	10.7	360.88	110.90
		2013	43.9	17.5	18.6	2.5	5.3	16.0	437.84	169.48
	Warrior	2012	62.5	15.0	9.5	1.9	2.3	10.1	368.32	94.61
		2013	46.5	18.3	16.6	2.4	4.6	15.1	329.17	124.81
	Distillery	2012	50.2	16.1	14.9	1.8	3.6	13.9	243.62	49.55
Wet distillers grains (WDG)	Claire	2012	73.3	15.2	5.7	0.8	1.3	6.7	335.01	56.70
		2013	72.7	18.3	5.0	0.5	1.1	6.6	393.98	69.49
	Istabraq	2012	73.2	16.5	4.9	0.8	1.1	6.6	291.36	58.11
		2013	71.2	20.4	4.6	0.5	1.1	6.8	283.83	60.00
	Viscount	2012	75.7	14.0	5.3	0.7	1.1	6.1	280.52	49.15
		2013	71.9	16.2	6.4	0.6	1.6	7.1	294.22	66.22
	Warrior	2012	75.0	14.1	5.1	0.8	1.1	6.3	319.73	54.96
		2013	72.5	18.5	4.8	0.5	1.2	6.6	290.04	58.98
	Distillery	2012	72.5	16.7	5.3	0.7	1.2	6.8	381.93	66.62
Distillers' dried grains with solubles (DDGS)	Claire	2012	73.8	15.5	5.4	0.7	1.1	6.6	312.26	52.26
		2013	72.6	16.9	6.0	0.6	1.2	6.8	428.53	89.93
	Istabraq	2012	76.1	16.4	4.6	0.7	0.9	6.0	295.94	57.69
		2013	72.7	16.9	5.7	0.7	1.2	6.8	330.11	74.70
	Viscount	2012	75.5	14.3	5.3	0.7	1.1	6.2	256.24	46.63
		2013	73.2	14.9	6.5	0.7	1.5	6.9	347.48	74.57
	Warrior	2012	75.4	14.5	4.9	0.6	1.0	6.1	284.28	50.25
		2013	72.8	16.2	5.9	0.7	1.3	6.8	373.90	77.96
	Distillery	2012	67.1	15.2	8.1	1.1	2.0	8.7	349.59	48.51
	Biofuel	2012	72.1	14.2	6.2	1.0	1.6	7.2	362.22	35.30

degree of modification with arabinose. By contrast, the percentage of arabinosylation of the WDG and DDGS fractions was lower by about half, than that of the starting material or TS, indicting a change in structure. The enzymatic fingerprinting also demonstrated that these fractions had increased proportions of xylose and xylobiose (X_2) fragments, which indicates that arabinose is lost during processing. This is consistent with the digestion pattern of oligosaccharides observed using the recombinant xylanase enzyme from the GH11 family, which hydrolyses only at unsubstituted xylan residues (Kolenová, Vršanská, & Biely, 2006; Biely, Vrsanska, Tenkanan, & Kluepfel, 1997).

3.6. Multivariate analyses

In order to visualise the similarities or differences between the samples and to identify which underlying variables may be responsible for the separation of them, Principal Coordinates Analysis (PCoA) was applied to the compositional dataset in Table 1, the fingerprinting data in Table 2 and the combined datasets. Although these three analyses showed broadly similar separations, the degree of discrimination is greater when the full dataset was used. This is shown in Fig. 2 with the analyses of the data in Table 1 ; Table 2 being shown in Supplementary Fig. S1. The PCoA plot in Fig. 2 shows wholemeal flour (WM, coloured black) and thin stillage (TS, coloured red) being separated from wet distillers grains (WDG, coloured green) and distillers' dried grains with solubles (DDGS, coloured blue) on the PCo1 axis, and TS and WDG being separated from WM and DDGS on the PCo2 axis with the exception of Claire 2013 DDGS. The grouping of the fractions from commercial and laboratory scale production together confirms that the laboratory scale system is a good model for commercial distilling and bioethanol production.

The F-statistics for correlations of variables with the PCo1 and PCo2 values from Fig. 2 were calculated in order to identify the most discriminatory variables. In the PCo1 direction the most discriminatory, top five variables are the percentage of 2-linked arabinose (F-PCo1 = 717.7), followed by xylobiose oligosaccharide (685.6) and $XA^{2+3}XX$ oligosaccharide as a percentage of total fingerprinted AX peak area (680.1), the percentage of unsubstituted xylose and percentage of arabinosylation (both 655.4) and finally the arabinose:xylose ratio (644.0). These variables are probably responsible for the separation of the WM and TS from the WDG. In the PCo2 direction the most discriminatory variables were the 'AX + MLG total peak area' (F-PCo2 = 136.0) and 'AX total peak area' (117.2). These variables are responsible for discriminating the TS and WDG from the WM and DDGS fractions (except for Claire 2013 DDGS), and also for the separation of the distillery TS from other TS fractions.

The PCoA also shows differences between the laboratory samples from the two growth years, particularly in the compositions of the TS (coloured red) and WM (coloured black) samples (Fig. 2). The distillery WM and TS samples also differ from the laboratory scale samples whereas the WDG sample clusters well with the laboratory scale samples from both years. The separation of the distillery DDGS from the other DDGS samples can be explained by slightly higher arabinose content that probably results from the TS (which contains WE-AX) being back sprayed onto the DDGS during processing. The separation of the 2013 Claire sample of DDGS from the other DDGS samples, on PCo2, is primarily due to the higher MLG content, shown as a greater MLG peak area in Table 2.

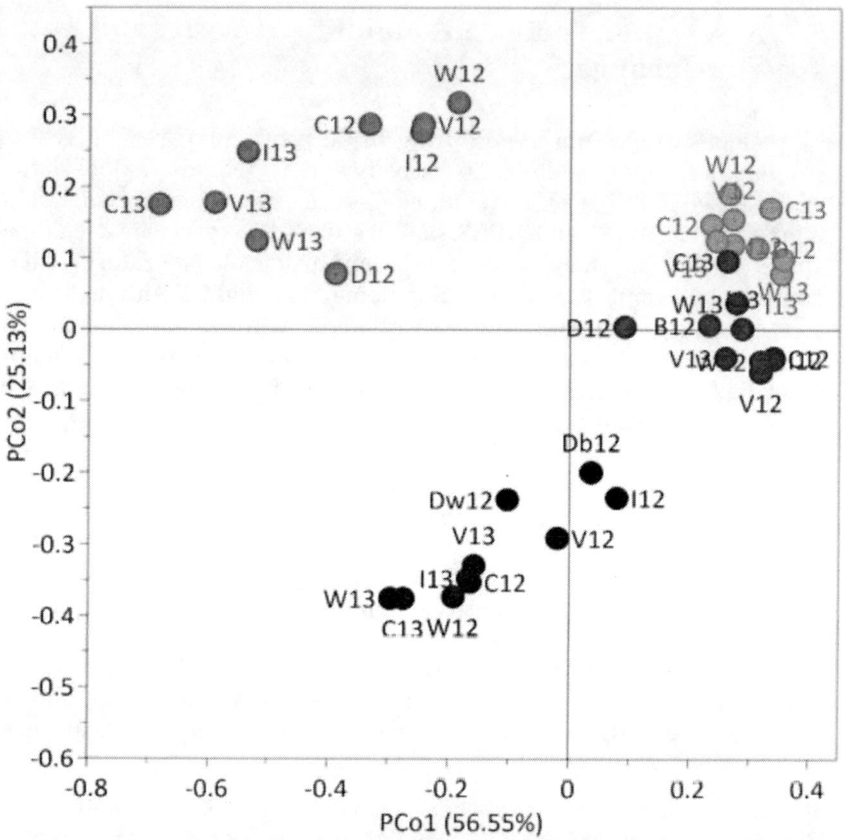

Fig. 2. Principal coordinates analysis (PCoA) plot for analyses of compositional data summarised in Table 1 ; Table 2 using the following coding of samples for cultivars: Claire (C), Istabraq (I), Viscount (V), Warrior (W), commercial distillery wheat (Dw) and barley (Db), and biofuel (B); followed by year: 2012 (12) and 2013 (13) and coloured by fraction type: wholemeal (WM, coloured black), thin stillage (TS, coloured red), wet distillers grains (WDG, coloured green) and distillers' dried grains with solubles (DDGS, coloured blue). The percentage of the variation in the distances between the samples accounted for by each PCo is shown in brackets. The first two PCos accounted for 81.68% of the variation in the distances between the samples and so only these two are retained for visualisation of the samples. (For interpretation of the references to color in this figure legend, the reader is referred to the web version of this article.)

3.7. Associations between solubility, viscosity and AX structure changes

The major changes observed during DDGS production were increased AX content and solubility, dynamic viscosity and decreased arabinosylation, as summarised in Fig. 1. These relationships are explored in more detail in Fig. 3 for the WM, WDG and DDGS samples from the four wheat cultivars; a shade plot of the matrix of correlations using all laboratory scale and commercial samples is given in Supplementary Fig. S2. Although dynamic viscosity was significantly correlated with the percentage of arabinosylation calculated from the enzyme mapping, ($r = 0.722, p < 0.001, n = 24$) overall, there were no significant ($p < 0.05$) correlations within the three types of grain sample: unprocessed wholemeal, WDG and DDGS (Fig. 3A). Whereas the wholegrain samples varied in both percentage arabinosylation and dynamic viscosity these were not related, while the WDG and DDGS samples each formed a tight group. These samples had similar low levels of arabinosylation but the DDGS samples had higher dynamic viscosity. A similar pattern was observed when dynamic viscosity was plotted against the percentage of the XA^3A^3XX arabinoxylan oligosaccharide (AXOS) released by endoxylanase digestion ($r = 0.730, p < 0.001, n = 24$) (Fig. 3C). This AXOS made a relatively minor contribution to the total AXOS released from wholegrain (between 1.6 and 2.3%) but showed the greatest proportional decrease during processing, to 0.6–0.7% in the DDGS.

Broadly similar plots were observed when WE-AX was plotted against dynamic viscosity ($r = 0.131, p = 0.541, n = 24$) (Fig. 3B) and percentage arabinosylation ($r = -0.292, p = 0.167, n = 24$) (Fig. 3D). Although the wholegrain samples ranged widely in both dynamic viscosity and percentage arabinosylation these were not correlated with WE-AX. The WDG samples varied widely in WE-AX but not in dynamic viscosity or percentage arabinosylation, while the DDGS samples formed tighter groups with WE-AX showing a linear correlation with dynamic viscosity. The latter was the only strong correlation ($r = 0.763, p = 0.028, n = 8$) observed between the parameters plotted for the three types of sample.

These results indicate that the degree of arabinosylation is not a major factor determining the extent of AX solubility and dynamic viscosity.

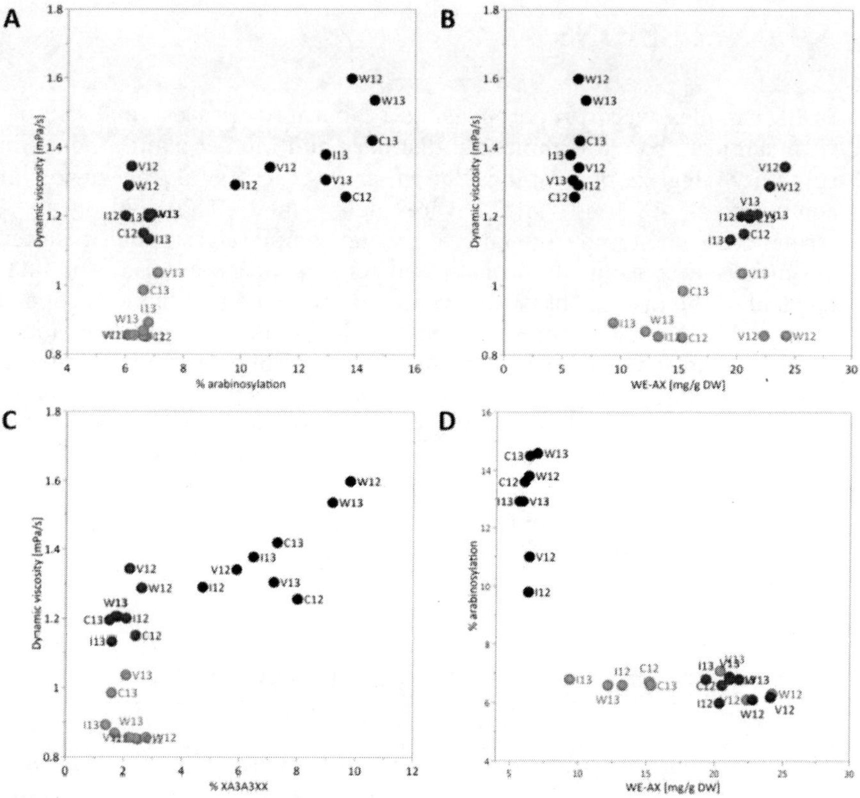

Fig. 3. Relationships between the AX structure, AX solubility of and dynamic viscosity of laboratory scale samples of wholemeal (WM, coloured black), wet distillers grains (WDG, coloured green) and DDGS (D, coloured blue). (A) dynamic viscosity *vs.* percentage arabinosylation; (B) dynamic viscosity *vs.* amount of WE-AX; (C) dynamic viscosity *vs.* percentage of XA3A3XX; (D) percentage of arabinosylation *vs.* amount of WE-AX. (For interpretation of the references to color in this figure legend, the reader is referred to the web version of this article.)

However, they also demonstrate two features which are of relevance to the utilisation of DDGS for high value products. These are that the AX present in DDGS is more uniform in composition and properties than that in WDG or the starting material, and that there is no clear relationship between the compositions of the starting material and DDGS. Hence, the selection of starting material is of little importance when producing DDGS for exploitation compared to the selection of the processing parameters.

4. CONCLUSIONS

DDGS samples were prepared using a laboratory process and compared with samples from a commercial distillery and biofuel plant. The amount, solubility, degree of arabinosylation and structure of the major fibre components, AX and MLG, were determined. The hydrolysis and fermentation of starch from laboratory and commercial samples resulted in an increased concentration of AX, but its structure was also modified. In particular, the process of DDGS production resulted in a higher proportion of soluble AX, contributing to increased dynamic viscosity of aqueous extracts than was found in the starting material. This study shows that laboratory scale production of DDGS can be used to predict the behaviour of grain in commercial distillery and biofuel processes. Furthermore, it also demonstrates that DDGS samples are more uniform in AX composition and properties than wholemeal or WDG samples, with little or no effect of variation in composition of the wholemeal on the final co-product, and that thin stillage and DDGS are good potential sources of AX for exploitation to produce novel food ingredients and prebiotics.

CONFLICT OF INTEREST STATEMENT

The authors of this paper have no affiliations with or involvement in any organization or entity with any financial or non-financial interest in the subject matter or materials discussed in this manuscript.

ACKNOWLEDGEMENTS

The work described here was supported by a BBSRC Integrated Biorefining Research & Technology club (IBTI) grant "Development of a process scheme for the production of high value functional products from DDGS" (BB/J019429/1 – University of Reading; BB/ J019380/1 – Rothamsted Research). Research at Rothamsted is funded by the Biotechnology and Biological Sciences Research Council (BBSRC) of the UK.

REFERENCES

1. Agu, R., Bringhurst, T., & Brosnan, J. (2006). Production of grain whisky and ethanol from wheat, maize and other cereals. Journal of The Institute of Brewing Institute of Brewing, 112(4), 314–323.
2. Anders, N., Wilkinson, M. D., Lovegrove, A., Freeman, J., Tryfona, T., Pellny, T. K., ... Mitchell, R. A. C. (2012). Glycosyl transferases in family 61 mediate arabinofuranosyl transfer onto xylan in grasses. Proceedings of the National Academy of Sciences of the United States of America, 109(3), 989–993.
3. Biely, P., Vrsanska, M., Tenkanan, M., & Kluepfel, D. (1997). Endo-P – 1,4-xylanase families: differences in catalytic properties. Journal of Biotechnology, 57, 151–166.
4. Bringhurst, T. A., & Brosnan, J. (2014). Chapter 6 – Scotch whisky: Raw material selection and processing. In: Whisky (pp. 49–122).
5. Broekaert, W. F., Courtin, C. M., Verbeke, K., Van de Wiele, T., Verstraete, W., & Delcour, J. A. (2011). Prebiotic and other health-related effects of cereal-derived arabinoxylans, arabinoxylan-oligosaccharides, and xylooligosaccharides. Critical Reviews in Food Science and Nutrition, 51(2), 178–194.
6. Buckeridge, M. S., Rayon, C., Urbanowicz, B., Tiné, M. A. S., & Carpita, N. C. (2004). Mixed linkage (1–3), (1–4)-b-D-glucans of grasses. Cereal Chemistry, 81(1), 115–127.
7. Bunzel, M., Ralph, J., Marita, J. M., Hatfield, R. D., & Steinhart, H. (2001). Diferulates as structural components in soluble and insoluble cereal dietary fibre. Journal of the Science of Food and Agriculture, 81(7), 653–660.
8. Chatzifragkou, A., Kosik, O., Prabhakumari, P. C., Lovegrove, A., Frazier, R. A., Shewry, P. R., & Charalampopoulos, D. (2015). Biorefinery strategies for upgrading Distillers' Dried Grains with Solubles (DDGS). Process Biochemistry, 50(12), 2194–2207.
9. Chatzifragkou, A., Prabhakumari, P. C., Kosik, O., Lovegrove, A., Shewry, P. R., & Charalampopoulos, D. (2016). Extractability and characteristics of proteins deriving from wheat DDGS. Food Chemistry, 198, 12–19.
10. Douglas, S. (1981). A rapid method for the determination of pentosans in wheat flour. Food Chemistry, 7(2), 139–145.
11. European Food Safety Authority (2010). Scientific Opinion on Dietary Reference Values for carbohydrates and dietary fibre. EFSA Journal, 8(3), 1–77.
12. Fauré, R., Courtin, C. M., Delcour, J. A., Dumon, C., Faulds, C. B., Fincher, G. B., ... O'Donohue, M. J. (2009). A brief and informationally rich naming system for oligosaccharide motifs of heteroxylans found in plant cell walls. Australian Journal of Chemistry, 62(6), 533–537.
13. Finnie, S. M., Bettge, A. D., & Morris, C. F. (2006). Influence of cultivar and environment on water-soluble and water-insoluble arabinoxylans in soft wheat. Cereal Chemistry, 83(6), 617–623.
14. Freeman, J., Lovegrove, A., Wilkinson, M. D., Saulnier, L., Shewry, P. R., & Mitchell, R. A. C. (2016). Effect of suppression of arabinoxylan synthetic genes in wheat endosperm on chain length of arabinoxylan and extract viscosity. Plant Biotechnology Journal, 14, 109–116.
15. Gibson, G. R., Probert, H. M., Van Loo, J., Rastall, R. A., & Roberfroid, M. B. (2004). Dietary modulation of the human colonic microbiota: updating the concept of prebiotics. Nutrition Research Reviews, 17, 259–275.
16. Han, J., & Liu, K. (2010). Changes in composition and amino acid profile during dry grind ethanol processing from corn and estimation of yeast contribution toward DDGS proteins. Journal of Agricultural and Food Chemistry, 58(6), 3430–3437.
17. Hopkins, M. J., Englyst, H. N., Macfarlane, S., Furrie, E., Macfarlane, G. T., & McBain, A. J. (2003). Degradation of cross-linked and non-cross-linked arabinoxylans by the intestinal microbiota in children. Applied and Environmental Microbiology, 69 (11), 6354–6360.

18. Huisman, M. M. H., Schols, H. A., & Voragen, A. G. J. (2000). Glucuronoarabinoxylans from maize kernel cell walls are more complex than those from sorghum kernel cell walls. Carbohydrate Polymers, 43(3), 269–279.
19. Jones, J. M. (2014). CODEX-aligned dietary fiber definitions help to bridge the "fiber gap". Nutrition Journal, 13(1), 34.
20. Kim, Y., Mosier, N. S., Hendrickson, R., Ezeji, T., Blaschek, H., Dien, B., ... Ladisch, M. R. (2008). Composition of corn dry-grind ethanol by-products: DDGS, wet cake, and thin stillage. Bioresource Technology, 99(12), 5165–5176.
21. Knudsen, K. E. B. (2014). Fiber and nonstarch polysaccharide content and variation in common crops used in broiler diets. Poultry Science, 93(9), 2380–2393.
22. Kolenová, K., Vršanská, M., & Biely, P. (2006). Mode of action of endo-b-1,4-xylanases of families 10 and 11 on acidic xylooligosaccharides. Journal of Biotechnology, 121(3), 338–345.
23. Krzanowski, W. (2000). Principles of multivariate analysis (2nd ed.). Oxford University Press. Liu, K. (2011). Chemical composition of distillers grains, a review. Journal of Agricultural and Food Chemistry, 59(5), 1508–1526.
24. Nantel, G. (1999). Carbohydrates in human nutrition. Journal of Food Composition and Analysis, 6–10. Nemeth, C., Freeman, J., Jones, H. D., Sparks, C., Pellny, T. K., Wilkinson, M. D., ... Shewry, P. R. (2010). Down-regulation of the CSLF6 gene results in decreased (1,3;1,4)-b-D-glucan in endosperm of wheat. Plant Physiology, 152(3), 1209–1218.
25. Nuez Ortín, W. G., & Yu, P. (2009). Nutrient variation and availability of wheat DDGS, corn DDGS and blend DDGS from bioethanol plants. Journal of the Science of Food and Agriculture, 89(10), 1754–1761.
26. Ordaz-Ortiz, J. J., Devaux, M. F., & Saulnier, L. (2005). Classification of wheat varieties based on structural features of arabinoxylans as revealed by endoxylanase treatment of flour and grain. Journal of Agricultural and Food Chemistry, 53(21), 8349–8356.
27. Pedersen, M. B., Dalsgaard, S., Arent, S., Lorentsen, R., Knudsen, K. E. B., Yu, S., & Lærke, H. N. (2015). Xylanase and protease increase solubilization of non-starch polysaccharides and nutrient release of corn- and wheat distillers dried grains with solubles. Biochemical Engineering Journal, 98, 99–106.
28. Pedersen, M. B., Dalsgaard, S., Knudsen, K. E. B., Yu, S., & Lærke, H. N. (2014). Compositional profile and variation of distillers dried grains with solubles from various origins with focus on non-starch polysaccharides. Animal Feed Science and Technology, 197, 130–141.
29. Rose, D. J., Patterson, J. A., & Hamaker, B. R. (2010). Structural differences among alkali-soluble arabinoxylans from Maize (Zea mays), Rice (Oryza sativa), and Wheat (Triticum aestivum) brans influence human fecal fermentation profiles. Journal of Agricultural and Food Chemistry, 58(1), 493–499.
30. Saulnier, L., Peneau, N., & Thibault, J.-F. (1995). Variability in grain extract viscosity and water-soluble arabinoxylan content in wheat. Journal of Cereal Science, 22 (3), 259–264.
31. Saulnier, L., Robert, P., Grintchenko, M., Jamme, F., Bouchet, B., & Guillon, F. (2009). Wheat endosperm cell walls: Spatial heterogeneity of polysaccharide structure and composition using micro-scale enzymatic fingerprinting and FT-IR microspectroscopy. Journal of Cereal Science, 50(3), 312–317.
32. Saulnier, L., Sado, P. E., Branlard, G., Charmet, G., & Guillon, F. (2007). Wheat arabinoxylans: Exploiting variation in amount and composition to develop enhanced varieties. Journal of Cereal Science, 46(3), 261–281.
33. Saulnier, L., & Thibault, J. F. (1999). Ferulic acid and diferulic acids as components of sugar-beet pectins and maize bran heteroxylans. Journal of the Science of Food and Agriculture, 79(3), 396–402.
34. Scientific Advisory Committee on Nutrition (2015). Carbohydrates and health. TSO The Stationary Office. Shewry, P. R., Hawkesford, M. J., Piironen, V., Lampi, A. M., Gebruers, K., Boros, D., ... Ward, J. L. (2013). Natural variation in grain composition

of wheat and related cereals. Journal of Agricultural and Food Chemistry, 61(35), 8295–8303.
35. Shewry, P. R., Piironen, V., Lampi, A. M., Edelmann, M., Kariluoto, S., Nurmi, T., ... Ward, J. L. (2010). The HEALTHGRAIN wheat diversity screen: Effects of genotype and environment on phytochemicals and dietary fiber components. Journal of Agricultural and Food Chemistry, 58(17), 9291–9298.
36. Toole, G. A., Le Gall, G., Colquhoun, I. J., Nemeth, C., Saulnier, L., Lovegrove, A., ... Shewry, P. R. (2010). Temporal and spatial changes in cell wall composition in developing grains of wheat cv. Hereward. Planta, 232(3), 677–689.
37. Verspreet, J., Damen, B., Broekaert, W. F., Verbeke, K., Delcour, J. A., & Courtin, C. M. (2016). A critical look at prebiotics within the dietary fiber concept. Annual Review of Food Science and Technology, 7(1), 167–190.
38. Widyaratne, G. P., & Zijlstra, R. T. (2007). Nutritional value of wheat and corn distiller's dried grain with solubles: Digestibility and digestible contents of energy, amino acids and phosphorus, nutrient excretion and growth performance of grower-finisher pigs. Canadian Journal of Animal Science, 87, 103–114.

CHAPTER 16

Oxidation and Complexation-Based Spectrophotometric Methods for Sensitive Determination of Tartrazine E102 in Some Commercial Food Samples

Magda M. S. Saleh*, Elham Y. Hashem, Najat O. A. Al-Salahi

Chemistry Department, Faculty of Science, Assiut University, Assiut, Egypt

ABSTRACT

Two new sensitive spectrophotometric methods are reported for determination of tartrazine (Tz) (E102) in some commercial food samples. The first method involves two coupled reactions, the reduction of Cu(II) to Cu(I) by the analyte in acetate buffer medium (pH = 5.9) at 30°C and the complexation reaction between Cu(I) and Tz oxidized form to yield Cu-Tz complex (method I). The other method is based on oxidation of Tz by alkaline $KMnO_4$. These reactions are monitored spectrophotometrically at maximum absorbances 332 and 610 nm for methods (I and II) respectively. Variables affecting these reactions are carefully studied and the conditions are optimized. The stability constants are calculated at 293, 303, 313 and

323 K. The thermodynamic parameters, Gibb's free energy change (DG), entropy change (DS), and enthalpy change (DH) associated with the complexation reaction are calculated and discussed. Under optimized conditions the proposed methods (I, II) obey Beer's law 10.69 - 85.50, 5.34 - 34.12 $\mu g \cdot ml^{-1}$ of Tz respectively. The molar absorptivity, sandel sensitivity, detection and quantification limits are calculated. Matrix effects are also investigated. The methods are successfully applied to the quantification of Tz in different commercially food samples. The results obtained are in good agreement with those obtained by the reported methods at the 95% confidence level.

Keywords: Tartrazine E102, Food Colorants, Complexation, $KMnO_4$, Spectrophotometric Analysis

1. INTRODUCTION

Food dyes are often added to foodstuffs and drinks in order to supply, intensify or restore their colour to create the desired coloured appearance [1] . Synthetic dyes are widely used as they show several advantages compared with natural dyes such as high stability to light, oxygen and pH, colour uniformity, low microbiological contamination and relatively lower production costs. Tartrazine is a synthetic organic food dye that can be found in common food products such as bakery products, dairy products, candies, and beverages. According to the limitations of European Union [EU] and Federal Food, Drug and Cosmetic Act [2] , the presence and content of Tz dye must be controlled in food products due to their potential harmfulness to human beings [3] . Also it appears to cause the most allergic and intolerance reaction of all the azo dyes, particularly among asthmatics and those with an aspirin intolerance [4] [5] . Therefore the determination of tartrazine in commercial food products is very important. Until now, different methods such as chromatography [6] - [12] , electroanalytical methods [13] - [15] and spectrophotometry [16] - [24] , have been reported for the determination of tartrazine. However, some of these methods are not suitable for routine monitoring as they are time consuming, complicated and have poor sensitivity and selectivity. To the best of our knowledge, no spectrophotometer methods based on complexation with copper have been reported for the quantification of Tz. This paper describes the development of a simple, and rapid two spectrophotometric methods for the assay of Tz in food samples. The first method is based on oxidation of tartrazine by Cu(II) in acetate-acetic acid medium (pH = 5.9) at

30°C followed by complex formation. The other method is based on oxidation of Tz with alkaline $KMnO_4$ under optimum conditions.

2. EXPERIMENTAL

2.1. Apparatus

An evolution 300 UV-Vis spectrophotometer with 1.0 cm matched cells fitted with vision pro software of Thermo Electronic Corporation (Cambridge, U.K.) was used for electronic spectral measurements. pH measurements were made with Jenway 3040 ion analyzer-pH meter, equipped with Jenway 924,005 combined glass electrode.

2.2. Chemicals

All chemicals were of analytical reagent grade. Deionized water was used to prepare all solutions. The tartrazine dye (E102) was supplied from Alfa Aesar (Germany). Stock standard solution containing 2.672 g $\cdot$ L^{-1} of the dye was prepared in deionized water. The working solutions were prepared daily by appropriate dilution. A 5×10^{-3} mol $\cdot$ L^{-1} of copper nitrate was prepared by dissolving the required amount in deionized water and standardized complexmetrically with EDTA [25] . The ionic strength of the solutions was maintained at a constant value of I = 0.1 mol $\cdot$ L^{-1}($NaClO_4$). Acetic acid (0.2 mol $\cdot$ L^{-1}) sodium acetate (0.2 mol $\cdot$ L^{-1}) buffer (pH = 5.9) was prepared in 100 ml volumetric flask. Potassium permanganate (Merck, Germany) 1×10^{-2} mol $\cdot$ L^{-1} was prepared in deionized water.

2.3. Preparation of Real Samples

Powdered gelatin samples (Lemon, pineapple, green apple, apricot, banana and peach) and powdered drinks (apple, orange and tamarind) were bought from local supermarket in Assiut city (Egypt). These samples contained sugars, fumaric and citric acids, beef gelatin, artificial flavour, colour (E102), tricalcium phosphate and aspartame. Each gelatin or drink samples were weighed exactly (about 1 g) and dissolved in 50 ml of deionized water. Then each solution was centrifuged (10 min, 3500 rpm) to remove the insoluble particles. The filtrate was collected in 100 ml volumetric flask and diluted to the mark with deionized water. For determination of Tz in the above real samples, 0.2 ml of sample solution was used and analyzed according to method I. Finally, the Tz content in food samples were determined using the calibration equation and standard addition calibration curves procedure.

2.4. General Procedures

2.4.1. Method I (Complexation Reaction)

Into a 10 ml volumetric flasks, transfer a suitable aliquot of standard solution in deionized water containing up to 181.6 μg of tartrazine and 4 ml of 2×10^{-3} mol · L^{-1} of Cu(II) solution. After mixing, the mixture was buffered to pH 5.9 with acetic acid-acetate buffer. Dilute the resulting solution to volume with deionized water and measure the absorbance at 332 nm, 303 K by using 1.0 cm quartz cell against a similarly prepared blank of the same pH. The calibration graph was constructed by plotting absorbance vs. Tz concentration.

2.4.2. Method II (Oxidation with $KMnO_4$)

Standard solutions in deionized water containing 16.031 - 187.03 μg · ml^{-1} Tz were transferred into individual 10 ml calibrated flasks, 4.0 ml of 1.0 mol · L^{-1} sodium hydroxide solution was added followed by 2.0 ml of 1×10^{-2} mol · L^{-1} $KMnO_4$ solution and it was diluted to the final volume with deionized water. After 60 min, the absorbance was measured at 610 nm against reagent blank treated similarly.

2.5. Interference from Matrix

Samples were prepared by mixing 0.5344 mg of Tz (method I) or 0.267 mg (in case of method II) with various amounts of common matrix cations, anions, sugar, gelatin, aspartame and dyes such as sunset yellow, and allura red. The procedure was continued as described under general procedures.

2.6. Determination of the Thermodynamic Parameters

The decomposition of stability constant into its enthalpy change (DH), Gibbs free energy change (DG) and entropic (DS) contributions is of fundamental importance to understand the various factors that may influence coordination [26] [27] . Enthalpy changes due to coordination reactions can always be obtained from the determination of stability constants at different temperatures according to the Van't Hoff [28] relationship (1, 2):

$$\Delta G = -2.303RT \log K_i = \Delta H - T\Delta S \qquad (1)$$

$$\log K_i = \left(\frac{-\Delta H}{2.303\,R}\right)\frac{1}{T} + \frac{\Delta S}{2.303\,R} \qquad (2)$$

where R is ideal gas constant (8.314 $JK^{-1} \cdot mol^{-1}$), K_i is the stability constant of the complex and T is the absolute temperature in Kelvin (K).

3. RESULTS AND DISCUSSION

3.1. Acid-Base Equilibria of Tartrazine

The absorption spectra of tartrazine salt (HL^{3-}) solution (5×10^{-5} mo $\cdot$ L^{-1}) in aqueous medium at I = 0.1 mol $\cdot$ L^{-1} ($NaClO_4$), 25°C were recorded at various pH values (Figure 1). The food additive tartrazine salt (HL^{3-}) yields four acid—base forms in solution of pH 0.86 - 12.4, H_5L^+, H_2L^{2-}, HL^{3-} and L^{4-} exhibiting the absorption maxima at 440, 430, 420 and 400 nm. The protonated acid form H_5L^+ at azo-nitrogen (l_{max} = 440 nm) predominates in strongly acidic medium (pH < 3). The solution spectra of food colorant display a symmetrical and highly intense band (l_{max}= 430 nm) at pH 3.2 - 4.3 corresponding to the protonated form of the dye salt (H_2L^{2-}). On decreasing, the acidity of the medium (4.2 - 6.5), a blue shift of the latter band at 420 nm is obtained resulting from dissociation of protonated nitrogen azo group (HL^{3-}). The electronic spectra of Tz in aqueous medium within pH 7.2 - 10.4 exhibit a broad band at 400 nm with a shoulder at 425 nm corresponding to tautomeric equilibria of monoanionic form of tartrazine salt (L^{4-}), Scheme 1. The absorbance versus pH graph were interpreted [29] assuming that a particular equilibrium established under selected conditions. Under our experimental condition $pKa_1[H_2L^{-2}/HL^{-3}$ (-HN=N-)] = 5.15 ± 0.10, pKa_2 [HL^{-3}/L^{-4} (OH)] = 9.25 ± 0.16.

3.2. Absorption Spectra of Tartrazine Reaction Products

Tartrazine likes Sudan dyes possess two reducible groups, a nitrogen-nitrogen double bond and a phenol group [30] . A reduction process, which corresponds to a spontaneous auto-reduction of the Cu^{2+} into Cu^+, was observed in cases of complexation of Cu^{2+} ions with sudan(I) [31] and sudan(II) [32] . Because of these features of sudan dyes, tartrazine is considered as chromogenic chelating and redox sensitive agent reacting with copper(II) to form Cu^+-Tz complex under optimum conditions. On the basis of redox sensitivity in the Cu(II)-tartrazine interaction in aqueous medium at I = 0.1 mol · L^{-1} ($NaClO_4$), pH = 5.9, 30°C, solution spectrum shows an absorption band with a maximum at 332 nm corresponding to Cu(I)-Tz complex (MLCT) (Figure 2(a)).

Figure 2(b) shows the absorption spectrum obtained for alkaline $KMnO_4$ solution (l_{max} = 530 nm). With addition of tartrazine solution, an absorption spectrum with two absorption bands at 390, 410 (double headed) and 610 nm was obtained. The latter band is attributed to the formation of manganate ion as a result of the oxidation of tartrazine with $KMnO_4$ in alkaline medium [33] [34] and the highly intense double headed band is related to Tz oxidized form.

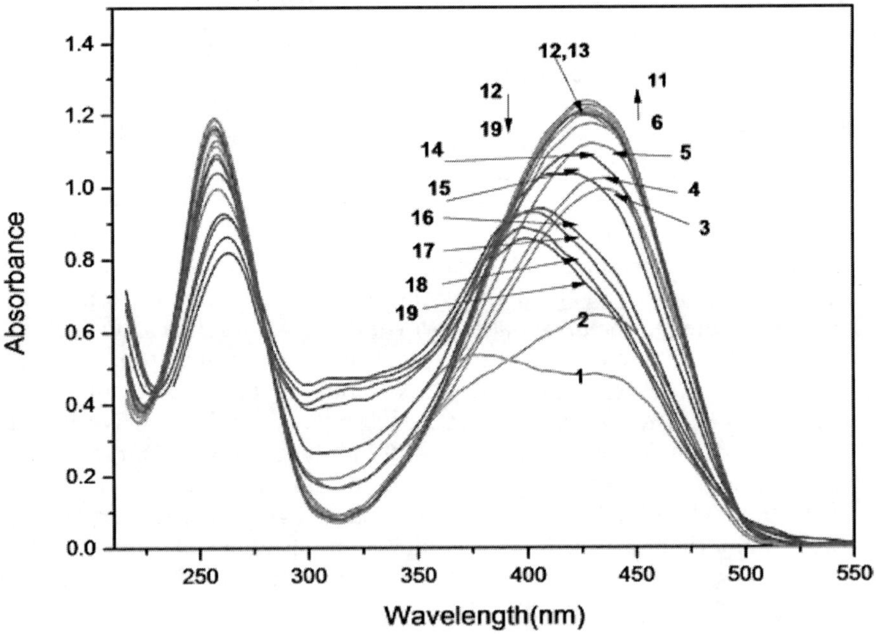

Figure 1. Absorption spectra of 5 × 10^{-5} mol · L^{-1} tartrazine in aqueous medium, I = 0.1 mol · L^{-1} ($NaClO_4$), 25°C at different pH values; 1, pH = 0.86; 2, pH = 0.9; 3, pH = 1.0; 4, pH 1.17; 5, pH = 1.6; 6, pH = 2.26; 7, pH = 2.73; 8, pH = 3.29; 9, pH = 3.73; 10, pH = 4.21; 11, pH = 5.7; 12, pH = 6.48; 13, pH = 7.19; 14, pH = 8.7; 15, pH = 9.2; 16, pH = 9.97; 17, pH = 10.58; 18, pH = 11.12; 19, pH = 12.83.

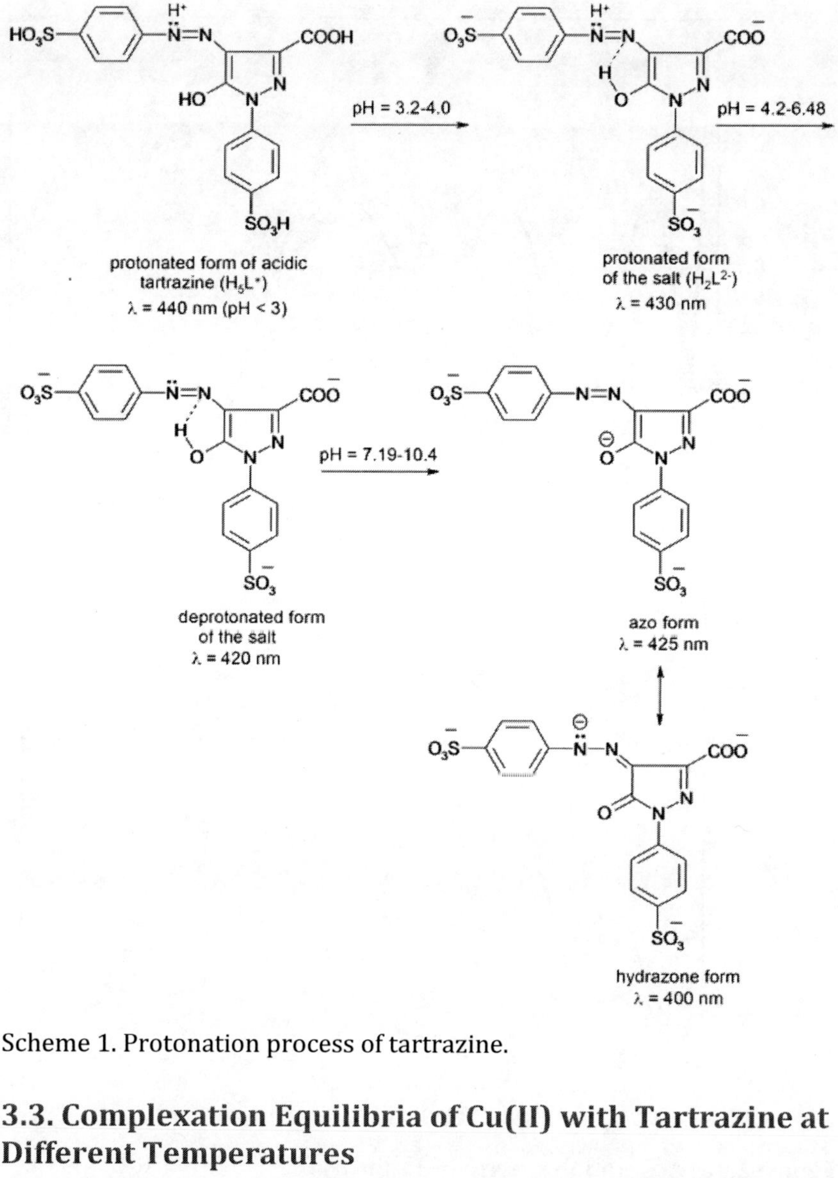

Scheme 1. Protonation process of tartrazine.

3.3. Complexation Equilibria of Cu(II) with Tartrazine at Different Temperatures

The complexation equilibria of Cu^{2+} with Tz were investigated at I = 0.1 mol · L^{-1} ($NaClO_4$) in aqueous medium over the pH range 3.5 - 10 at 293, 303, 313 and 323 K. The solution spectra were recorded in presence of an excess of the metal ion and in equimolar solutions. The absorption spectra for both metal/Tz ratios at various pH are analogous at the same temperature and exhibit an absorption band at 332 nm.

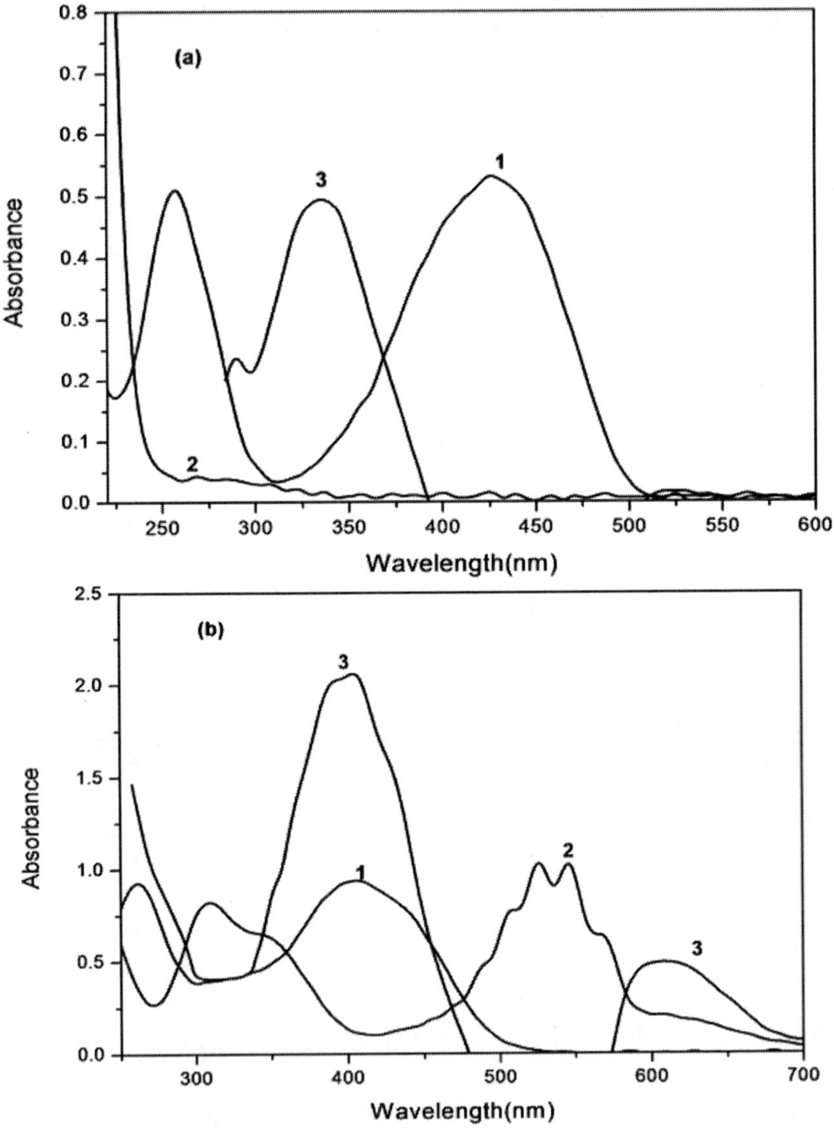

Figure 2. (a) Absorption spectra of Tz-Cu(II) (pH = 5.9), I = 0.1 mol · L^{-1} (1) Tz 2 × 10^{-5}mol · L^{-1}; (2) Cu(II) 2 × 10^{-4} mol · L^{-1}; (3) 1:1 Tz-Cu(II); [Cu^{2+}] = [Tz] = 2 × 10^{-4} mol · L^{-1}. (b) Absorption spectra of Tz-$KMnO_4$ reaction (in alkaline medium), (1) Tz 5 × 10^{-5} mol · L^{-1}; (2) alkaline $KMnO_4$ (5 × 10^{-4} mol · L^{-1}); (3) Tz-$KMnO_4$ reaction product [Tz] = 7 × 10^{-5}mol · L^{-1}, [$KMnO_4$] = 2 × 10^{-3} mol · L^{-1}, [NaOH] = 0.4 mol · L^{-1}.

The absorbance versus pH graphs for the above solutions at different temperatures show the range of complex formation (3.0 - 6.5) and existence of one complex equilibria within the pH range studied (Figure 3(a)). All graphs exhibit a similar descending branch above pH 6.5 which is due to the hydrolysis of the complexed Tz. The absorbance versus pH graphs for copper-Tz system at different temperatures were interpreted using relations derived earlier by Sommer et al. [35] , Idriss et al. [36] [37] and Saleh et al. [34] . The following Equations (3, 4) were valid for equimolar and solutions with an excess of metal ion respectively:

$$\log\left[\Delta A/(\varepsilon_1 C_L - \Delta A)^2\right] = q\ \mathrm{pH} + \log \varepsilon_1 + \log K^* \tag{3}$$

$$\log\left[\Delta A/(\varepsilon_1 C_L - \Delta A)\right] = q\ \mathrm{pH} + \log C_M + \log K^* \tag{4}$$

All symbols have their usual meanings. The logarithmic transformation are linear with a slope of q = 2, indicating the release of two protons during complexation and an intercept including K^*. By considering the values of the dissociation constants of Tz under our experimental conditions, one can assume the following complexation equilibria with Cu(II) in the pH range 3.0 - 6.5.

$$\mathrm{Cu(II)} + \mathrm{H_2L^{-2}} \xrightarrow{\mathrm{pH=3.0-6.5}} \left[\mathrm{Cu^+ + H_2L^-}\right] \xrightarrow{\text{Intermediate state}} \left[\mathrm{Cu^+L^{3-}}\right] + 2\mathrm{H^+}, K^* \tag{A}$$

The mean equilibrium constant K^* was determined at different temperatures by considering equilibrium A. the stability constant of Cu(I) complex is related to the equilibrium constant K^* by the expression $K_1 = K^*/K_{a1}\cdot K_{a2}$. The calculated values of log K_1 for the complex at 293, 303, 313 and 323 K are given in Table 1. The proposed binding model of Tz-copper complex is shown below.

Table 1. Stability constants at different temperatures and thermodynamic parameters DG, DH and DS of copper(I)-Tz complex.

Complex species	Log K_1 at temperature, K				$-\Delta G$ $\mathrm{Kj\cdot mol^{-1}}$	ΔH $\mathrm{Kj\cdot mol^{-1}}$	ΔS $\mathrm{J\cdot mol^{-1}\cdot K^{-1}}$
	293	303	313	323			
$[\mathrm{Cu^-L^{3-}}]^{2-}$	6.922	7.02	7.172	7.322	41.319	27.354	142.65

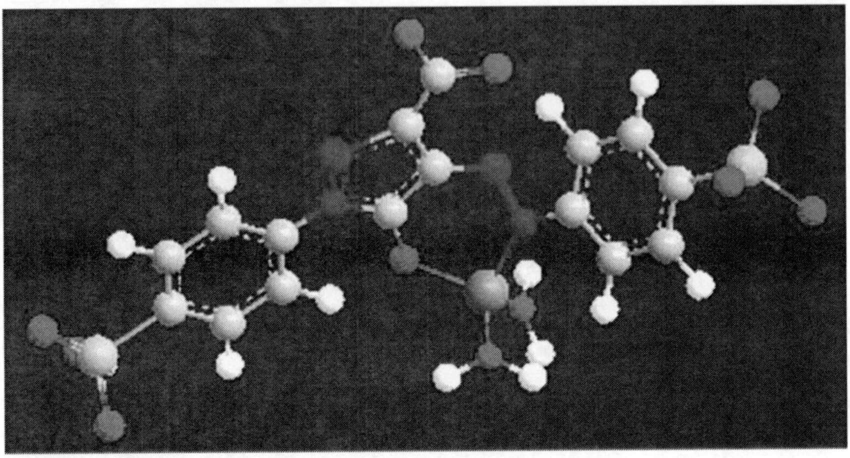

The proposed binding model of Tz with copper(I).

3.4. Thermodynamic Functions of Tartrazine-Copper(II) Reaction

According to Van't Hoff equation the values of overall thermodynamic parameters DG, DH and DS accompanying complex formation have been determined (Figure 3(b)). The DH and DS values can then be considered as a sum of three contributions, Cu(II)-Tz redox reaction, release of H_2O molecules from octahedral copper(II) ions and metal-ligand bond formation. From Table 1, the -ve value of DG for the complexation process Cu(I)-Tz suggests a spontaneous nature of such process [38] . The +ve value of DH means that this reaction is endothermic favourable at high temperature. The +ve value of DS for the complexation process confirm that the Cu(I)-Tz complex formation is entropically favourable [39] .

3.5. Optimization of Variables

3.5.1. Complexation Method

In order to optimize the condition we have investigated a number of parameters such as pH, reagent concentration, the temperature and time.

Effect of pH

With other conditions fixed the effect of pH on absorbance of Cu(I)-Tz complex at l_{max} = 332 nm was investigated from pH 3.5 - 10.0. From Figure 3(a), the quantitative determination of Tz was achieved at pH = 5.9.

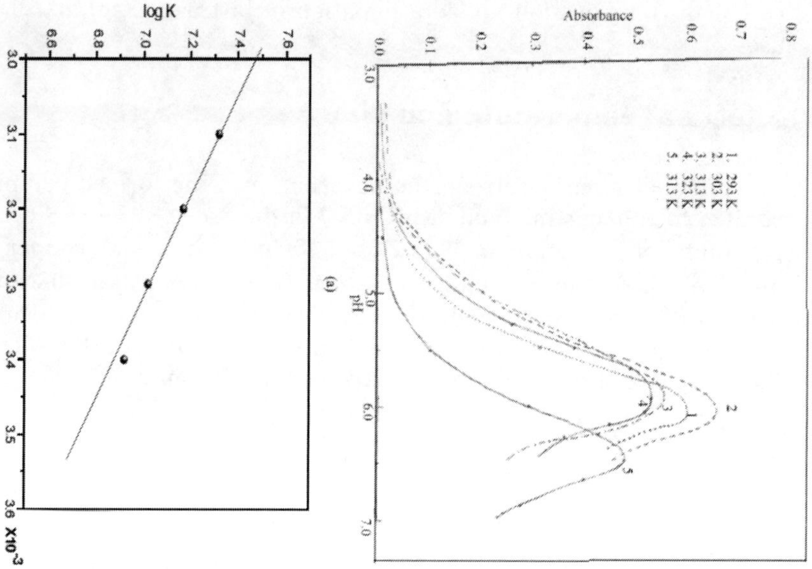

Figure 3. (a) Absorbance vs pH graph for Tz-Cu(I) complex in aqueous medium, I = 0.1 mol · L^{-1} ($NaClO_4$), l_{max} = 332 nm at various temperature, (1 - 4) [Tz] = 2 × 10^{-4} mol · L^{-1}, [Cu^{2+}] = 6 × 10^{-4} mol · L^{-1} at 293, 303, 313, and 323 K respectively; (5) [Tz] = [Cu^{2+}] = 2 × 10^{-4} mol · L^{-1} at 313 K. (b) Van't Hoff plot of log K_1 of Tz-Cu(I) complex against 1/T.

In order to determine the best buffer solution at pH = 5.9, several buffer systems at isomolar concentration of 0.2 mol · L^{-1} were studied. The best analytical sensitivity was obtained in presence of acetate-acetic acid buffer. Next, the effect of buffer volume on analytical sensitivity was also studied in the range of 0.2 - 2.5 ml, and the maximum absorbance was obtained at a buffer volume of 0.5 ml. the results are shown in Figure 4(a).

Effect of Copper Concentration

The effect of copper(II) concentration on the absorbance of the complex was investigated by varying the reagent concentration under the optimum conditions ([Tz] = 2 × 10^{-4} mol · L^{-1}), pH = 5.9, T = 293 K. It is clear from Figure 4(a) that the maximum absorbance was attained with 3.0 ml; above this volume, the absorbance remained unchanged. To ensure the complete complexation for determination of tartrazine, 4.0 ml Cu(II) was used.

Effect of Temperature and Time

The effect of temperature on the absorbance of the formed complex was studied in aqueous medium at pH = 5.9 in the range 393 - 323 K, keeping constant concentration of Tz (2.0×10^{-4}mol · L^{-1}) and copper (8.0×10^{-4} mol · L^{-1}). The maximum absorbance was obtained at 303 K (Figure 4(b)). Under optimum conditions, the reaction time was determined by following the absorbance of the complex at different time intervals. Complete complex formation was attained after 5 min. at 303 K.

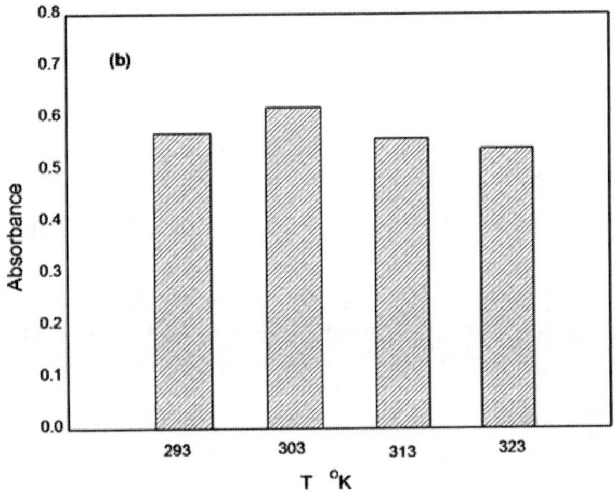

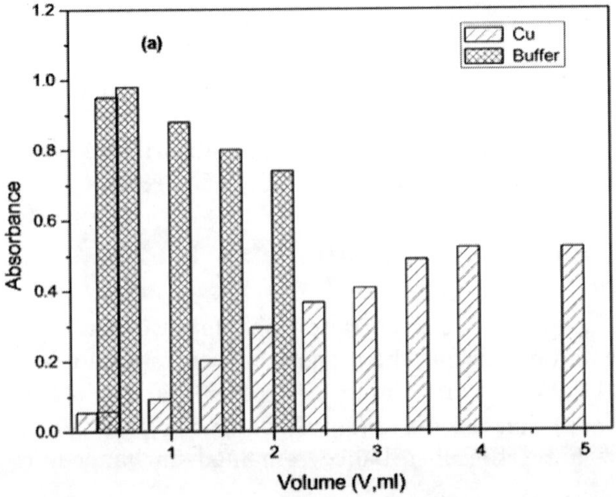

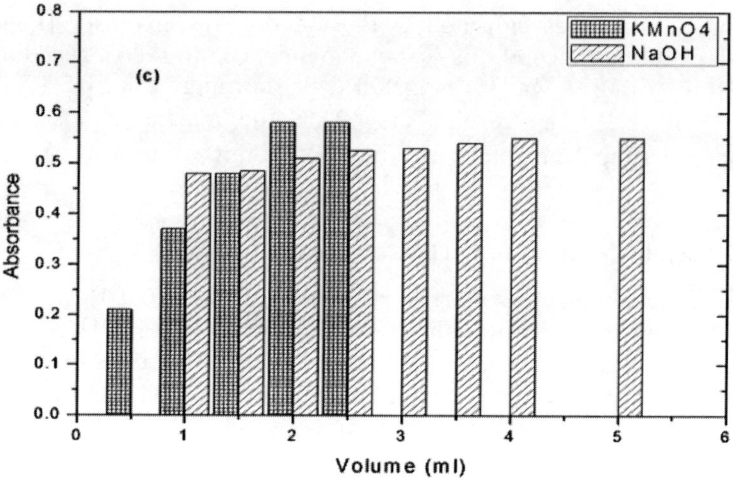

Figure 4. (a) Reaction conditions for complexation reaction at 293 K, [Tz] = 2 × 10^{-4} mol · L^{-1}, l_{max} = 332 nm, pH = 5.9. (1) Effect of different volumes of 2 × 10^{-3} mol · L^{-1} Cu^{2+} on absorbance of Cu^{+}-Tz^{3-} complex, acetate buffer volume = 1 ml. (2) Effect of different acetate buffer volume on absorbance of Cu-Tz complex, [Cu^{2+}] = 8 × 10^{-4} mol · L^{-1}. (b) Effect of absolute temperature on absorbance of Tz complex, [Tz] = 2 × 10^{-4} mol · L^{-1}, [Cu^{2+}] = 8 × 10^{-4} mol · L^{-1}; buffer volume = 0.5 ml. (c) Reaction conditions for oxidation reaction of Tz with alkaline $KMnO_4$ at 293 K, [Tz = 1 × 10^{-4} mol · L^{-1}]. 1) Effect of $KMnO_4$ (1 × 10^{-2} mol · L^{-1}) (by volume), [NaOH] = 0.4 mol · L^{-1}, t = 60 min. 2) Effect of NaOH (1.0 mol · L^{-1}) (by volume), [$KMnO_4$] = 2 × 10^{-3} mol · L^{-1}, t = 60 min.

3.5.2. $KMnO_4$ Oxidation Method

Redox reactions have been used as the basis for the development of simple and sensitive spectrophotometric methods for the determination of many compounds [33] [40] . Tartrazine reacts with $KMnO_4$ in strongly alkaline medium producing green manganate (l_{max} = 610 nm) (Figure 2(b)). During the current study the produced colour intensity increased gradually with time to reach maximum after 60 min and was stable for at least 24 h. The factors affecting the formation of manganate ions were further studied and optimized.

Effect of NaOH and $KMnO_4$

The dependence of redox reaction on the volume of NaOH (1.0 mol · L^{-1}) was investigated in the range 1.0 - 5.0 ml. The results show that maximum absorbance was obtained using 4.0 ml of 1.0 mol · L^{-1} NaOH with no significant changes indicated as the volume was increased (Figure 4(c)). Under optimum conditions, the dependence of redox reaction on the volume of $KMnO_4$ was investigated over the range from 0.5 - 3.0 ml (1×10^{-2} mol · L^{-1}). The results show that the maximum absorbance was obtained using 2.0 ml of 1×10^{-2} mol · L^{-1} $KMnO_4$ (Figure 4(c)).

3.5.3. Stoichiometry of Tartrazine Reactions

The stoichiometryt of tartrazine reaction with Cu(II) or $KMnO_4$ was established by the continuous variation method [34] [41] . The solutions having $C_o = C_R + C_{Tz} = 6.0 \times 10^{-4}$ mol · L^{-1} at pH 5.9 or > 11.0. The plot of absorbance versus mole fraction of reagent at l_{max} 332 or 610 nm, reveals the formation of 1:1 (Tz:copper) complex and 1:2 (Tz:$KMnO_4$).

3.6. Quantification

3.6.1. Validation of the Proposed Methods

For our two methods (I, II) the ranges of linearity of absorbance as a function of food additive Tz concentration provide a satisfactory measure of the sensitivity of the method. Under the optimum conditions the absorbance of the complex (Figure 5) or manganate obeys Beer's law in Tz concentration range of 10.69 - 85.50 or 5.34 - 34.12 μg · ml^{-1} respectively. For the two proposed methods (I, II), the molar absorptivity and Sandell sensitivity [42] values are 0.361×10^4, 0.80×10^4 L · mol^{-1} · cm^{-1} and 0.1480, 0.0668 μg · cm^{-2} respectively. The regression equations, correlation coefficient, limit of detection, and limit of quantification [43] were also calculated and summarized in Table 2.

3.6.2. Accuracy and Precision

The accuracy and precision of the proposed spectrophotometric methods were determined at three different concentration levels of additive food colour-Tz by analyzing five replicate samples of each concentration. The relative standard deviation (R.S.D%) obtained for the analytical results did not exceed 2% (Table 3) which proved a high reproducibility of the results and precision of the methods.

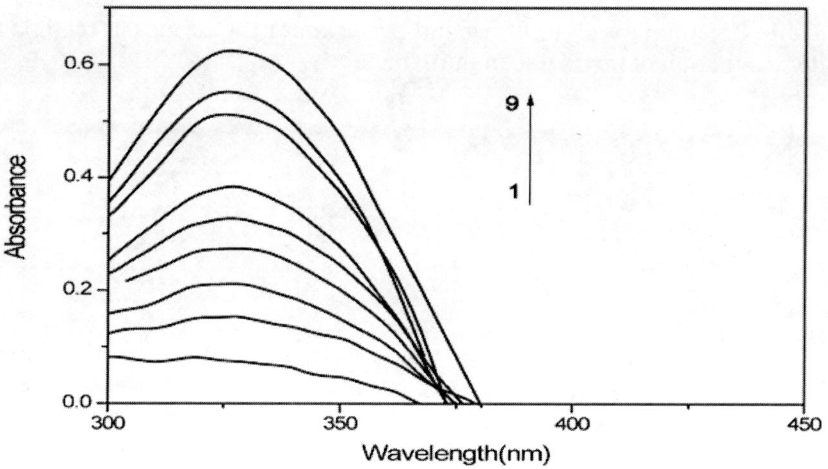

Figure 5. Absorption spectra of tartrazine-copper complex, pH = 5.9 (acetate buffer), tartrazine concentration range 2 × 10^{-5} (1) to 2.0 × 10^{-4}mol · L^{-1} (9) with regular successive additions in presence of 8.0 × 10^{-4}mol · L^{-1}Cu(II) at 303 K.

Table 2. Summary of optical and regression characteristic of the proposed methods for determination of tartrazine.

Method/reagent Parameter	Copper(II)	$KMnO_4$
Colour	Colourless	Green
λ_{max} (nm)	332	610
Beer's law limits (μg·ml^{-1})	10.69 - 85.50	5.34 - 34.12
Ringborn limts (μg·ml^{-1})	12.59 - 81.28	7.08 - 31.62
Molar absorptivity (L·mol^{-1}·cm^{-1})	0.361×10^4	0.80×10^4
Sandel's sensitivity (μg·cm^{-2})	0.1480	0.0668
Regression equation A = a + bc	$A = 6.55 \times 10^{-3}C - 8.62 \times 10^{-4}$	$A = 0.0147C + 6.79 \times 10^{-3}$
Slope (b)	6.55×10^{-3}	0.0147
Intercept (a)	8.62×10^{-4}	6.79×10^{-3}
Correlation coefficient	0.9999	0.9969
Limit of detection (LOD) (μg·ml^{-1})	0.101	0.355
Limit of quantification (LOQ) (μg·ml^{-1})	0.305	1.075

Table 3. Summary of accuracy and precision of the proposed methods for determination of tartrazine in pure form.

Method/Reagent	Amount μg·ml^{-1} Taken	Found ± SD[a]	RSD %	Recovery %	C.L[b]
Cu(II) Intraday assay	32.062	31.929 ± 0.11	0.34	99.585	±0.137
	53.437	51.777 ± 0.13	0.25	96.894	±0.162
	64.124	63.532 ± 0.20	0.31	99.076	±0.249
Intreday assay	32.062	31.166 ± 0.12	0.39	97.205	±0.149
	53.437	52.54 ± 0.15	0.29	98.321	±0.186
	64.124	63.838 ± 0.19	0.30	99.554	±0.236
$KMnO_4$ Intraday	16.031	16.001 ± 0.05	0.31	99.813	±0.062
	26.719	26.205 ± 0.07	0.27	98.076	±0.087
	32.062	32.395 ± 0.11	0.34	101.039	±0.137

a) Mean for 5 independed analysis. b) C.L., confidence limit at 95% confidence level at 4 degree of freedom (t = 2.776).

3.6.3. Intraday and Interday Precision and Accuracy

Under optimum conditions the intraday precision was carried out for our procedures through replicate analysis (n = 5) for Tz corresponding to 32.062, 53.437, 64.124 μg · ml^{-1} (method I) and 16.031, 26.719, 32.062 μg · ml^{-1} (method II). The interday precision was also evaluated through replicate analysis of the standard Tz for three consecutive days at the same concentration levels as in within day precision. The results of these assays are reported in Table 3. For intraday and interday precision, the recovery and RSD values were in the range of 96.89% - 101.04% and 0.25% to 0.39% respectively. The green manganate ions produced from Tz-$KMnO_4$ reaction in alkaline medium was stable only for 24 hours.

3.6.4. Analytical Recovery and Matrix Effects

To investigate the selectivity of the proposed methods, the effect of various substances on the determination of Tz (53.437 and 26.719 mg · L^{-1}) for method I and II respectively was tested under optimum conditions. Several representative potential interferences such as inorganic cations, anions, molecular species and dyes were investigated individually for their effect on Tz recovery. Tolerance limits were defined by the concentration of interferents which caused on <5% error in the determination of Tz. The obtained mean recoveries and standard deviation ranged between 99.0% - 96.0% and ± 1.0 - 2.0 respectively are shown in Table 4. These results proved the accuracy of the proposed methods and absence of interference from common matrix. The accuracy of our method was also checked by studying the influence of various dyes on the recovery of Tz. Tartrazine determination

Table 4. Matrix effects on the recovery of tartrazine (E = 102).

Matrix components	Complexation method		Oxidation method	
	Concentration mg·L^{-1}	Recovery %	Concentration mg·L^{-1}	Recovery %
1) Na^{+}, K^{+}, Mg^{2+}, Ca^{2+}, Cl^{-}, SO_4^{2-}, PO_4^{3-}	3000	99.0 ± 1.0	2500	98.0 ± 1.5
2) Citrate, stearat, fumarate, aspartame, gelatin, sugar, L-ascopric acid.	1000	98.0 ± 1.0	8.0	98.5 ± 1.0
3) Sunset yellow	4.5	97.5 ± 1.5	3.5	97.0 ± 1.5
4) Tropaeolin ooo	6.1	96.5 ± 2.0	4.5	96.0 ± 2.0
5) Allura red Ac	5.0	97.0 ± 2.0	4.0	96.8 ± 1.0

Table 5. Statistical analysis of results obtained by the complexation method for tartrazine in powdered drink samples.

Drink[a] sample	Present µg·ml^{-1}	Added µg·ml^{-1}	Found µg·ml^{-1}	±RSD %	Recovery %	Mean		t-value[b]	F-value[b]	Reported method[(41)]	
						Recovery %	±RSD %			Recovery %	±RSD %
		10.690	13.893	0.68	96.48						
Apple	3.741	21.375	24.578	0.69	97.86	98.121	0.70	1.98	1.15	97.5	0.75
		32.070	35.800	0.73	100.02						
		10.690	11.756	0.55	95.632						
Orange	1.603	21.375	23.510	0.62	102.320	98.75	0.60	-	-	-	-
		32.070	33.130	0.63	98.390						
		10.690	12.825	0.60	99.97						
Tamarind	2.1375	21.375	22.440	0.63	95.44	98.46	0.65	-	-	-	-
		42.760	34.199	0.72	99.96						

a) From local markets of Assiut city, Egypt. b) Theoretical t-value ($\upsilon = 4$) and F-value ($\upsilon = 4.4$) at 95% confidence level are 2.78 and 6.39.

Table 6. Statistical analysis of results obtained by the complexation method for tartrazine in powdered gelatin.

Gelatin[a] sample	Present µg·ml^{-1}	Added µg·ml^{-1}	Found µg·ml^{-1}	±RSD %	Recovery %	Mean		t-value[b]	F-value[b]	Reported method[(21)]	
						Recovery %	±RSD %			Recovery %	±RSD %
		10.690	12.82	0.54	99.92						
Lemon	2.137	21.375	22.44	0.62	95.44	97.40	0.60	1.49		97.0	–
		32.070	33.13	0.64	96.84						
		10.690	12.820	0.66	92.25						
Apricot	3.206	21.375	23.512	0.72	95.57	95.93	0.71	–	–	–	–
		32.070	35.270	0.75	99.98						
		10.690	12.820	0.59	99.94						
Banana	2.1375	21.375	22.440	0.70	95.45	98.45	0.64	–	–	–	–
		32.070	34.199	0.63	99.96						
		10.690	13.894	0.66	99.95						
Pineapple	3.206	21.375	23.512	0.69	95.66	97.52	0.68	–	–	–	–
		32.070	34.199	0.69	96.94						
		10.690	11.75	0.67	99.93						
Green apple	1.068	21.375	21.37	0.69	95.23	97.29	0.70	–	–	–	–
		32.070	32.06	0.74	96.67						
		10.690	10.687	0.66	90.89						
Peach	1.068	21.375	22.440	0.69	99.98	96.95	0.69	–	–	–	–
		32.070	33.130	0.72	99.97						

a) From local markets of Assiut city, Egypt. b) Theoretical t-value ($\upsilon = 4$) and F-value ($\upsilon = 4.4$) at 95% confidence level are 2.78 and 6.39.

was possible in presence of 0.045 mg/10 ml sunset yellow (0.035 mg in case of method II), 0.05 mg Allura red (0.04 mg in case of method II) and 0.061 mg Tropaeolin 000 (0.045 mg in case of method II). This was attributed to the great sensitivity of the methods that necessitated dilution for the food sample and consequently the matrix beyond their interference capability.

3.7. Applications

The complexation method was successfully applied to the determination of Tz in two different commercial food products (powered drink and powdered gelatin samples). To investigate the applicability of the proposed method, recovery experiments were performed using multiple points standard addition method. For this purpose, a known amount of Tz was spiked to the formulated preparations and the total amount of the dye was estimated. The results are summarized in Table 5 and Table 6. The t- and F-tests indicated no significant differences between the calculated and theoretical values of both the proposed and the reported methods [24] [44] at 95% confidence level, which indicate good precision and accuracy in the analysis of investigated Tz in commercial food products.

4. CONCLUSION

In this paper, two spectrophotometric methods for the determination of Tz with Cu(II) and $KMnO_4$were proposed and successfully validated. The methods were based on the redox reaction with Cu(II) followed by complex formation and oxidation with strongly alkaline $KMnO_4$. The positive value of change in enthalpy (DH) for complexation reaction suggested that the reaction was endothermic favourable at high temperature. The negative value of free energy change indicated that copper-Tz redox reaction, dissociation of the ligand and complexation process were spontaneous. It was found that the complexation method was selective and sensitive enough to enable the determination of lower amounts of Tz and could be applied to the food quality control.

CITE THIS PAPER

Magda M. S. Saleh, Elham Y. Hashem, Najat O. A. Al-Salahi (2016) Oxidation and Complexation-Based Spectrophotometric Methods for Sensitive

Determination of Tartrazine E102 in Some Commercial Food Samples. *Computational Chemistry*,**04**,51-64. doi: 10.4236/cc.2016.42005

REFERENCES

1. Argetino, C.A. (2010) Chapter XII. http://www.anmat.gov.ar/alimentos/codigea/CAPITULO_II.pdf
2. FDA (2007) Summary of Color Additives for Use in United States in Foods, Drugs, Cosmetics and Medical Devices. FDA.
3. Tanaka, T. (2006) Reproductive and Neurobehavioural Toxicity Study of Tartrazine Administered to Mice in the Diet. Food and Chemical Toxicology, 44, 179-187.?
4. Michel, O., Naeije, N., Bracamonte, M., Duchateau, J. and Sergysels, R. (1984) Decreased Sensitivity to Tartrazine after Aspirin Desensitization in an Asthmatic Patient Intolerant to Both Aspirin and Tartrazine. Annals of Allergy, 52, 368-370.?
5. Moutinho, I.L.D., Bertges, L.C. and Assis, R.V.C. (2007) Prolonged Use of the Food Dye Tartrazine (FD&C Yellow n? 5) and Its Effects on the Gastric Mucosa of Wistar Rats. Brazilian Journal of Biology, 67, 141-145.?
6. Al-Degs, Y.S. (2009) Determination of Three Dyes in Commercial Soft Drinks Using HLA/GO and Liquid Chromatography. Food Chemistry, 117, 485-490.?
7. Alves, S.P., Brum, D.M., de Andrade, é.C.B. and Netto, A.D.P. (2008) Determination of Synthetic Dyes in Selected Foodstuffs by High Performance Liquid Chromatography with UV-DAD Detection. Food Chemistry, 107, 489-496.?
8. Garc?a-Falcón, M.S. and Simal-Gandara, J. (2005) Determination of Food Dyes in Soft Drinks Containing Natural Pigments by Liquid Chromatography with Minimal Clean-Up. Food Control, 16, 293-297.?
9. Husain, A., Sawaya, W., Al-Omair, A., Al-Zenki, S., Al-Amiri, H., Ahmed, N. and Al-Sinan, M. (2006) Estimates of Dietary Exposure of Children to Artificial Food Colours in Kuwait. Food Additives and Contaminants, 23, 245-251.?
10. Ma, M., Luo, X., Chen, B., Su, S. and Yao, S. (2006) Simultaneous Determination of Water-Soluble and Fat-Soluble Synthetic Colorants in Foodstuff by High-Performance Liquid Chromatography-Diode Array Detection-Electrospray Mass Spectrometry. Journal of Chromatography A, 1103, 170-176.?
11. Minioti, K.S., Sakellariou, C.F. and Thomaidis, N.S. (2007) Determination of 13 Synthetic Food Colorants in Water- Soluble Foods by Reversed-Phase High-Performance Liquid Chromatography Coupled with Diode-Array Detector. Analytica Chimica Acta, 583, 103-110.
12. Poul, M., Jarry, G., Elhkim, M.O. and Poul, J.M. (2009) Lack of Genotoxic Effect of Food Dyes Amaranth, Sunset Yellow and Tartrazine and Their Metabolites in the Gut Micronucleus Assay in Mice. Food and Chemical Toxicology, 47, 443-448.?
13. Ghoreishi, S.M., Behpour, M. and Golestaneh, M. (2012) Simultaneous Determination of Sunset Yellow and Tartrazine in Soft Drinks Using Gold Nanoparticles Carbon Paste Electrode. Food Chemistry, 132, 637-641.?
14. KAPOR, M.A., Yamanaka, H., Carneiro, P.A. and Zanoni, M.V.B. (2001) Electroanalysis of Food Dyes: Determination of Indigo-Carmine and Tartrazine. Eclética Química, 26, 53-68.
15. Silva, M.L.S., Garcia, M.B.Q., Lima, J.L. and Barrado, E. (2007) Voltammetric Determination of Food Colorants Using a Polyallylamine Modified Tubular Electrode in a Multicommutated Flow System. Talanta, 72, 282-288.?
16. López-de-Alba, P.L., Michelini-Rodriguez, L.I., Wróbel, K. and Amador-Hernández, J. (1997) Extraction of Sunset Yellow and Tartrazine by Ion-Pair Formation with Adogen-464 and Their Simultaneous Determination by Bivariate Calibration and Derivative Spectrophotometry. Analyst, 122, 1575-1579.

17. Nevado, J.B., Flores, J.R., Cabanillas, C.G., Llerena, M.V. and Salcedo, A.C. (1998) Resolution of Ternary Mixtures of Tartrazine, Sunset Yellow and Ponceau 4R by Derivative Spectrophotometric Ratio Spectrum-Zero Crossing Method in Commercial Foods. Talanta, 46, 933-942.?
18. Berzas, J.J., Flores, J.R., Llerena, M.V. and Farinas, N.R. (1999) Spectrophotometric Resolution of Ternary Mixtures of Tartrazine, Patent Blue V and Indigo Carmine in Commercial Products. Analytica Chimica Acta, 391, 353-364.?
19. Vidotti, E.C., Cancino, J.C., Oliveira, C.C. and Rollemberg, M.D.C.E. (2005) Simultaneous Determination of Food Dyes by First Derivative Spectrophotometry with Sorption onto Polyurethane Foam. Analytical Sciences, 21, 149-153.?
20. Vidotti, E.C. and Rollemberg, M.D.C.E. (2006) Derivative Spectrophotometry: A Simple Strategy for Simultaneous Determination of Food Dyes. Química Nova, 29, 230-233.?
21. Llamas, N.E., Garrido, M., Di Nezio, M.S. and Band, B.S.F. (2009) Second Order Advantage in the Determination of Amaranth, Sunset Yellow FCF and Tartrazine by UV-Vis and Multivariate Curve Resolution-Alternating Least Squares. Analytica Chimica Acta, 655, 38-42.?
22. Ni, Y., Wang, Y. and Kokot, S. (2009) Simultaneous Kinetic Spectrophotometric Analysis of Five Synthetic Food Colorants with the Aid of Chemometrics. Talanta, 78, 432-441.?
23. Schenone, A.V., Culzoni, M.J., Marsili, N.R. and Goicoechea, H.C. (2013) Determination of Tartrazine in Beverage Samples by Stopped-Flow Analysis and Three-Way Multivariate Calibration of Non-Linear Kinetic-Spectrophotome- tric Data. Food chemistry, 138, 1928-1935.?
24. Sahraei, R., Farmany, A. and Mortazavi, S.S. (2013) A Nanosilver-Based Spectrophotometry Method for Sensitive Determination of Tartrazine in Food Samples. Food chemistry, 138, 1239-1242.?
25. Vogel, A.I. (1973) A Text Book of Quantitative Inorganic Analysis. 3rd Edition, Longman, London.
26. Nair, V.S.K. and Parthasarathy, S. (1971) Studies on Metal Complexes in Aqueous Solution—VII: 4-Nitro and 4-Methyl Phthalates of Some Transition Metals. Journal of Inorganic and Nuclear Chemistry, 33, 3019-3024.?
27. Dickerson, R.E., Geis, I. and Benjamin, I.W.A. (1976) Chemistry, Matter and the Universe. California, USA.
28. Ives, D.J.G. (1972) Chemical Thermodynamics, University Chemistry, Macdonald Technical and Scientific.
29. Kuban, V. and Havel, J. (1973) Some 2-(2-Thiazolylazo)-4-Methoxyphenol (TAMP) Complex Equilibria, Acid-Base Properties of TAMP in Water and in Various Mixed Solvents. Acta Chemica Scandinavica, 27, 528-540.
30. Wu, L.P., Li, Y.F., Huang, C.Z. and Zhang, Q. (2006) Visual Detection of Sudan Dyes Based on the Plasmon Resonance Light Scattering Signals of Silver Nanoparticles. Analytical Chemistry, 7, 5570-5577.
31. Huo, J., Guo, Y., Meng, S., Wang, M. and Wang, Y. (2010) Complex Formation of Sudan I with Cu (II) and Its Identification from Chilli Species. 4th International Conference on Bioinformatics and Biomedical Engineering (iCBBE), Chengdu, 18-20 June 2010, 1-4.
32. Gürkan, R. and Altunay, N. (2013) A Reliable Method of Quantification of Trace Copper in Beverages with and without Alcohol by Spectrophotometry after Cloud Point Extraction. Química Nova, 36, 1146-1154.??
33. Hassan, E.M. and Belal, F. (2002) Kinetic Spectrophotometric Determination of Nizatidine and Ranitidine in Pharmaceutical Preparations. Journal of Pharmaceutical and Biomedical Analysis, 27, 31-38.?
34. Saleh, M.M., Hashem, E.Y., Youssef, A.K. and Abdel-Kader, D.A. (2015) UV-Visible Spectrophotometric Methods for Direct Determination of Sulfasalazine Antibiotic Drug in Its Pharmaceutical Formulations. World Journal of Phar- macy and Pharmaceutical Sciences, 4, 205-226.

35. Voznica, P., Havel, J. and Sommer, L. (1980) The Reactions of Gallium, Indium and Thallium with 2-(2-Pyridylazo)- 1-Naphthol-4-Sulphonic Acid and Their Spectrophotometric Determination. Collection of Czechoslovak Chemical Communications, 45, 54-79.?
36. Idriss, K.A., Seleim, M.M., Saleh, M.S., Abu-Bakr, M.S. and Sedaira, H. (1988) Spectrophotometric Study of the Complexation Equilibria of Zirconium (IV) with 1-Amino-4-Hydroxyanthraquinone and the Determination of Zirconium. Analyst, 113, 1643-1647.?
37. Idriss, K.A. and Saleh, M.M. (1993) Acid-Base Characteristics of Naphthazarin and Solution Equilibria of Yttrium (III) Chelates. Analyst, 118, 223-227.?
38. Gupta, K.K., Kaur, T. and Dadhich, E. (2013) Polarographic Study of Mixed Ligand Complexes of Pb(II) and Tl(I) with Thio Disuccinic Acid and Some Amino Acids in Aqueous Medium. Journal of Ultra Chemistry, 9, 249-256.
39. Al-Sarawy, A.A., El-Bindary, A.A., El-Sonbati, A.Z. and Mokpel, M.M. (2006) Potentiometric and Thermodynamic Studies of Azosulfonamide Drugs. Polish Journal of Chemistry, 80, 289-295.?
40. Rahman, N., Khan, N.A. and Azmi, S.N.H. (2004) Extractive Spectrophotometric Methods for the Determination of Nifedipine in Pharmaceutical Formulations Using Bromocresol Green, Bromophenol Blue, Bromothymol Blue and Eriochrome Black T. Il Farmaco, 59, 47-54.?
41. Job, P. (1928) Formation and Stability of Inorganic Complexes in Solution. Annali di Chimica, 9, 113-203.
42. Sandell, E.B. (1959) Colorimetric Determination of Traces of Metals.? 3rd Edition, Interscience Publishers, Inc., New York.
43. Narade, S., Patil, S., Surve, S., Shete, D. and Pore, Y. (2010) Simultaneous UV Spectrophotometric Method for the Determination of Diacerein and Aceclofenac in Tablets. Journal of Pharmaceutical Science and Research, 2, 137-142.?
44. Li, R., Jiang, Z.T. and Liu, Y.H. (2008) Direct Solid-Phase Spectrophotometric Determination of Tartrazine in Soft Drinks Using β-Cyclodextrin Polymer as Support. Journal of Food and Drug Analysis, 16, 91-96.?

CHAPTER 17

A Comprehensive Strategy to Detect the Fraudulent Adulteration of Herbs: The Oregano Approach

Connor Black, Simon A. Haughey , Olivier P. Chevallier, Pamela Galvin-King, Christopher T. Elliott

Institute for Global Food Security, Advanced ASSET Centre, School of Biological Sciences, Queen's University Belfast, Northern Ireland, United Kingdom

ABSTRACT

Fraud in the global food supply chain is becoming increasingly common due to the huge profits associated with this type of criminal activity. Food commodities and ingredients that are expensive and are part of complex supply chains are particularly vulnerable. Both herbs and spices fit these criteria perfectly and yet strategies to detect fraudulent adulteration are still far from robust. An FT-IR screening method coupled to data analysis using chemometrics and a second method using LC-HRMS were developed, with the latter detecting commonly used adulterants by biomarker identification. The two tier testing strategy was applied to 78 samples obtained from a variety of retail and on-line sources. There was 100% agreement between the two tests that over 24% of all samples tested had some form of adulterants present. The innovative strategy devised could potentially be used for testing the global supply chains for fraud in many different forms of herbs.

Keywords: Oregano; Authenticity; Adulteration; Fourier transform infrared; High resolution mass spectrometry; Biomarkers

1. INTRODUCTION

Globally, herbs and spices play a significant part in the diets of many as they are important ingredients in a multitude of foods, beverages, medicines and cosmetics. With consumers having greater access and a desire to use these products, the demand has increased vastly over the last thirty years making it a multibillion dollar industry (Furth & Cox, 2004). Marieschi, Torelli, Poli, Sacchetti, and Bruni (2009) stated that the global herb and spice trade was worth $2.97 billion with the EU market amounting to 520 thousand tonnes and a value of €1.8 billion. Sales in 2014 at all United Kingdom (UK) supermarkets were £173 million for dried herbs and spices, and £107 million for fresh herbs (spices not included) (Nielsen, 2015). Economically motivated adulteration (EMA) of food is a common concept, which has occurred within the food industry since trading began (Spink & Moyer, 2013). As is the case with any food commodity, there is a greater possibility of food adulteration when the demand and prices increase and when complex supply chains are involved. Herbs and spices fulfil all this criteria.

Herbs and spices are two very different food items with spices tending to be bright vibrant colours emanating often from warm climates in a diverse form e.g. cumin and turmeric and herbs are usually green leaved and derived from plants in cooler environments e.g. parsley and thyme. Supply and demand is a fundamental economic principle which determines the price of all commodities. However, as well as this, the price of spices is also dictated by the intensity of their colours and therefore, common adulteration of spices had been the addition of illegal dyes such as Sudan dyes, which are group 3 genotoxic carcinogens (Cornet *et al.*, 2006 ; Ruf *et al.*, 2012). However, since this issue was highlighted in 2003, the addition of Sudan dyes as a food additive has been banned worldwide. Herbs are not traded on colour and so there is no economic advantage gained from adding dyes. Instead, the price of herbs is dependent on how compact the product is and therefore bulking agents have been used as most commercially sold herbs tend to be either chopped or ground. Consequently, it is relatively easy to add other cheaper ground bulking agents without the supply chain and indeed the consumer noticing.

Oregano is a culinary herb most commonly associated with pizzas and other Mediterranean dishes. The main producers of oregano reside in the United States of America, Mexico, Greece and Turkey. Compared to most herbs, oregano has a complicated history as the true identity of it is very difficult to define. This is partly due to the large heterogeneity of the *Origanum* genus, but also due to the grouping of different botanical genera; *Origanum* (*lamiaceae*) from the Mediterranean and *Lippia* (*verbenaceae*) from Mexico (Marieschi et al. 2009). Due to the confusion, this led to a clear market distinction between Mediterranean and Mexican oregano, with both having different cleanliness specifications such as the addition of sumac leaves (ASTA, 2015). Mexican oregano has a much stronger and robust flavour compared to Mediterranean oregano, which could be due to the varying percentages of essential oils within the leaves. The essential oil percentage in Mexican oregano leaves is around 3–4%, whilst the percentage within Mediterranean oregano leaves is around 2–2.5%.

Even with the clear commercial distinction between Mexican and Mediterranean oregano, there are still several different definitions regarding Mediterranean oregano. The European Pharmacopoeia (PhEur) and the European Spice Association only allow *Origanum vulgare L.* ssp. *hirtum* and *Origanum onites L.*, to be marketed as true oregano with impurities of extraneous materials of up to 2% being considered tolerable (ESA, 2015a; ESA, 2015b; Marieschi *et al.*, 2011a; Marieschi *et al.*, 2011b ; European Pharmacopoeia Journal, 2005). However, ISO/FDIS 7925 (2015) allow leaves of all *Origanum* genus, species and subspecies except *Origanum majorana L*, to be marketed as oregano. Impurities of up to 1% are considered tolerable, which is in line with the value accepted by American Spice Trade Association (ASTA, 2015 ; Nielsen, 2015).

There has been a range of different detection methods developed and used in the determination of the authenticity of food. In a review by Reid, O'Donnell, and Downey (2006) many methods of detections were scrutinised including spectroscopy (ultraviolet-visible (UV), near infrared (NIR), mid infrared (MIR), Raman), isotopic analysis, chromatography, electric nose, polymerase chain reaction, enzyme-linked immunosorbent assay and thermal analysis – all of which are techniques that have been applied to food authentication since 2001. Food fingerprinting is of particular interest as a method of detection. Ellis et al. (2012) published a review of some of the fingerprinting technologies, with particular interest paid to NIR, MIR and Raman spectroscopic techniques. Few analytical methods are available to screen a large number of dried plant materials quickly and the detection of oregano adulteration has been limited to a few techniques. Marieschi *et al.*, 2009; Marieschi *et al.*, 2010; Marieschi *et al.*, 2011a ; Marieschi *et al.*, 2011b applied a random amplified polymorphic

DNA (RAPD) method which has led to the development of sequence-characterized amplified region makers (SCARs) for a number of potential adulterants; *Rhus coriaria L.*, *Cistus incanus L.*, *Olea europaea L.*, *Rubus coriaria L.*, which lack a clearly detectable essential oil profile and *Satureja montana L.*, *O. majorana L.*, which belong to the lamiaceae family. Work has also been carried out by Bononi *et al.*, 2010 ; Bononi and Tateo, 2011 who have utilized liquid-chromatography mass spectrometry (LC-MS) and gas-chromatography mass spectrometry (GC-MS) to identify the presence of olive leaves in ground oregano, using oleuropein as a marker.

The aim of this study was to develop and fully validate a two-tier approach utilising Fourier-Transform Infrared spectroscopy (FTIR) and Liquid Chromatography High Resolution Mass spectrometry (LC-HRMS) to screen for and confirm oregano adulteration. When these two techniques are combined with multivariate data analysis software they have the ability to process a large number of samples (Gao *et al.*, 2014 ; Osorio *et al.*, 2014). By applying FTIR and LC-HRMS the ability to produce the world's first comprehensive testing system of oregano adulteration was trialled. A survey of commercially available oregano samples in the UK were then undertaken to determine the current level of abuse.

2. MATERIALS AND METHODS

2.1. Sample collection and preparation

Samples of oregano with full provenance and traceability and a number of previously identified adulterants (olive leaves, myrtle leaves, sumac leaves, cistus leaves, hazelnut leaves), were sourced from different parts of the world. Commercially available oregano samples were purchased at various retailers including convenience shops, supermarkets and market places in the UK and Ireland. In addition samples were also purchased from online retailers a small number that were obtained from EU and non EU countries. The samples were milled to a homogeneous powder on a PM-100 Retsch Planetary Ball Mill (Haan, Germany) by weighing approximately 5 g into grinding jars and milling at 500 rpm for 5 min.

For LC-HRMS analysis, milled homogenate herb sample (0.05 g) was extracted in 2 mL of methanol/water solution (1:1, v/v), mixed for 10 min, sonicated for 15 min at maximum frequency in a water bath at room temperature, centrifuged at 10,000*g* for 10 min at 4 °C and the supernatant collected (1 mL). The supernatant was dried and reconstituted in 1.5 mL of ultra-pure water. Subsequently, the extract was filtered through a 0.22 µm Costar Spin-X Centrifuge Tube Filter (10,000*g* at 4 °C for 10 min). Filtered extracts were immediately transferred into Waters maximum recovery vials for UPLC-QTof-MS analysis.

2.2. Spectral data acquisition using Fourier-Transform Infrared (FTIR)

For FTIR, the milled samples were placed in the ATR sample area of a Thermo Nicolet iS5 spectrometer (Thermo Fisher Scientific, Dublin, Ireland) equipped with ATR iD5 diamond crystal and ZnSe lens and DTGS KBr detector. The slip-clutch pressure tower is applied to the sample and tightened until the correct pressure was utilised which gives more reproducible results. Each spectrum was acquired in the 550–4000 cm^{-1} range. The acquisition parameters were: number of sample scans: 32; collection length: 47 s; resolution: 4.000; levels of zero filling: 0, number of scan points: 12415; laser frequency: 11742.96 cm^{-1}; apodization: N-B Strong; phase correction: mertz; number of background scans: 32; background gain: 4.0. The acquisition was repeated 3 times. Spectral data for each sample was averaged before further data processing.

2.3. Chromatographic and mass spectrometry conditions

Analyses were carried out on a Waters Acquity UPLC I-Class system (Milford, MA, USA) coupled to a Waters Xevo G2-S QTof mass spectrometer (Manchester, UK) with an electrospray ionisation source operating in positive or negative mode with lock-spray interface for real time accurate mass correction. Instrument settings were as follow: source temperature was set at 120 °C, cone gas flow at 50 L h^{-1}, desolvation temperature at 450 °C, and desolvation gas flow at 850 L h^{-1}. The capillary voltage was set at 1.0 kV in positive mode and 1.5 kV in negative mode, respectively. Source offset was 60 (arbitrary unit). Mass spectra data were acquired in continuum mode using MS^Efunction (low energy: 4 eV; high energy: ramp from 15 to 30 eV) over the range *m/z* 50–1200 with a scan time of 0.08 s. A lock-mass solution of Leucine Enkephalin (1 ng μL^{-1}) in methanol/water containing 0.1% formic acid (1:1, v/v) was continuously infused into the MS via the lock-spray at a flow rate of 5 μL min^{-1}.

The chromatographic separation was conducted on an Acquity HSS T3 column (100 mm × 2.1 mm, 1.8 μm). The column oven temperature was set at 45 °C, injection volume at 5 μL and flow rate at 0.4 mL min^{-1}. Mobile phase consisted of (A) water with 0.1% formic acid and (B) methanol with 0.1% formic acid. The gradient was set as follows: 1.50 min of 99% (A) followed by a linear increase from 1% to 99% (B) over 15 min, isocratic cleaning step at 99% (B) for 2 min, then returned to initial conditions 99% (A) over 0.25 min and column equilibration step at 99% (A) for 1.25 min.

Each sample was injected three times in order to assure reproducibility. Prior to all analyses 10 pooled conditioning samples were injected. For quality control pooled samples were injected at intervals of every 10 samples throughout the entire experiment to determine the chromatographic reproducibility of retention times and peak intensities (Chevallier *et al.*, 2015 ; Graham *et al.*, 2013).

2.4. Data processing and data analysis

Principal Components Analysis (PCA), an unsupervised technique, and Orthogonal Partial Least Squares Discriminate Analysis (OPLS-DA), a supervised technique, were used for building the qualitative models in this investigation as previously described (Haughey, Galvin-King, Ho, Bell, & Elliott, 2015).

The generation of calibration models was carried out using similar methodology as previously published (Haughey, Galvin-King, Malechaux, & Elliott, 2014). The data pre-processing included standard normal variate technique (SNV), which compensates for differences in pathlengths due to scattering effects, 2nd order derivative and Pareto scaling using the SIMCA 14 chemometric software. Spectral data was analysed between the wavenumber ranges 550–1800 cm^{-1} and 2800–3999 cm^{-1}. The data generated include R2 which is an estimate of the fit of the model and Q2 which is an estimate of the predictive ability of the model and it is calculated by cross-validation. The latter is calculated by removing each 1/7th of the data in succession and building a new model on the remaining data with the omitted data predicted using this method of cross validation. Predicted Residual Sum of Squares (PRESS) is calculated by comparison with the original data with the best predictability of the model indicated by a low value. The SIMCA 14 chemometric software automatically converts PRESS into Q2 to resemble the scale of the R2 with good predictions having high Q2.

Raw data generated by the mass spectrometer were imported to Progenesis QI 2.0 software (Waters, Newcastle, UK). After data conversion to the appropriate format using a filter set at 2, data were aligned to the best pool sample selected and peak picking from 0.5–17.5 min was carried out with sensitivity set at automatic and chromatographic peak width to 0.08. The analysed spectral data were then exported to SIMCA 14 for multivariate analysis. As a quality control measure all the spectral data were Center Scaled and analysed using PCA. All pooled samples (QC) were found to be tightly clustered within the center of each representative scores plot which indicates good reproducibility of the data. Following this, all data were mean centred, Pareto scaled and grouped into Adulterant and Oregano prior to OPLS-DA. R2 (cumulative), Q2 (cumulative) and Root Mean

Squared Error of cross validation (RMSECV) were used to determine the validity of the model. R2 (cum) indicates the variation described by all components in the model and Q2 is a measure of how accurately the model can predict class membership.

3. RESULTS AND DISCUSSION

3.1. FT-IR spectroscopic analysis

FTIR is a technique that is based on the absorbance of light at particular wavelengths and has been a popular methodology in detecting food adulteration. In a review by Rodriguez-Saona and Allendorf (2011), examples of using FTIR in conjunction with multivariate analysis has been applied for authentication of herbal products, fruit juices, agricultural products, edible oils, dairy, and numerous other food products. In this study, the powdered oregano and adulterant samples (olive leaves, myrtle leaves, hazelnut leaves, cistus leaves, sumac leaves) were analysed on the FT-IR spectrometer. Fig. 1 shows the FT-IR spectra of pure oregano, olive leaves and myrtle leaves. Although there was some overlap observed between the peaks in the spectrum of the oregano and the peaks in the spectra of the adulterants, visually there are observable differences in the fingerprint region from 900 to 1800 cm^{-1}. FTIR peaks are attributed for stretching and bending vibrations that characterize the functional groups. The regions of interest included: (i) 1100–1400 cm^{-1}, generally the most prominent peak, is due to the vibration peak of CO in alcohol hydroxyl group (ii) 1400–1500 cm^{-1} corresponding to CO and CC stretching vibrations specific to phenyl groups; (ii) 1500–1600 cm^{-1} corresponds to aromatic vibrations and N-H bending and (iv) 1600–1740 cm^{-1} corresponding to bending NH, CO stretching (aldehydes, ketones, esters, free fatty acids and glycerides). To further observe the influence of adulterants on the spectrum of oregano, the latter was adulterated in 10% additions (0–100%) of olive leaves and the spectra recorded. Fig. 2 shows the resulting spectra indicating the monotonic increase intensity exemplified by the peak shown in the inset. Due to these differences identified in the spectral data it was possible to apply chemometric modelling for discriminant analysis. The chemometric software (SIMCA 14) was used to generate a qualitative model using PCA (unsupervised) and supervised OPLS-DA (supervised) algorithms with Pareto scaling to determine if it was possible to differentiate pure oregano from its adulterants. The data was pre-processed using SNV, 2nd order derivative algorithm with Savitzky-Golay smoothing (11 point window and 2nd order polynomial). For the unsupervised PCA model, the first four principal components describe most of the variation (84.5%) as follows: PC1 43.3%; PC2 19.4%; PC3 14.5%; PC4 7.3%. Separation was achieved mainly along PC1 and PC2 with the positive scores related to oregano samples and negative scores associated with the adulterants. The measure of fit (R2) of this PCA model was 94% and the measure of predictive ability (Q2), based on cross validation was 86%. For the supervised chemometric model, OPLS-DA was used with the same pre-processing parameters used for the PCA plot. The OPLS-DA model generated one predictive components and one orthogonal components which explained 34.8% and 23.6% of the differences respectively. The measure of prediction (Q2), based on cross-validation, was 95.9% and RMSECV = 9.7%, indicating very good predictability of the data. Fig. 3A shows PCA and OPLS-DA scores plots, and indicated that this method could be used for discriminant analysis and this approach to rapidly screen for adulteration of oregano is sufficiently robust and thus fit for purpose.

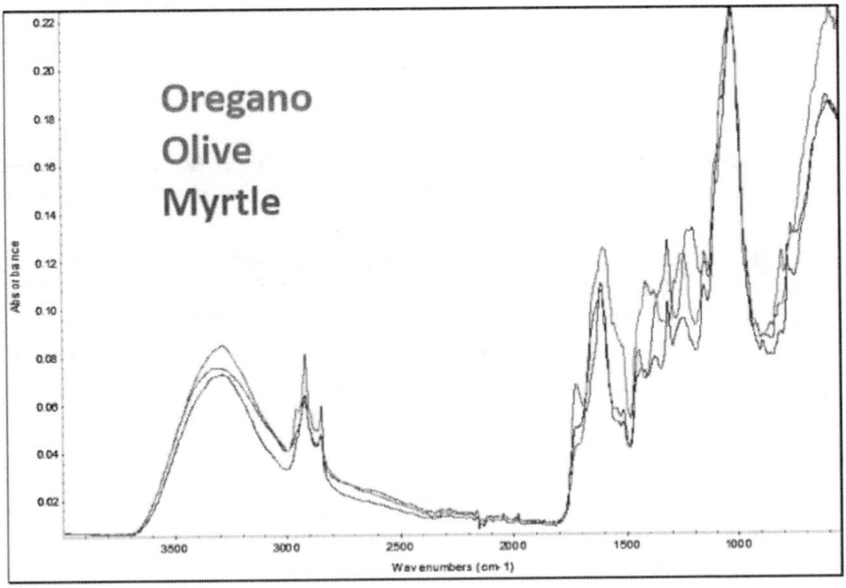

Fig. 1. FT-IR spectra of oregano and the adulterants Olive leaves and Myrtle leaves.

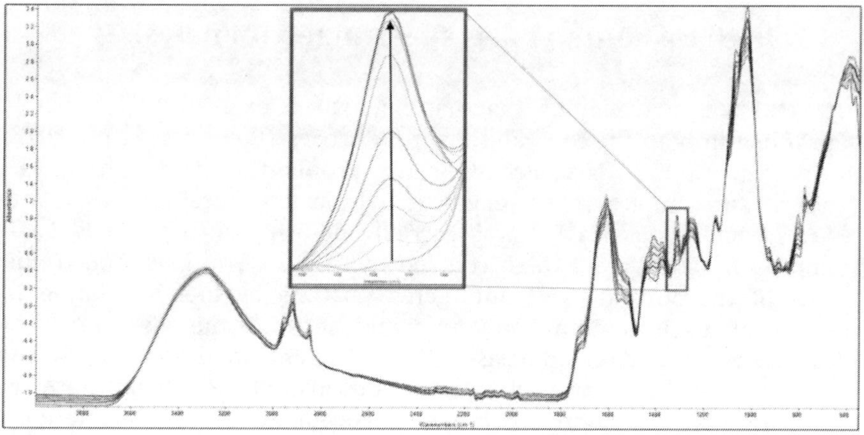

Fig. 2. FT-IR spectra of Oregano adulterated with olive leaves in 10% additions (0–100%) showing a monotonic increase in intensity exemplified by the inset with the arrow indicating the increase in olive leaf adulteration.

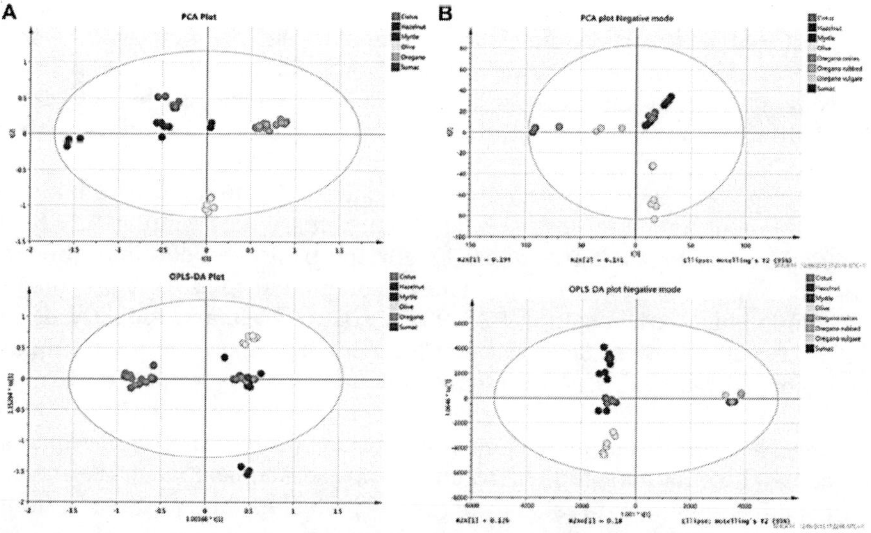

Fig. 3. (A) Unsupervised PCA and Supervised OPLS-DA scores plots from FTIR spectral data; (B) Unsupervised PCA and Supervised OPLS-DA scores plots from LC-HRMS data in positive ionisation mode.

3.2. High resolution mass spectrometric analysis

LC-HRMS is a technique that has been used extensively for metabolic profiling in both the food and plant industries (Commisso *et al.*, 2013; De Vos *et al.*, 2007 ; Masson *et al.*, 2010). Profiling analysis of oregano and adulterant samples (olive leaves, myrtle leaves, hazelnut leaves, sumac leaves, and cistus leaves) was carried out using an untargeted analysis approach on a Waters UPLC coupled to a G2-S QT of mass spectrometer. One of the advantage of untargeted analysis by high resolution mass spectrometry in combination with chemometrics is the possibility to build models on an untargeted basis, with subsequent exploration of the data to discover the characteristic markers that contribute most significantly to the classification. Furthermore, recent improvements in instrumentation and processing software allow a faster, more reproducible and more comprehensive data analysis. In our case, up to 4500 ions in each ionisation mode were reliably detected along the chromatographic gradient (Fig. 4). The extracted data were then exported to chemometric software to be subjected to similar data treatment as was the spectroscopic data presented above. The PCA score plot generated (Fig. 3B) showed clear discrimination between the pure oregano and the adulterants, with the oregano samples clustered together on one side of the plot and the adulterant samples scattered on the other side of the plot. Additionally, there was clear separation between the two oregano species which can be marketed as Mediterranean oregano; *Origanum vulgare* and *Origanum onites*, with one sample which contained both species situated in-between the two groups. OPLS-DA was then performed and a model was generated with one latent component and three orthogonal components with resulting R2 = 99.7%, Q2 = 94.4% and RMSECV of 10.3% for the positive mode ionisation data and another one with one latent component and three orthogonal components with resulting R2 = 99.3%, Q2 = 94.9% and RMSECV of 9.4% for negative mode. Several other individual models were subsequently produce by comparing Oregano to each individual adulterant (olive, myrtle, hazelnut, cistus, sumac) with their respective S-plots to enable the potential identification of markers to each adulterant (Table 1). This methodology allowed the identification of 16 unique markers in positive mode and 12 in negative mode, with all adulterant samples having at least 4 unique markers. This data will be used for the future development of a targeted method using MS/MS analysis.

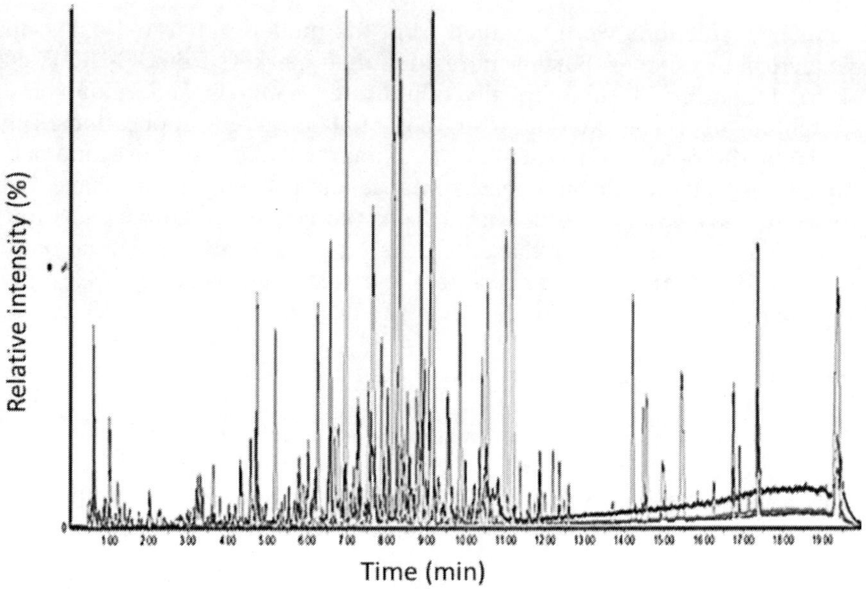

Fig. 4. Overlay full scan chromatograms of oregano, olive leaves and myrtle leaves.

Table 1. Values of the statistical parameters obtained for different OPLS-DA models generated using UPLC-QT of MS data for both ionisation modes.

	Ionisation mode	Latent component	Orthogonal component	R2 (cum)	Q2 (cum)	RMSECV
Myrtle Vs Oregano	ESI –	1	1	0.994	0.984	0.062
	ESI +	1	3	0.999	0.994	0.039
Sumac Vs Oregano	ESI –	1	0	0.99	0.95	0.102
	ESI +	1	1	1	0.997	0.022
Olive Vs Oregano	ESI –	1	2	0.996	0.934	0.127
	ESI +	1	1	0.997	0.957	0.102
Cistus Vs Oregano	ESI –	1	2	0.997	0.982	0.066
	ESI +	1	5	1	0.961	0.097
Hazelnut Vs Oregano	ESI –	1	1	0.991	0.936	0.105
	ESI +	1	1	0.994	0.953	0.089

3.3. Survey of commercial samples

To test the models developed for both analytical methodologies, a survey of oregano was carried out which included samples from retailers, service sector, internet sources and a few bought from commercial outlets in countries outside UK/Ireland. The spectral data generated for these samples from the FTIR and LC-HRMS were predicted as unknowns using the relevant OPLS-DA model produced earlier.

Similar predictions were obtained from the models generated using both analytical techniques. Furthermore, identified markers from LC-HRMS data were found in all commercially adulterated samples and reinforce the potential of these markers for potential targeted applications. Table 2 shows the results of the survey of commercially available oregano based on the spectroscopic and spectrometric data. The samples have been broken down into those procured from the retail and service sectors in UK/Ireland and those purchased on the internet or at commercial outlets outside of UK/Ireland. The results show that approximately 24% of the oregano samples tested were adulterated and the scale of adulteration ranged from 30% to over 70%, indeed two samples had virtually no oregano present. The scale and level of the adulteration uncovered was not expected. In addition the similar figures for 'store' bought and 'internet' bought were surprising as one would have thought the retail trade would have better systems of control in place. The most common adulterants found in the samples were olive leaves and myrtle leaves. The results of the survey were passed to the regulator to alert the affected companies to the economically motivated fraud ongoing in the sector. Based on the results of this survey the samples which were indicated to be oregano by both methods can be included in the calibrations models to increase the robustness of the test.

Table 2. Results from the oregano survey.

Oregano Survey	UK/Ireland[a]	Internet/Other[b]
Samples Tested	53	25
Samples Adulterated	13	6
Samples Adulterated %	24.5	24
Level of Adulteration[c]	~30 to >70%	~30 to >70%
Most Common Adulterants	1. Olive leaves 2. Myrtle leaves	1. Olive leaves 2. Myrtle leaves

[a] Includes Retail and Service Sector.
[b] Includes Amazon, Ebay and Purchases made abroad.
[c] Based on scores from chemometric analysis.

4. CONCLUSIONS

The detection of fraud in foods and food ingredients has become an even more important topic since the horsemeat scandal of 2013. Many consumers lost faith in the food they were purchasing and the food industry recognised that more robust measures in terms of auditing and testing had to be put in place. Often fraud is perpetrated in high value food commodities and those which come via complex supply chains. Probably herbs and spices fit these characteristics more than any other food ingredients and are thus highly vulnerable. Testing methods for the food industry must be easy to use, rapid and low costs. Our two tier system of testing provides not only a cost effective means of testing but one also that will survive rigours of a legal process. The survey data presented is disturbing in the level of adulteration found. It is clear that a serious level of fraud is being perpetrated and that bona fide businesses and consumers are being financially harmed. It is likely similar (if not worse) levels of fraud are occurring in many global regions. We believe the system we have developed and validated for oregano should be expanded to cover all herbs sold in the market. Only then will there be a sufficient deterrent in place to stop fraudulent activity in these widely consumed food ingredients.

ACKNOWLEDGEMENTS

The research undertaken was funded by a UK BBSRC CASE studentship BB/M0503162/1, Invest NI Proof of Concept Project PoC452 and EU FP7 FOODINTEGRITY project (agreement no: 613688).

REFERENCES

1. ASTA 2015. (2015). American spice trade. Association, ASTA cleanliness specifications for spices, seeds and herbs, http://www.ina.or.id/knoma-hpsp/herb/HPSP-12-Ollop. Accessed 10/10.
2. Bononi, M., Fiordaliso, I., & Tateo, F. (2010). Rapid GC/MS test for identification of Olea Europaea L. leaves in ground oregano. Italian Journal of Food Science, 22, 479–483.
3. Bononi, M., & Tateo, F. (2011). LC/MS/MS-ESI (–) identification of oleuropein as marker of Olea europaea L. leaves used as a bulking agent in ground oregano and sage. Italian Journal of Food Science, 23, 145–151.
4. Chevallier, O. P., Graham, S. F., Alonso, E., Duffy, C., Silke, J., Campbell, K., Botana, L. M., et al. (2015). New insights into the causes of human illness due to consumption of azaspiracid contaminated shellfish. Scientific Reports, 5, 1–8.
5. Commisso, M., Strazzer, P., Toffali, K., Stocchero, M., & Guzzo, F. (2013). Untargeted metabolomics: An emerging approach to determine the composition of herbal products. Computational and Structural Biotechnology Journal, 4, 1–7.
6. Cornet, V., Govaert, Y., Moens, G., Van Loco, J., & Degroodt, J. M. (2006). Development of a fast analytical method for the determination of sudan dyes in chili- and curry-containing foodstuffs by high-performance liquid chromatographyı photodiode array detection. Journal of Agricultural and Food Chemistry, 54, 639–644.
7. De Vos, R. C. H., Moco, S., Lommen, A., Keurentjes, J. J. B., Bino, R. J., & Hall, R. D. (2007). Untargeted large-scale plant metabolomics using liquid chromatography coupled to mass spectrometry. Nature Protocols, 2, 778–791.

8. European Pharmacopoeia Journal 2005, 2155–2156.
9. Ellis, D. I., Brewster, V. L., Dunn, W. B., Allwood, J. W., Golovanov, A. P., & Goodacre, R. (2012). Fingerprinting food: Current technologies for the detection of food adulteration and contamination. Chemical Society Reviews, 41, 5706–5727.
10. ESA (2015a). European Spice Association, Quality Minima Documenthttp://www.esa-spices.org/index-esa.html/publications-esa. Accessed 10.10.2015. ESA (2015b). European spice association, list of culinary herbs and spiceshttp:// www.esa-spices.org/index-esa.html/publications-esa. Accessed 10.10.2015. Furth, P., & Cox, D. (2004). The spice market expands. Food Technology, 58, 30.
11. Gao, B., Qin, F., Ding, T., Chen, Y., Lu, W., & Yu, L. (2014). Differentiating organically and conventionally grown oregano using ultraperformance liquid chromatography mass spectrometry (UPLC-MS), headspace gas chromatography with flame ionization detection (headspace-GC-FID), and flow injection mass spectrum (FIMS) fingerprints combined with multivariate data analysis. Journal of Agricultural and Food Chemistry, 62, 8075–8084.
12. Graham, S. F., Chevallier, O. P., Roberts, D., Hoelscher, C., Elliott, C. T., & Green, B. D. (2013). Investigation of the human brain metabolome to identify potential markers for early diagnosis and therapeutic targets of Alzheimer's disease. Analytical Chemistry, 85, 1803–1811.
13. Haughey, S. A., Galvin-King, P., Ho, Y.-C., Bell, S. E. J., & Elliott, C. T. (2015). The feasibility of using near infrared and Raman spectroscopic techniques to detect fraudulent adulteration of chili powders with Sudan dye. Food Control, 48, 75–83.
14. Haughey, S. A., Galvin-King, P., Malechaux, A., & Elliott, C. T. (2014). The use of handheld near-infrared reflectance spectroscopy (NIRS) for the proximate analysis of poultry feed and to detect melamine adulteration of soya bean meal. Analytical Methods, 7, 181–186.
15. ISO 7925 (2015). Dried oregano (Origanum vulgare L.) – Whole or ground leaves – Specification. Accessed 10.10.2015. Marieschi, M., Torelli, A., Bianchi, A., & Bruni, R. (2011a). Detecting Satureja montana L. and Origanum majorana L. by means of SCAR-PCR in commercial samples of Mediterranean oregano. Food Control, 22, 542–548.
16. Marieschi, M., Torelli, A., Bianchi, A., & Bruni, R. (2011b). Development of a SCAR marker for the identification of Olea europaea L.: A newly detected adulterant in commercial mediterranean oregano. Food Chemistry, 126, 705–709.
17. Marieschi, M., Torelli, A., Poli, F., Bianchi, A., & Bruni, R. (2010). Quality control of commercial Mediterranean oregano: Development of SCAR markers for the detection of the adulterants Cistus incanus L., Rubus caesius L. and Rhus coriaria L. Food Control, 21, 998–1003.
18. Marieschi, M., Torelli, A., Poli, F., Sacchetti, G., & Bruni, R. (2009). RAPD-based method for the quality control of Mediterranean oregano and its contribution to pharmacognostic techniques. Journal of Agricultural and Food Chemistry, 57, 1835–1840.
19. Masson, P., Alves, A. C., Ebbels, T. M. D., Nicholson, J. K., & Want, E. J. (2010). Optimization and evaluation of metabolite extraction protocols for untargeted metabolic profiling of liver samples by UPLC-MS. Analytical Chemistry, 82, 7779–7786.
20. Nielsen (2015). Accessed 10.10.2015. Osorio, M. T., Haughey, S. A., Elliott, C. T., & Koidis, A. (2014). Evaluation of methodologies to determine vegetable oil species present in oil mixtures: Proposition of an approach to meet the EU legislation demands for correct vegetable oils labelling. Food Research International, 60, 66–75.
21. Reid, L. M., O'Donnell, C. P., & Downey, G. (2006). Recent technological advances for the determination of food authenticity. Trends in Food Science & Technology, 17, 344–353.
22. Rodriguez-Saona, L. E., & Allendorf, M. E. (2011). Use of FTIR for rapid authentication and detection of adulteration of food. In M. P. Doyle & T. R.

Klaenhammer (Eds.). Annual review of food science and technology (Vol. 2, pp. 467–483).
23. Ruf, J., Walter, P., Kandler, H., & Kaufmann, A. (2012). Discovery and structural elucidation of the illegal azo dye Basic Red 46 in sumac spice. Food Additives & Contaminants, 29, 897–907.
24. Spink, J., & Moyer, D. C. (2013). Understanding and combating food fraud. Food Technology, 67, 30–35.

CHAPTER 18

Fast Ion Chromatographic Method for the Determination of Formates in Alcoholic Drinks

Maria Balcerzak[1] & Dawid Kapica[1]

[1] *Department of Analytical Chemistry, Warsaw University of Technology, Noakowskiego 3, 00-664 Warsaw, Poland*

ABSTRACT

Monitoring of human exposure to formic acid is important due to its high biological activity. Elevated amounts of formic acid in the human body can cause serious damage to optical nerve, respiratory failure, renal failure, deep metabolic acidosis, coma, and eventual death from cardiovascular arrest. Such symptoms are characteristic for methanol poisoning owing to rapid enzymatic conversion of methanol into formic acid of substantially higher than methanol alone toxicity. Early diagnosis of methanol consumed and formic acid produced is essential for successful medical treatment. Small amounts of methanol can be introduced into the human body with alcoholic drinks (methanol can be a by-product of distillation and fermentation processes). A fast method for the determination of formates in commercial alcoholic drinks by the use of suppressed ion chromatography (IC) with conductometric detection has been developed. The applicability of two anion-exchange columns, Metrohm Metrosep A Supp 7 and Dionex IonPac AS9-HC, for selective detection of formates in mixtures with acetates and common inorganic ions occurring in such kind of samples was examined. Quantitative isolation of formates from the matrix can be reached in less than 10 min using 3.6 mmol/L Na_2CO_3 as an eluent. Preliminary minimizing of alcoholic matrix by the evaporation (under IR source) allows to improve the quality of chromatographic results. The evaluated amounts of formates in two different Polish commercial products, "Absolwent vodka" and "Golden Rum," tested in the work were in the range of 0.2–3.1 mg/L. The 96–107% recoveries of the formates from the examined samples were found out.

Keywords: Ion chromatography Formates Alcoholic drinks Anionic profile

INTRODUCTION

High biological activity of formic acid generates the interest in monitoring of its sources for humans. The elevated level of the acid occurring in the human body can cause significant fatal effects such as serious damage to optical nerve, respiratory failure, hepatic and renal failure, and coma. Blood concentrations above 10 mmol/L (0.5 mg/mL) can induce deep metabolic acidosis and lead to death (Kinoshita et al. 1998; Tanaka et al. 1991). Formic acid can enter the human body through diet, production by intestinal microflora, and, as a result, methanol ingestion or inhalation of its vapors. Methanol ingestion is the most important source of high concentrations of formates in body fluids and tissues. Methanol is rapidly, within 2–24 h after ingestion, metabolized (dehydrogenated) by alcohol dehydrogenase (ADH) to formaldehyde and further to formic acid of substantially higher than methanol toxicity (Hovda et al. 2005; Kinoshita et al. 1998; Kruse 1992; Wallage and Watterson 2008). The content of formate in blood is a marker of methanol intoxication. Its rapid screening in the human body is of great significance for clinical diagnosis. High concentration of formic acid, corresponding to those related to fatal methanol poisonings, can be found in putrefied postmortem physiological fluids, presumably due to bacterial action and decomposition of lipids, carbohydrates, and proteins (Viinamäki et al. 2011).

Small amounts of methanol can occur in commercial alcoholic drinks as a by-product of fermentation and distillation steps (5.2–85.6 mg/L in vodkas (Arbuzov and Savchuk 2002) and 3.1–4.6 mg/100 mL ethanol in rums (Lachenmeier et al. 2003)). Wide consumption of alcoholic beverages can make a source of formic acid for humans (Kapur et al. 2007). Until now, data on the content of formates in such products are scarce. Analytical control of the sources of formic acid for humans is of great value for the general food industry. Organic acids, including formic acid, can affect sensory properties of food items as well as their stability.

Gas chromatography (Lee et al. 2012; Lee et al. 2013; Ligor et al. 2008; Ryhl-Svendsen and Glastrup 2002; Sokoro et al. 2007), capillary electrophoresis (Klampfl et al. 2000; Kubáň and Boček 2013; Kubáň et al. 2013; Kubáň et al. 2014; Pantučkova et al. 2013; Suárez-Luque et al. 2006; Travassos Lemos et al. 2015), and ion chromatography (Brudin et al. 2010; Käkölä et al. 2008; Kuo 1998; Lodi and Rossin 1995; Magne et al. 1998; Rantakokko et al. 2004; Walford 2002; Wojtczak et al. 2013) are techniques most often used for the determination of low-molecular-mass carboxylic acids in various materials, including food products. Complications arising from the major sample components, e.g., proteins, lipids, and salts, in the examination of the samples of biological matrices, can generate the necessity of coupling of the technique with a suitable sample preparation step for their elimination.

Ion chromatography (IC) is a technique widely applied in the investigations of ionic components of various materials (Weiss 2004). Fast, simple, and selective determination of inorganic and small organic ionic species in one run makes the technique competitive to the other instrumental techniques. Little sample preparation steps and small samples required for analysis are the advantages of IC analytical procedures. The possibility of the use of ion chromatography for rapid identification and quantification of small organic acids, among them formic acid, in various food and beverages samples was documented (Lodi and Rossin 1995; Magne et al. 1998; Walford 2002; Brudin et al. 2010; Wojtczak et al. 2013). The content of organic acids in foods can serve as a marker of degradation and fermentation processes occurring during production and storage steps. The IC technique is particularly suitable for direct examination of the occurrence of small organic acids in drinking waters, among them those being by-products of ozonation processes (Kuo 1998; Rantakokko et al. 2004). Such compounds can be consumed by microorganisms stimulating their regrowth. Simultaneous inorganic (cations and anions) multicomponent analytical results can make the basis of the differentiation in the originality of alcoholic products (Arbuzov and Savchuk 2002; Lachenmeier et al. 2003). Reliable determination of formates and acetates in the presence of inorganic anions requires careful optimization of the IC analytical procedure for effective baseline separation of formate, acetate, and fluoride signals exhibiting close retention times (Kontozova-Deutsch et al. 2008; Kontozova-Deutsch et al. 2011; Janiszewska and Balcerzak 2013).

In the present work, a rapid method for the determination of formates in vodka and rum products, differing in production processes, by the use of suppressed anionic IC with conductometric detection was developed. Vodka is produced by rectifying ethyl alcohol, the product of fermentation of grains (wheat, rye, corn), potatoes, sugar beets, and fruits. The distillation products (of 96–97% alcohol content) are diluted with water usually to alcohol level below 40% prior to bottling. Rum is a distilled alcoholic beverage made from sugarcane by-products (molasses or syrup) or sugarcane juice. The distilled spirits are diluted with water to 60–70% of alcohol and aged (usually for a year) in oak barrels. During aging, the rum acquires golden color.

MATERIALS AND METHODS

Apparatus

The 761 Compact IC System from Metrohm AG (Herisau, Switzerland) with a MSM II (Metrohm Suppressor Module) suppressor and a conductometric detector was used. Anion separations were investigated with the use of two different chromatographic columns, Metrosep A Supp 7 (250 mm × 4 mm) from Metrohm, Switzerland, and IonPac AS9-HC (250 mm × 4 mm) from Dionex, USA (the volume of a sample injection loop was 20 μL). Data acquisition and evaluation of chromatograms were carried out with the IC Net 2.3 Metrodata (Metrohm) software.

An infrared (IR) radiation lamp was used for sample evaporation.

Filters of 0.45 μm pore size (MCE, Fisherbrand) were used for sample filtration throughout experiments.

Reagents

All reagents used were of analytical grade. Water purified in a Millipore Elix3/Simplicity UV system (a specific resistance >18.2 M'Ω cm) was used in all experiments.

Standard solutions of formate and acetate were prepared from high-purity acids (from Fluka) dedicated to IC and containing 1000 ± 4 mg/L of each anion. Standard solutions of inorganic anions (fluoride, chloride, bromide, nitrate(V), phosphate, and sulfate) were prepared from high-purity sodium salts (from Fluka) containing 1000 ± 4 mg/L of each anion and dedicated to IC.

The eluent of 3.6 mmol/L of sodium carbonate prepared by dissolution of 381.6 mg of Na_2CO_3(from Fluka) in 1 L of deionized water was used.

Alcoholic Drinks Examined

Two different Polish industrial alcoholic products, pure vodka (Absolwent) and rum (Golden Rum), were bought on the local market. The samples taken from commercial bottles were directly analyzed by the analytical procedure described below.

Analytical Procedure

Alcohol samples of 5–50 mL were transformed into a 50-mL beaker and evaporated to ca. 1 mL under IR lamp (20 cm distance between the lamp and the sample surface). Each evaporated sample was diluted to 5.0 mL with deionized water and submitted to chromatographic analysis after filtration through a 0.45-μm filter.

RESULTS AND DISCUSSION

Formate is a small organic anion of ion chromatographic signal registered under the conditions commonly used for the determination of inorganic anions. The ion chromatogram of the mixture of formate and some inorganic anions (F^-, Cl^-, NO_2^-, Br^-, NO_3^-, HPO_4^{2-}, and SO_4^{2-}) registered when using a very popular anion-exchange A Supp 5 Metrohm column, Na_2CO_3 + $NaHCO_3$ eluent, and conductometric detection is given in Fig. 1. Acetate and oxalate are organic acids which signals can also be registered under such conditions. Close retention times of formate, acetate, and fluoride ions can provide problems with quantification of particular anions in the samples containing all of them.

A kind of chromatographic column and an eluent applied limit the effectiveness of the separation of formate ion signal from those of acetate and inorganic anions in case of their occurrence in the examined sample. In this work, the effectiveness of two anion-exchange columns, Metrosep A Supp 7 from Metrohm and IonPac AS9-HC from Dionex, in the separation of F^-, CH_3COO^-, and $HCOO^-$ ions and quantification of formates was examined. Both columns are dedicated to the determination of inorganic and small organic anions occurring in the examined solutions. The experiments carried out have shown that complete resolution of fluoride, acetate, and formate signals is possible with both columns and Na_2CO_3as an eluent. Figure 2 presents the ion chromatograms of the mixture of the ions obtained with the use of the examined columns and the same (3.6 mmol/L Na_2CO_3) eluent. Both columns exhibited similar effectiveness in the separation of F^-, CH_3COO^-, and $HCOO^-$ ions. The

Metrosep A Supp 7 column was chosen for further studies owing to shorter analysis time of the samples containing the other inorganic anions (Cl^-, NO_2^-, Br^-, NO_3^-, HPO_4^{2-}, SO_4^{2-}) usually occurring in food products and giving IC signals under the conditions used. The separation of all ions was possible in shorter, ca. 35 min, time on the Metrosep A Supp 7 column as compared with ca. 60 min when using the IonPac AS9-HC column. The analysis time of the mixture of all ions on the IonPac AS9-HC column decreases in case of more concentrated eluent (ca. 30 min for 9 mmol/L Na_2CO_3). The isolation of formate signal from the acetate is, however, worse under such conditions.

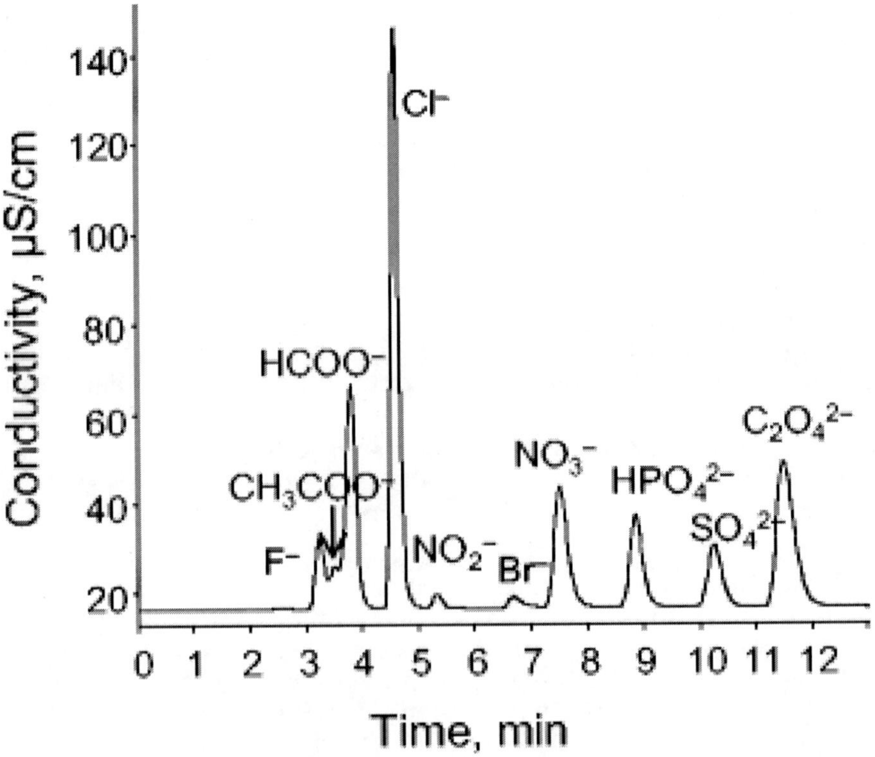

Fig. 1. The ion chromatogram of synthetic mixture of formate, acetate, and common inorganic anions registered with Metrohm Metrosep A Supp 5 chromatographic column (3.2 mmol/L Na_2CO_3 + 1.0 mmol/L $NaHCO_3$ eluent)

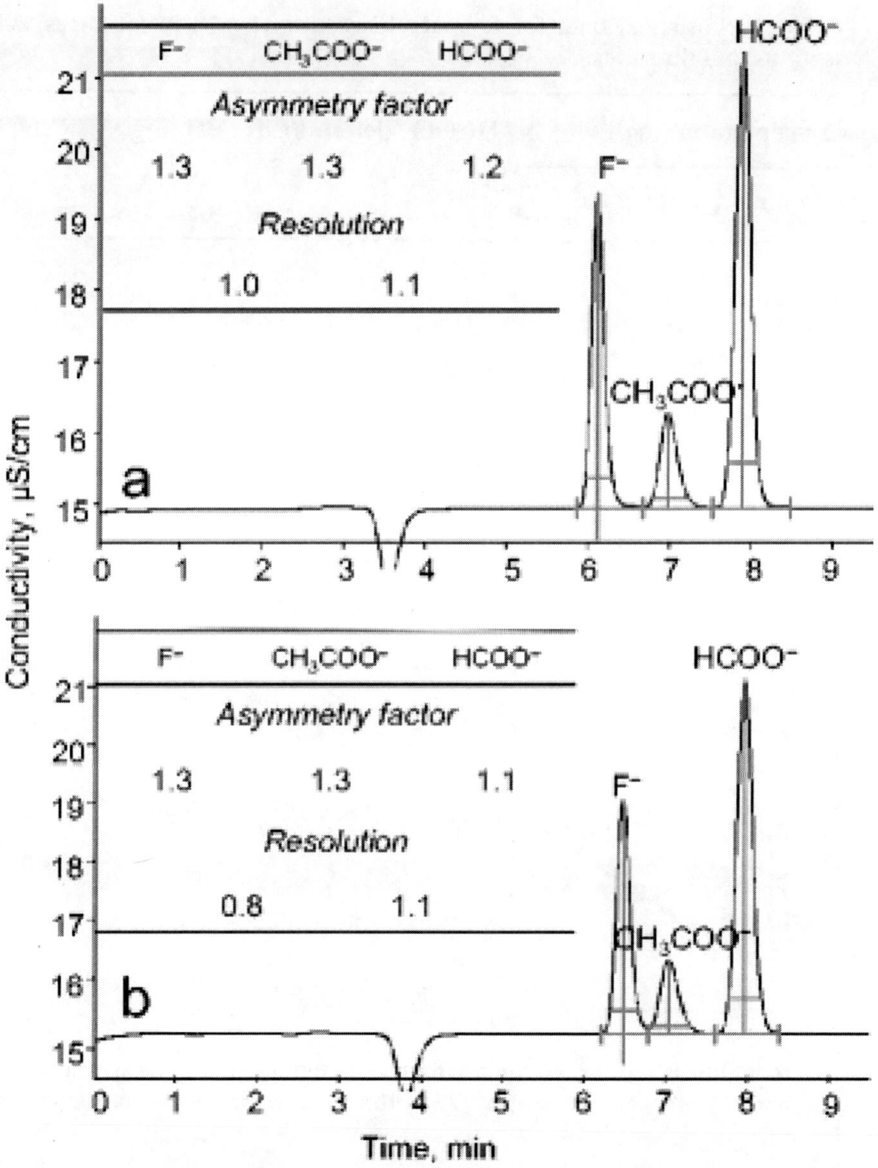

Fig. 2. The chromatograms of synthetic mixture of formate (10 mg/L), acetate (10 mg/L), and fluoride (2 mg/L) ions, and the values of asymmetry factors and resolutions obtained with the use of **a** Metrohm Metrosep A Supp 7 and **b** Dionex IonPac AS9-HC (3.6 mmol/L Na_2CO_3 eluent, 0.7 mL/min)

Table 1. Statistical data for formate IC results (n = 3) in mixtures with acetates and fluorides

Concentration, mg/L			$HCOO^-$ determined ± SD, mg/L	RSD, %
$HCOO^-$	CH_3COO^-	F^-		
0.5	0.5	0.1	0.50 ± 0.02	0.4
1.0	1.0	0.2	0.98 ± 0.01	0.9
2.0	2.0	0.4	1.98 ± 0.01	0.4
5.0	5.0	1.0	5.02 ± 0.02	0.3
10.0	10.0	2.0	10.01 ± 0.03	0.3

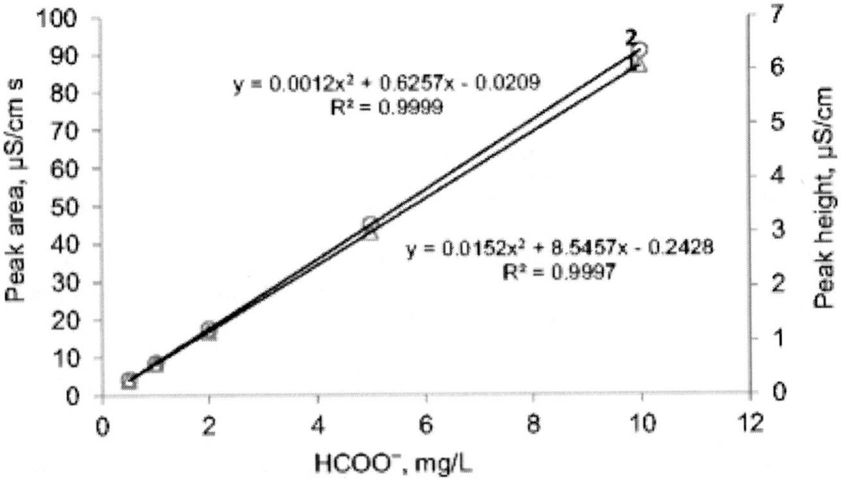

Fig. 3. Calibration curves for formate ions using *1* peak areas and *2* peak heights (Metrosep A Supp 7 column, 3.6 mmol/L Na_2CO_3 eluent, 0.7 mL/min)

Five standard solutions containing the mixtures of formates with acetates and fluorides were used for the calibration procedure (Table 1). Each of the calibration solution was run triplicate. Two calibration curves were generated by plotting (1) the peak areas and (2) the peak heights against the concentrations of formate in the injected standards (Fig. 3). Both parameters, peak areas and peak heights, can make the basis of the quantitation procedure. Statistical data of formate results obtained using peak areas for quantitation are presented in Table 1. Detection limit (DL) for formate ions was evaluated in the analysis of synthetic mixtures containing their minimum concentrations being detected: 0.25 mg/L ($HCOO^-$) in the mixture with 0.25 mg/L (CH_3COO^-) and 0.05 mg/L (F^-) ions and calculated as a triple standard deviation (SD) of the results ($n = 10$). DL and quantitation limit (QL) were evaluated as 0.014 and 0.042 mg/L, respectively. Calibration range corresponded to 0.04–10.0 mg/L of formate ions.

Analysis of Alcoholic Drink Samples

Two typical Polish alcoholic products, vodka (Absolwent) and rum (Golden Rum), available on the local market were analyzed for the confirmation of the applicability of the described chromatographic procedure for the determination of formates in such kind of samples. The samples of 5–50 mL taken from the commercial bottles were preliminary evaporated to ca. 1 mL to minimize the amount of alcoholic matrix. The evaporation was carried out under IR lamp (located ca. 20 cm above the surface of the sample), and the residue was diluted with deionized water to 5.0 mL of the final volume prior to submitting to chromatographic analysis. The experiments have shown that such approach improves baseline separation of formate signal. Figure 4 shows the IC chromatograms of the commercial rum sample directly submitted to IC analysis and after preliminary evaporation of ethanolic matrix. Lower amount of alcoholic matrix provides better adjustment of formate signal to the baseline. Changing the matrix from pure water to ethanol-water (1:4) does not practically affect quantification results. The regression curves for formate ions in pure aqueous and ethanolic-aqueous matrices calculated on the basis of peak areas are presented in Fig. 5.

The ion chromatograms of the examined spirit products under the conditions used are presented in Fig. 6. The evaluated contents of formate in both examined spirit drinks are presented in Table 2. Higher amounts of formates were determined in rum (3.11 ± 0.04 mg/L) as compared with vodka (0.21 ± 0.03 mg/L) samples. Spike recovery of formates (standard solution added to the commercial sample prior to the evaporation) was in the range of 96.0–107.0%. Figure 7 presents the chromatograms of the vodka sample spiked with formate (0.2 mg/L), fluoride (0.04 mg/L), and acetate (0.60 mg/L) ions.

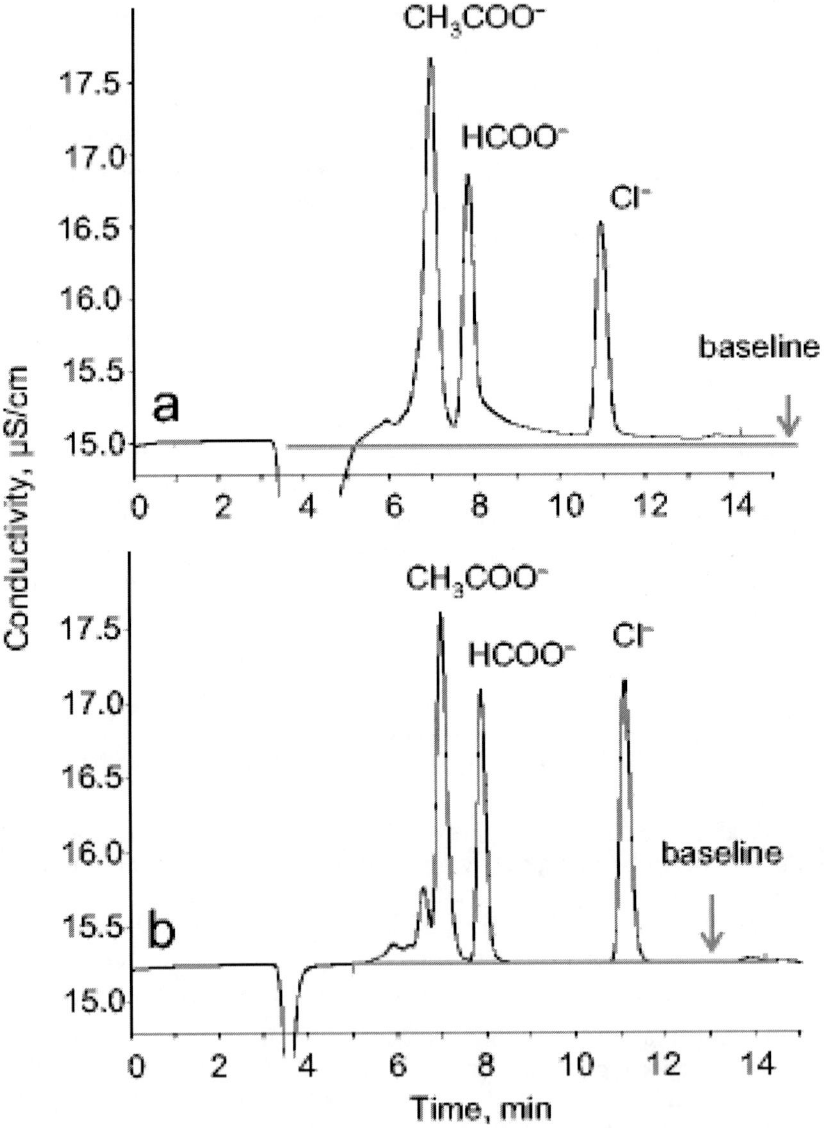

Fig. 4. he IC chromatograms of commercial Golden Rum. **a** Directly submitted to IC analysis. **b** After preliminary evaporation of ethanolic matrix (Metrosep A Supp 7 column, 3.6 mmol/L Na_2CO_3 eluent, 0.7 mL/min)

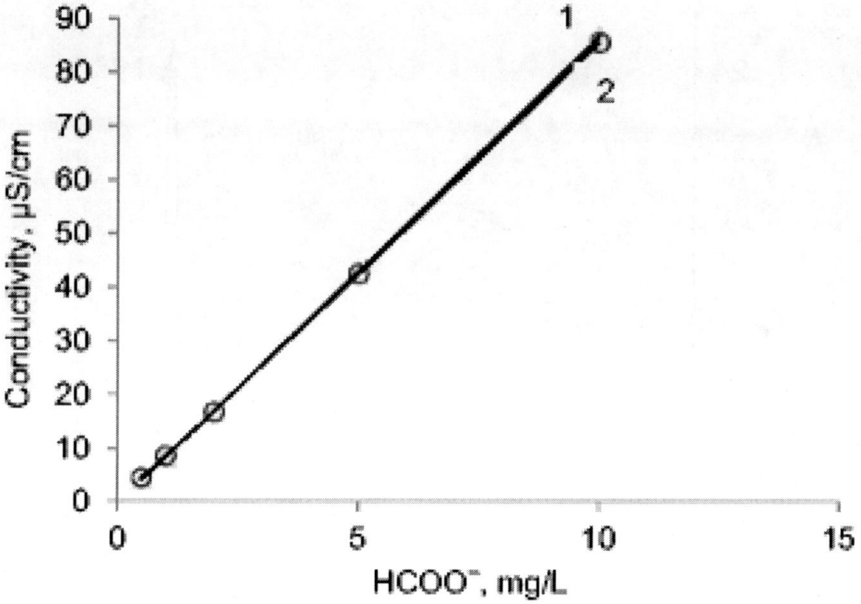

Fig. 5. Calibration curves for formate ions in *1* aqueous and *2* ethanolic-aqueous (1:4) solutions

Table 2. The results for formate in the examined commercial spirit drinks (n = 3)

Golden Rum			Absolwent vodka		
Sample, mL	Examined solution, mg/L	Original product, mg/L	Sample, mL	Examined solution, mg/L	Original product, mg/L
5	3.10	3.10	10	0.42	0.21
10	6.29	3.15	25	1.05	0.21
25	15.39	3.08	50	2.19	0.22
Statistical data (original products)					
$\bar{x}$ = 3.11 mg/L, SD = 0.04 mg/L, CV = 1.1%			$\bar{x}$ = 0.21 mg/L, SD = 0.03 mg/L, CV = 1.8%		

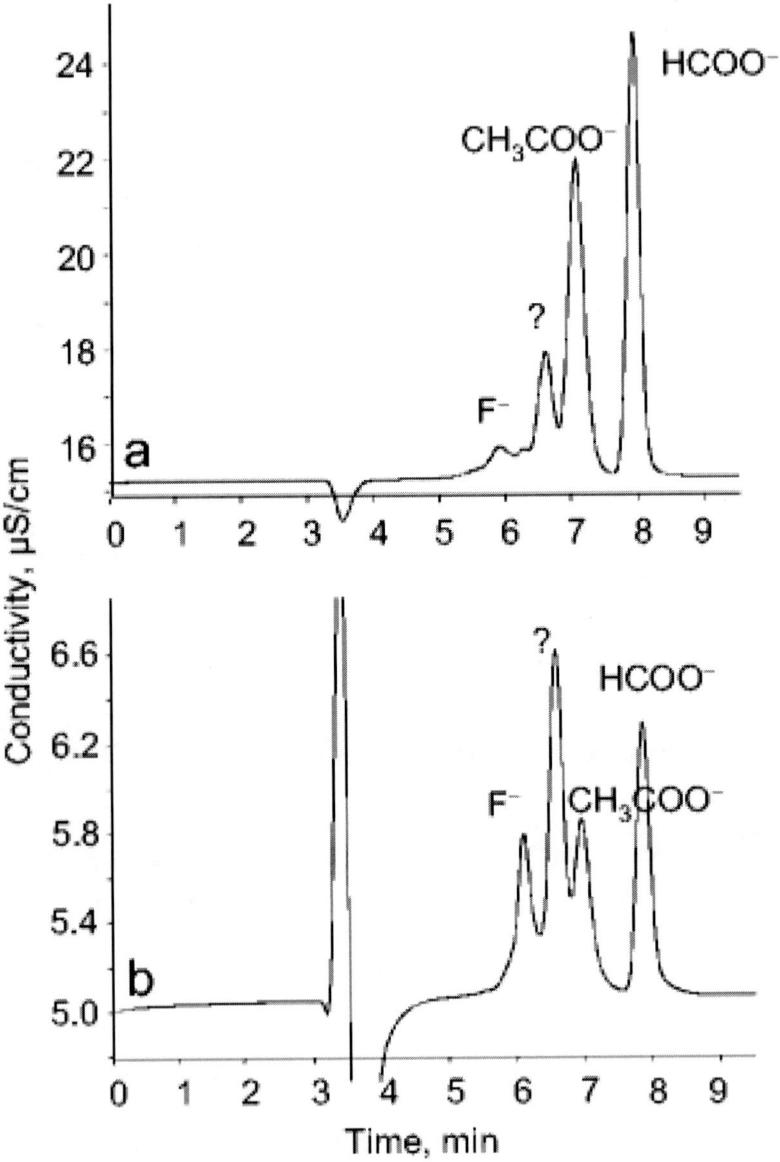

Fig. 6. The IC chromatograms of commercial. **a** Rum and **b** vodka samples (A Supp 7 column, 3.6 mmol/L Na_2CO_3 eluent, 0.7 mL/min) (?—unidentified peak)

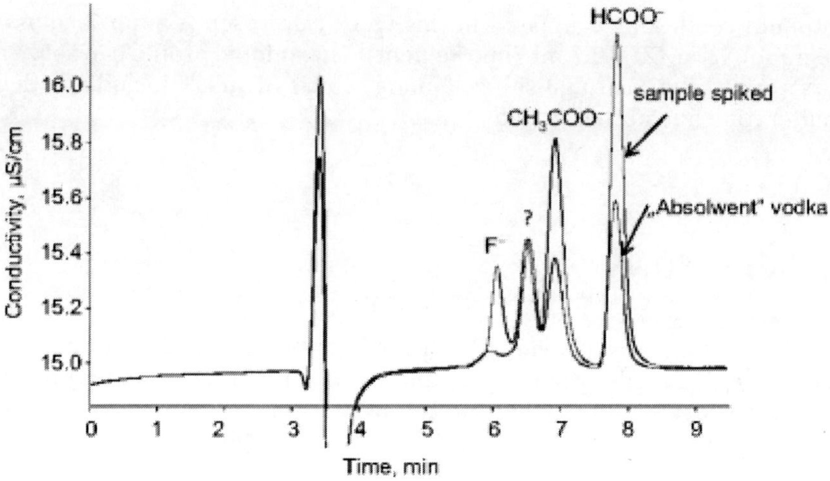

Fig. 7. The IC chromatograms of the original and spiked (formate (0.2 mg/L), fluoride (0.04 mg/L), and acetate (0.60 mg/L)) Absolwent vodka sample (?—unidentified peak)

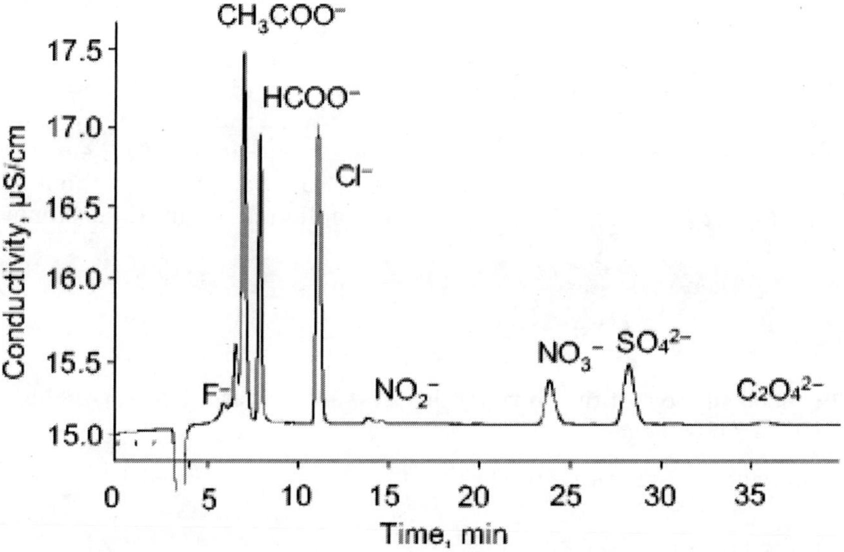

Fig. 8. The multi-anionic profile of Golden Rum sample (A Supp 7 column, 3.6 mmol/L Na_2CO_3eluent, 0.7 mL/min)

The developed method offers the possibility of rapid evaluation of the multi-anionic profile of alcoholic drinks. Figure 8 presents the whole anionic chromatogram of the examined Golden Rum sample under optimum conditions evaluated in this work (Metrosep A Supp 7 column, 3.6 mmol/L Na_2CO_3 (0.7 mL/min) eluent). The anionic profile can serve as a criterion for distinguishing among spirit drinks, including their adulteration and authentication.

CONCLUSIONS

Ion chromatography provides the possibility of rapid evaluation of the multi-anionic (small organic acids and common inorganic anions) profile of alcoholic drinks. Two analytical columns examined, Metrosep A Supp 7 from Metrohm and IonPac AS9-HC from Dionex, allow effective isolation of formate signals from such kind of samples. The analysis time of the mixture of the examined organic ($HCOO^-$, CH_3COO^-, and $C_2O_4^{2-}$) and common inorganic (F^-, Cl^-, NO_2^-, Br^-, NO_3^-, HPO_4^{2-}, and SO_4^{2-}) ions is shorter (ca. 35 min) when using the Metrosep A Supp 7 column as compared with the IonPac AS9-HC column (ca. 60 min) (3.6 mol/L Na_2CO_3 eluent in both cases). The described analytical procedure is applicable to direct analysis of alcoholic commercial products for the content of formates. Minimizing of the amount of alcoholic matrix by preliminary careful evaporation of the sample under IR light source improves baseline separation of formate ion signal. The evaluated amounts of formate ions in the examined rum samples were about 15 times higher (ca. 3 mg/L) than in vodka samples (ca. 0.2 mg/L) examined.

ACKNOWLEDGEMENTS

Financial support of the work by the Warsaw University of Technology is kindly acknowledged.

Compliance with Ethical Standards

CONFLICT OF INTEREST

Maria Balcerzak declares that she has no conflict of interest. Dawid Kapica declares that he has no conflict of interest.

ETHICAL APPROVAL

This article does not contain any studies with human participants or animals performed by any of the authors.

REFERENCES

1. Arbuzov VN, Savchuk SA (2002) Identification of vodkas by ion chromatography and gas chromatography. J Anal Chem 57:428–433
2. Brudin SS, Shellie RA, Haddad PR, Schoenmakers PJ (2010) Comprehensive two-dimensional liquid chromatography: ion chromatography × reversed-phase liquid chromatography for separation of low-molar-mass organic acids. J Chromatogr A 1217:6742–6746
3. Hovda KE, Urdal P, Jacobsen D (2005) Increased serum formate in the diagnosis of methanol poisoning. J Anal Toxicol 29:586–588
4. Janiszewska J, Balcerzak M (2013) Analytical problems with the evaluation of human exposure to fluorides from tea products. Food Anal Methods 6:1090–1098
5. Käkölä JM, Alén RJ, Isoaho JP, Matilainen RB (2008) Determination of low-molecular-mass aliphatic carboxylic acids and inorganic anions from kraft black liquors by ion chromatography. J Chromatogr A 1190:150–156
6. Kapur BM, Vandenbroucke AC, Adamchik Y, Lehotay DC, Carlen PL (2007) Formic acid, a novel metabolite of chronic ethanol abuse, causes neurotoxicity, which is prevented by folic acid. Alcohol Clin Exp Res 31:2114–2120
7. Kinoshita H, Ijiri I, Ameno S, Tanaka N, Kubota T, Tsujinaka M, Watanabe R, Ameno K (1998) Combined toxicity of methanol and formic acid: two cases of methanol poisoning. Int J Legal Med 111:334–335
8. Klampfl CW, Buchberger W, Haddad PR (2000) Determination of organic acids in food samples by capillary zone electrophoresis. J Chromatogr A 881:357–364CrossRefGoogle Scholar
9. Kontozova-Deutsch V, Krata A, Deutsch F, Bencs L, Van Grieken R (2008) Efficient separation of acetate and formate by ion chromatography: application to air samples in a cultural heritage environment. Talanta 75:418–423
10. Kontozova-Deutsch V, Deutsch F, Bencs L, Krata A, Van Grieken R, De Wael K (2011) Optimization of the ion chromatographic quantification of airborne fluoride, acetate and formate in the Metropolitan Museum of Art, New York. Talanta 86:372–376
11. Kruse JA (1992) Methanol poisoning. Intensive Care Med 18:391–397
12. Kubáň P, Boček R (2013) Direct analysis of formate in human plasma, serum and whole blood by in-line coupling of microdialysis to capillary electrophoresis for rapid diagnosis of methanol poisoning. Anal Chim Acta 768:82–89
13. Kubáň P, Foret F, Boček R (2013) Capillary electrophoresis with contactless conductometric detection for rapid screening of formate in blood serum after methanol intoxication. J Chromatogr A 1281:142–147
14. Kubáň P, Ďurč P, Bittová M, Foret F (2014) Separation of oxalate, formate and glycolate in human body fluid samples by capillary electrophoresis with contactless conductometric detection. J Chromatogr A 1325:241–246
15. Kuo CY (1998) Improved application of ion chromatographic determination of carboxylic acids in ozonated drinking water. J Chromatogr A 804:265–272
16. Lachenmeier DW, Attig R, Willi F, Athanasakis C (2003) The use of ion chromatography to detect adulteration of vodka and rum. Eur Food Res Technol 218:105–110

17. Lee XQ, Huang DK, Lou DW, Pawliszyn J (2012) Needle trap extraction for GC analysis of formic and acetic acids in aqueous solution. J Sep Sci 35:1675–1681
18. Lee XQ, Zhang LK, Huang DK, An N, Yang F, Jiang W, Fang B (2013) Analysis of the stable carbon isotope composition of formic and acetic acids. Anal Biochem 436:178–186
19. Ligor M, Jarmalaviciene R, Szumski M, Maruška A, Buszewski B (2008) Determination of volatile and non-volatile products of milk fermentation processes using capillary zone electrophoresis and solid phase microextraction coupled to gas chromatography. J Sep Sci 31:2707–2713
20. Lodi S, Rossin G (1995) Determination of some organic acids in sugar factory products. J Chromatogr 706:375–383
21. Magne V, Mathlouthi M, Robilland B (1998) Determination of some organic acids and inorganic anions in beet sugar by ionic HPLC. Food Chem 61:449–453
22. Pantučkova P, Kubáň P, Boček P (2013) Supported liquid membrane extraction coupled in-line to commercial capillary electrophoresis for rapid determination of formate in undiluted blood samples. J Chromatogr A 1299:33–39
23. Rantakokko P, Mustonen S, Yritys M, Vartiainen T (2004) Ion chromatographic method for the determination of selected inorganic anions and organic acids from raw and drinking waters using suppressor current switching to reduce the background noise. J Liquid Chromatogr & Related Technol 27:829–842
24. Ryhl-Svendsen M, Glastrup J (2002) Acetic acid and formic acid concentrations in the museum environment measured by SPME-GC/MS. Atm Environ 36:3909–3916
25. Sokoro AR, Lehotay D, Eichhorst J, Treble R (2007) Quantitative endogenous formate analysis in plasma using headspace gas chromatography without a headspace analyzer. J Anal Toxicol 31:342–346
26. Suárez-Luque S, Mato I, Huidobro JF, Simal-Lozano J, Sancho MT (2006) Capillary zone electrophoresis method for the determination of inorganic anions and formic acid in honey. J Agric Food Chem 54:9292–9296
27. Tanaka E, Honda K, Horiguchi H, Misawa S (1991) Postmortem determination of the biological distribution of formic acid in methanol intoxication. J Forensic Sci 36:936–938
28. Travassos Lemos MA, Cassella RJ, de Jesus DP (2015) A simple analytical method for determining inorganic anions and formate in virgin olive oils by capillary electrophoresis with capacitively coupled contactless conductivity detection. Food Control 57:327–332
29. Viinamäki J, Rasanen I, Vuori E, Ojanperä I (2011) Elevated formic acid concentrations in putrefied post-mortem blood and urine samples. Forensic Sci Int 208:42–46
30. Weiss J (2004) Handbook of ion chromatography. Weinheim, Wiley-VCH
31. Walford SN (2002) Applications of ion chromatography in cane sugar research and process problems. J Chromatogr A 956:187–199
32. Wallage HR, Watterson JH (2008) Formic acid and methanol concentrations in death investigations. J Anal Toxicol 32:241–247
33. Wojtczak M, Antczak A, Lisik K (2013) Contamination of commercial cane sugars by some organic acids and some inorganic anions. Food Chem 136:193–198

CHAPTER 19

Determination of Omega Fatty Acid Profiles in Egg Yolk by HILIC-LC-MS and GC-MS

Justyna Walczak[1] & Szymon Bocian[1] & Tomasz Kowalkowski[1,2] & Tadeusz Trziszka[3] & Bogusław Buszewski [1,2]

[1]*.Department of Environmental Chemistry and Bioanalytics, Faculty of ChemistryNicolaus Copernicus UniversityTorunPoland*

[2]*.Interdisciplinary Centre of Modern TechnologyNicolaus Copernicus UniversityTorunPoland*

[3]*.Department of Animal Products Technology and Quality ManagementWroclaw University of Environmental and Life SciencesWroclawPoland*

ABSTRACT

The paper presents the analysis of the profile composition of fatty acids in the molecules of phosphatidylcholine and phosphatidylethanolamine, by using hydrophilic interaction liquid chromatography and gas chromatography coupled with mass spectrometry. The profiles of 15 phosphatidylcholine and 8 phosphatidylethanolamine species were analyzed with a newly developed hydrophilic interaction liquid chromatography (HILIC)–electrospray ionization (ESI)–tandem mass spectroscopy (MS/MS) method, by using a new stationary bonded phase. The application of the new method in control and experimental groups of egg yolk revealed significant differences in the composition of phospholipid species containing mainly polyunsaturated fatty acids. Additionally, using GC-MS, the profile of fatty acids in four groups with different dietary supplementation of hens was analyzed and 20 fatty acids in egg yolks were determined. Monounsaturated fatty acids were found in higher amounts than saturated fatty acids and polyunsaturated fatty acids in egg yolks. Oleic acid (18:1) was the major monounsaturated fatty acid in egg yolk while palmitic acid (16:0) was the major saturated fatty acid. Linoleic acid (18:2), arachidonic acid (20:4), and docosahexaenoic acid (22:6) reached the highest levels among the polyunsaturated fatty acids.

Keywords: Lipidomics Phospholipids Fatty acids HILIC Stationary bonded phase

INTRODUCTION

Fats are the most concentrated source of energy in the human diet. They are a source of fatty acids (FAs), including unsaturated essential fatty acids (EFAs), and fat-soluble vitamins A, D, and E. They are also the source from which the body derives components to build the cells, tissues, organs, and materials for the synthesis of some biologically active substances (eicosanoids), in particular, prostaglandins belonging to tissue hormones such as prostacyclin (PGI_2) (Fetterman and Zdanowicz 2009). Depending on the presence of double bonds and their number, fatty acid molecules can be divided into the following: saturated fatty acids (SFAs), monounsaturated fatty acids (MUFAs), and polyunsaturated fatty acids (PUFAs). The position of the first double bond counted from the methyl group occurs between series of n-3 (ω-3) or n-6 (ω-6) acids. FAs of the family ω-3 (α-linolenic acid ALA—C18:3) and ω-6 (linoleic acid LA—C18:2) are not synthesized in the human body and most animals, due to the lack of desaturases which are responsible for introducing the double bond into acid molecule between 3–4 and 6–7 carbon atoms (counting from the methyl group), and therefore must be supplied with food (Behr 1994; Flachs et al. 2009). PUFAs delivered with food may undergo enzymatic transformations that rely on the introduction of further double bonds under the influence of a specific desaturase (Δ6, Δ5) and elongation of the hydrocarbon chain with the participation of elongase enzyme (Fig. 1). Transformation of EFAs takes place in the endoplasmic reticulum of cells, where, in the renaturing and elongation process, they are converted into long-chain LC PUFA—having an essential biological activity (Newton 1996). The sources of PUFAs are plants (Behr 1994), fish (especially cold water fish), clams, oysters, and shrimps (Friedman and Moe 2006), but also plant products, such as nuts (especially English walnuts), sesame seeds (Namiki 2007), and linseed, and vegetable oils such as soybean oil and rapeseed oil (Whelan and Rust 2006).

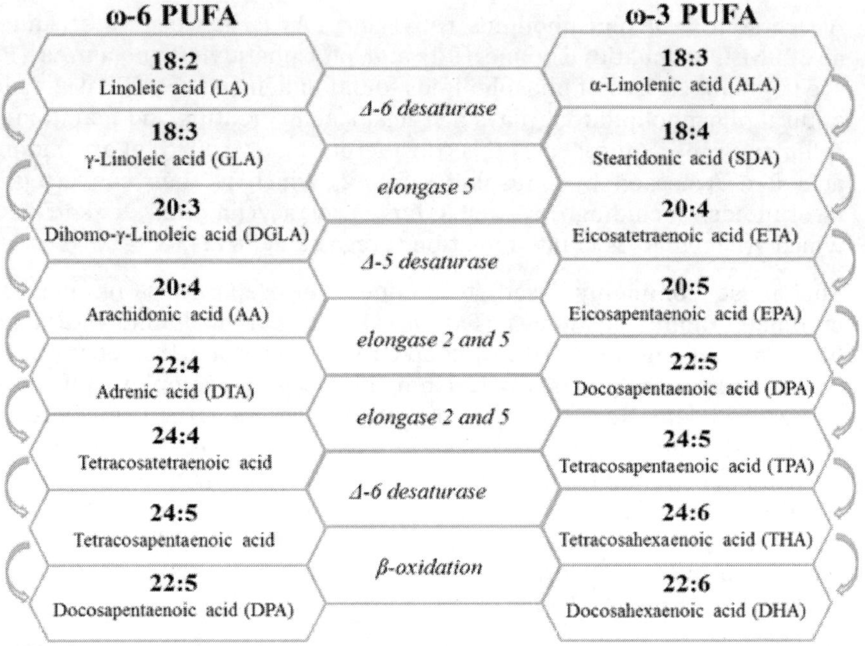

Fig. 1. Diagram of metabolism process of omega-3 and omega-6 fatty acids

LC PUFAs are deemed necessary in the human body, because they are essential for the proper development and functioning of the brain and retina. The omega-3 and omega-6 acids are the building components of phospholipidic membrane, and their proportion in the diet to a large extent reflects their relative proportions in the tissues of the body. Long-chain PUFAs are the components of myelin sheath of the peripheral nerves, synaptic membrane, and neurotransmitters (Hibbeln and Salem 1995). Omega-3 with omega-6 acids affects the normal construction and functioning of the skin. Omega-3 fatty acids prevent excessive drying of the skin and prevent its damage. A number of omega-6 acids are responsible for wound healing and epidermis regeneration (Bojarowicz and Wozniak 2008). Eicosapentaenoic acid (EPA) mainly stimulates the cardiovascular system, while docosahexaenoic acid (DHA) is an important structural component of highly active nerve tissue. DHA represents up to 60 % of the total fatty acids in the phospholipids of neurons and plays a key role in the development of the nervous system, occurring during gestation and early childhood. Too low level of DHA in the diet of women will shorten the gestation and cause low birth weight of children (Hansen 1994).

A rich source of phospholipids (PLs) and FAs is hen egg yolk (Ahn et al. 2006). Phosphatidylcholine (PC) and phosphatidylethanolamine (PE) are the main classes of phospholipids found in hen egg yolk. PC is a major natural phospholipid that plays a crucial role in creating and maintaining cell membranes. Moreover, PC is considered to be a source of arachidonic acid (AA) released by phospholipase A2, which is then converted to eicosanoids, including prostaglandins, prostacyclin, and leukotrienes, which are mediators of inflammation (Frega et al. 2012).

The most commonly used techniques for separation of PLs are chromatography techniques (Pari et al. 2016). Hydrophilic interaction liquid chromatography (HILIC) coupled with mass spectrometry (MS) is undoubtedly a helpful analytical tool in assays of complex substances (Buszewski and Noga 2012). Many papers mention the application of HILIC in the separation analysis of PLs. Zhao et al. (2010) proposed a very interesting method of separation and qualitative and quantitative analysis of PC from soy lecithin using TiO_2 column in HILIC mode. Holčapek and coworkers determined individual polar and nonpolar lipid classes in human plasma, erythrocytes, and biological tissues using HILIC (Holčapek et al. 2015; Lisa et al. 2011).

The aim of this work was to develop a new method for the separation of phospholipids extracted from egg yolk. For this purpose, a new stationary bonded phase was used, containing two amino groups and alkyl chain C12. Furthermore, hydrophilic interaction liquid chromatography coupled with tandem mass spectrometry (HILIC-MS/MS) was used for the analysis of individual species of PC and PE. This analysis was performed for the control group and the experimental group (by supplementation of diet). Additionally, the effect of hens' dietary supplementation on the content of PUFAs and the composition of the fatty acid profile of lipid yolk were determined.

EXPERIMENTAL

Chemicals

Acetonitrile and methanol (all LC-MS grade) were purchased from Sigma-Aldrich (Steinheim, Germany). Ultrapure water was obtained from the Milli-Q water system (Millipore, Bedford, MS, USA). The phospholipid standards of PC (≥99 %), PE (≥97 %), lysophosphatidylethanolamine (LPE) (≥97 %), lysophosphatidylcholine (LPC) (≥98 %), sphingomyelin (SM) (≥95 %), phosphatidylglycerol (PG) (≥97 %), and lysophosphatidylglycerol (LPG) (≥97 %) were purchased from Larodan

Lipids (Malmö, Sweden). All reagents used in the lab procedure were high-performance liquid chromatography (HPLC) grade; methanol, hexane, and dichloromethane were obtained from Sigma-Aldrich (Steinheim, Germany). Sodium chloride, sulfuric acid, and anhydrous potassium carbonate were obtained from POCh (Gliwice, Poland). Standards for 37 fatty acid methyl ester (FAME) mix (Supelco 37 Component FAME Mix C4-C24) used to establish calibration curves were purchased from Supelco/Sigma-Aldrich (Steinheim, Germany). Helium and argon, purity 99.999 %, were purchased from B.O.C. (Bydgoszcz, Poland).

A 1.5-mg/mL stock solution of all phospholipid standards for HILIC analysis was prepared in pure methanol. Working solutions from 0.2 to 1 mg/mL were obtained by diluting the stock solution with methanol and stored at −20 °C.

Samples

Pure PL fraction from egg yolk (Gladkowski et al. 2011) was obtained from Wroclaw University of Environmental and Life Sciences. Raw eggs were obtained from Japanese quail (*Coturnix japonica*). The control group of birds were given complete feeding stuff Farm line of G-090 for quail, which included soybean meal, wheat, corn, wheat bran, vegetable oils, calcium carbonate, monocalcium phosphate, sodium chloride, macro and trace elements, vitamins, amino acids, and biologically active compounds. The experimental group received feed Farm line of G-090 enriched with selenium yeast *Saccharomyces cerevisiae* containing 500 mg/kg selenium and a standard mix of feed enriched with linseed and fish oils, both in the amount of 1 to 6 %, and dried algae in the amount of 0.5 to 1.5 %. The diet was also supplemented by Humokarbowit and Humobentofet formulations in the amount of 0.5 to 2 %, and 0.01 % vitamin E. Four samples from egg yolk were analyzed. The first one was the control group (standard, not enriched diet); the second, third, and fourth were egg yolk from the experimental group (rich diet) in a 5-, 9-, and 13-week consuming diet for laying hens, respectively. More details on the composition of the diet are described by Gladkowski et al. (2014).

HILIC-LC-MS/MS Analysis

Phospholipid separation was performed by liquid chromatography using a homemade alkyl-amide stationary bonded phase that contains two amine groups and organic dodecanoyl chains. It is a modification of a commonly known alkyl-amide packing (Bocian et al. 2015; Buszewski et al. 1994). The stationary phase was synthesized on Kromasil 100 silica gel with specific surface area of 320 m^2/g, average pore diameter of 110 Å, and specific pore volume of 0.9 mL/g. Adsorbents were packed into 125 mm × 4.6 mm i.d. stainless steel columns, with the use of the slurry method, by means of laboratory-made equipment and a Haskel (Burbank, CA, USA) packing pump. Approximately 1.5 g of modified silica was made into slurry with 15 mL of chloroform and placed in the packing apparatus. Methanol was used as a packing pressurizing solvent during the filling process. Columns were packed at a constant pressure of 40 MPa. The obtained stationary phases exhibit high carbon content that equals 15.6 %, which provide the surface coverage density of 3 μmol/m^2. The application of the support with two amine groups allows to obtain higher coverage density of organic chains as well as better stability in the mobile phase that contains significant amount of water. Due to the composition of hydrophobic and polar groups, this stationary phase offers unique selectivity in the separation of analytes with polar functionalities, e.g., PLs. The structure of the chemically bonded stationary phase is shown in Fig. 2.

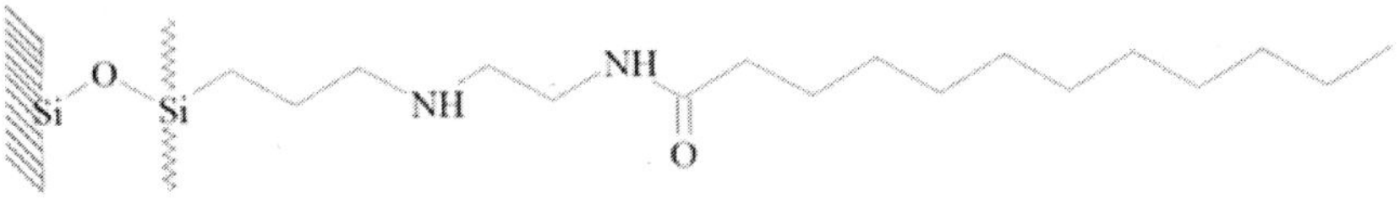

Fig. 2. The structure of the bonded stationary phase

The apparatus was a Shimadzu LC-MS 8050 system (Tokyo, Japan). The absorbance was measured at 206 nm using a diode array detector (SPD-M20A). The instrument includes a binary solvent delivery system (LC-30AD), a controller (CBM 20A), an autosampler (SIL-30A), and a column thermostat (CTO-20AC). Instrument control, data acquisition, and processing were performed with LabSolution 5.8 software for HPLC. The injection volume was 1 μL. The column was maintained at 25 °C. Isocratic mobile phase consisted of 80 % acetonitrile and 20 % water. The flow rate was 0.5 mL/min. MS/MS analysis was performed in the positive ionization mode on a triple quadrupole (Shimadzu, Tokyo, Japan) equipped with an electrospray ionization (ESI) source. The ESI settings were as follows: nebulizing gas flow 3 L/min, heating gas flow 10 L/min, temperature of the drying gas 400 °C, DL temperature 250 °C, and interface temperature 300 °C. PL species were monitored in the scheduled multiple reaction monitoring (MRM) mode. Product ion mode m/z 184 was used for identification of PC, and neutral loss mode of 141 was used for identification of PE. Fractions of PC and PE were collected manually to Eppendorf tubes, and each fraction was evaporated using Labconco CentriVap DNA concentrator (Kansas City, USA). Next, the fraction was dissolved in pure methanol. The collected fractions were analyzed using

Kinetex C18 column 150 × 4.6 mm with pore size of 100 Å (Phenomenex, Torrance, CA, USA). The flow rate of the mobile phase was 0.4 mL/min, the injection volume was 1 μL, and the separation temperature was 25 °C. Ninety percent acetonitrile and 10 % water were used as a mobile phase.

GC-MS Analysis of Fatty Acids

The GC-MS analysis was performed on 6890N gas chromatograph (Agilent Technologies, Waldbronn, Germany) coupled with mass spectrometer Agilent 5975 Inert XL MSD equipped with Zebron ZB-WAX (30 m length, 0.25 mm film, 0.25 μm internal diameter; Phenomenex, Torrance, CA, USA) capillary column. The oven temperature program was as follows: initially 60 °C held for 2 min, then ramped at 13 °C/min to 150 °C, and then ramped at 2 °C/min to 230 °C and held for 6 min. The temperature of the splitless injector was 240 °C. The MS analyses were carried out in full-scan mode, with scan mass range of m/z 35–450. The spectra were collected at electron ionization (EI) of 70 eV; both ion source and line transfer temperatures were set at 200 °C. FAMEs were identified by comparison of retention times and MS spectra with authentic standards. The acquisition of chromatographic data was performed by using the ChemStation 3 software (Agilent).

Derivatization of Fatty Acids

FAs were methylated according to the modified procedure described by Christie (1984). Ten milligrams of PL fraction from each sample of egg yolk was transferred into the test tube and dissolved in 4 mL of pure methanol. Four milliliters of 1 % solution of sulfuric acid (VI) in methanol was added, and the test tubes were placed in a water bath at 80 °C for 20 min. After cooling, 4 mL of 6 % NaCl was added. The samples were mixed, and the mixture was extracted three times with 4-mL portions of hexane. Hexane extracts were connected and washed with 6 mL of 4 % potassium carbonate solution and subsequently dried. Dried extracts were concentrated by evaporation of the solvent, then dissolved in dichloromethane and analyzed by GC-MS system. Each fraction was analyzed three times.

Statistical Analysis

Significant differences among the investigated groups in terms of the content of ∑ SFA, ∑ MUFA, ∑ PUFA, total ω-3, total ω-6, and ω-6/ω-3 ratio were analyzed by means of ANOVA followed by Newman–Keuls post hoc test ($p = 0.05$). Additionally, the factor analysis (FA) with Varimax rotation was also utilized to classify the fatty acid composition in different groups of breeds. All the calculation was performed on Statistica software (Statistica ver.7, StatSoft).

RESULT AND DISCUSSION

HILIC-ESI-MS/MS Methods for Determination of Species of PC and PE

In Fig. 3, a chromatogram of egg yolk PLs using a homemade stationary bonded phase in hydrophilic interaction liquid chromatography conditions is presented. The sample components were eluted within 20 min, with the following elution order: PG, LPG, PE, LPE, PC, SM, and LPC. Analysis of PLs from egg yolk using the new stationary phase allows to separate PLs with a polar "head" of the molecule.

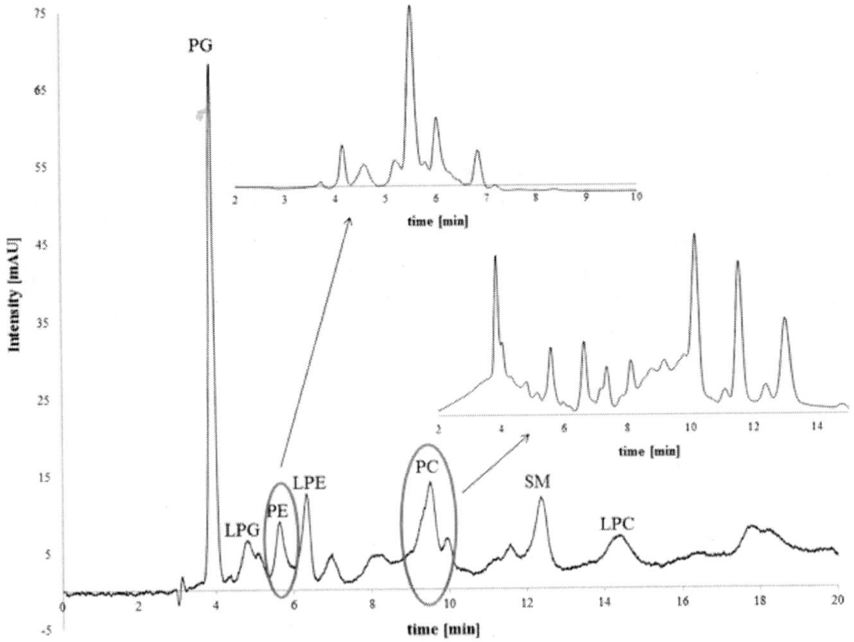

Fig. 3. Separation of the phospholipid classes in the egg yolk samples: *PG* phosphatidylglycerol, *LPG* lysophosphatidylglycerol, *PE* phos phatidylethanolamine, *LPE* lysophosphatidylethanolamine, *PC* phosphatid ylcholine, *SM* sphingomyelin, *LPC* lysophosphatidylcholine. Mobile phases are 80 % acetonitrile and 20 % water; flow rate 0.5 mL/min

The new stationary phase has a good selectivity in the designation of PLs. It is believed that the structure of the stationary phase (amino groups and the dodecanoyl chain) determines good selectivity against the analyzed compounds. Compared with other works (Anesi and Guella 2015; Schwalbe-Herrmann et al. 2010; Viidanoja 2015), in which the HILIC chromatography was used, the PLs were separated isocratically. Analysis of the hydrophilic interaction liquid chromatography–electrospray ionization–tandem mass spectroscopy (HILIC-ESI-MS/MS) was used to determine the fatty acid composition of the molecules of PC and PE (fractions 3 and 5, respectively, in Fig. 3). Fractions of PC and PE were separated using C18 column. The separation mechanism of PC and PE on the octadecyl stationary phase was based on hydrophobic interactions. Lipids which are more hydrophobic are strongly retained and eluted later from the column. Elution order of PLs in reversed phase depends on the differences in the chain lengths and the number of double bonds of acyl residues. The elution order of lipids in reversed phase is presented using the coefficient equivalent carbon number (ECN) for fatty acid (Perona and Ruiz-Gutierrez 2003). ECN is defined as $ECN = CN - 2n$ (where CN is the sum of the total carbon atom number of fatty acyl groups and n is the number of the sum of double bonds). The retention time of phospholipids increases proportionally to the ECN coefficient (Table 1). PC and PE are the predominant components of PLs in egg yolk. Due to the fact that egg is a complex biological matrix, it is very hard to separate the combination of FAs. In order to determine the molecule PC and PE, a specific scan mode MRM using positive ESI coupled to MS/MS was applied, since these lipids lose their headgroups during collision-induced dissociation (CID). PC can be identified by the appearance of the spectrum in positive ion mode with the value of fragment m/z 184, derived from phosphocholine headgroup $C_5H_{15}NO_4P^+$. In the case of PE, a loss of neutral molecule m/z 141 corresponds to the loss of the polar head $C_2H_9NO_4P^+$ (phosphoethanolamine) (Brugger et al. 1997). Table 1 presents all identified PUFAs contained in PC and PE for the two groups of egg yolks. In this study, we have found the following phospholipids, PC 44:12, PC 38:7, PC 38:6, PC 38:4, PC 36:6, PE 38:6, and PE 38:5, which occur in the experimental group (13 weeks of consuming the diet for laying hens), and they are not present in the control group.

GC-MS Analysis of Fatty Acids

The total relative fatty acid profile was analyzed in the egg yolk samples using GC-MS. Since the metabolism of FAs is dependent on the relative cellular fatty acid composition, rather than the absolute concentrations, we focused on the comparison of fatty acid profiles in the four studied groups. As shown in Tables 2, 3, 4, and 5, we were able to verify the results obtained with the metabolomic approach by targeted GC-MS analysis.

Table 1. Phospholipid profile of PC and PE species in control and experimental groups of egg yolk

PL species	Fatty acids	MRM	ECN	Control group	Experimental group
PC 44:12	22:6/22:6	878.7 → 184.1	20	–	+
PC 44:11	22:5/22:6	880.7 → 184.1	22	+	–
PC 42:10	20:4/22:6	854.6 → 184.1	22	+	+
PC 42:6	20:0/22:6	862.8 → 184.1	30	+	+
PC 40:7	20:3/20:4	832.7 → 184.1	26	+	–
PC 38:7	18:3/20:4	804.6 → 184.1	24	+	–
PC 38:7	16:1/22:6	804.7 → 184.1	24	–	+
PC 38:6	16:0/22:6	806.6 → 184.1	26	–	+
PC 38:5	18:2/20:3	808.8 → 184.1	28	+	–
PC 38:4	18:0/20:4	810.6 → 184.1	30	–	+
PC 36:6	16:1/20:5	778.7 → 184.1	24	–	+
PC 36:5	18:2/18:3	780.8 → 184.1	26	+	+
PC 36:4	16:0/20:4	782.6 → 184.1	28	+	+
PC 36:3	18:1/18:2	784.7 → 184.1	30	+	+
PC 36:2	18:0/18:2	786.6 → 184.1	32	+	+
PE 40:7	18:1/22:6	790.6 → 141	26	+	+
PE 40:6	18:0/22:6	792.6 → 141	28	+	+
PE 38:6	16:0/22:6	764.7 → 141	26	–	+
PE 38:5	16:0/20:5	766.5 → 141	26	–	+
PE 36:4	16:0/20:4	740.6 → 141	28	+	+
PE 38:3	18:3/20:0	770.6 → 141	32	+	–
PE 36:3	18:1/18:2	742.5 → 141	30	+	+
PE 36:2	18:0/18:2	744.6 → 141	32	+	+

Individual and total saturated fatty acid concentrations are given in Table 2. In the test samples of eggs, saturated fatty acids constitute about 44.80 % of the total amount of FAs. This percentage was at the same level for group II but had significantly decreased for groups III and IV. The most abundant of saturated fatty acids was palmitic acid, constituting about 31.37 % of the total fatty acid composition. The highest value of palmitic acid was recorded for experimental group number II. FAs with an odd number of carbon atoms in the molecule represented only 0.14 % of the total FA composition.

Table 2. Saturated fatty acid composition (% of total fatty acids) in total lipids of egg yolks (n = 3)

Fatty acids	I	II	III	IV
C14:0	0.23	0.55	0.61	0.48
C15:0	0.06	0.08	0.10	0.36
C16:0	33.04	33.06	29.47	29.90
C17:0	0.11	0.16	0.14	0.16
C18:0	11.33	12.19	10.50	9.99
C20:0	0.03	0.04	0.05	0.17
∑ SFA	44.80^{a}	46.09^{a}	40.73^{b}	41.07^{b}

I is the control group; II, III, and IV (experimental groups) are the direct 5, 9, and 13 weeks of consuming diet with enriched feed for hens, respectively. The values within a row marked with different letters are significantly different (at $p < 0.05$)

Table 3. Monounsaturated fatty acid composition (% of total fatty acids) in total lipids of egg yolks (n = 3)

Fatty acids	I	II	III	IV
C14:1	0.13	0.11	0.13	0.24
C16:1	4.47	3.11	3.42	3.22
C17:1	0.06	0.08	0.11	0.40
C18:1	35.44	37.31	40.91	42.74
C20:1	0.45	0.61	0.43	0.40
∑ MUFA	40.57^{a}	41.21^{a}	40.87^{b}	46.99^{c}

The values within a row marked with different letters are significantly different (at $p < 0.05$)

Table 4. Polyunsaturated fatty acid composition (% of total fatty acids) in total lipids of egg yolks (n = 3)

Fatty acids	I	II	III	IV
C18:2 n-6	11.31	8.86	8.95	9.02
C18:3 n-6	0.05	nd	nd	nd
C18:3 n-3	nd	0.04	0.05	nd
C20:3 n-6	0.13	0.03	nd	nd
C20:3 n-3	0.03	nd	nd	nd
C20:4 n-6	2.47	0.86	0.60	0.46
C20:5 n-3	nd	0.28	0.35	0.39
C22:5 n-6	0.12	0.23	0.62	nd
C22:6 n-3	0.52	2.39	3.56	2.07
$\sum$ PUFA	14.63^{a}	12.74^{b}	14.13^{c}	11.94^{d}

The values within a row marked with different letters are significantly different (at $p < 0.05$)

Table 5. Mean of total ω-3, total ω-6, and ω-6/ω-3 in eggs (n = 3)

Fatty acids	I	II	III	IV
ω-3	0.55^{a}	2.72^{b}	3.95^{c}	2.46^{d}
ω-6	14.09^{a}	9.98^{b}	10.18^{c}	9.48^{d}
ω-6/ω-3	25.72^{a}	3.68^{b}	2.57^{c}	3.86^{b}

The values within a row marked with different letters are significantly different (at $p < 0.05$)

The composition of monounsaturated fatty acids is presented in Table 3. The MUFA content was at the same level for groups I and II and significantly decreased for groups III and IV. The highest value of palmitoleic acid (C16:1) was recorded in the control group followed by experimental groups III and IV, while lowest concentration was recorded for II group (Table 3). Oleic acid, with an average content of 39.10 %, dominated in the overall composition.

Polyunsaturated fatty acids play a crucial role among unsaturated acids mainly due to their physiological functions. The total content of PUFA was significantly different between all investigated groups. Among the PUFAs, the most abundant acid was linoleic acid, which constituted 9.54 % of the total amount of polyunsaturated fatty acids (Table 4). The biggest difference in fatty acid composition was determined for α-linolenic acid (ALA), which increased to 0.05 % in the experimental group III. The contents of ALA and the DHA were 0.05 and 0.52 %, respectively, in the control group of PL fraction, but the presence of only DHA (2.39, 3.56, and 2.07 %) was observed in the experimental groups. This phenomenon can be explained by the metabolic pathway of PUFA, wherein DHA is synthesized from ALA by the elongation and desaturation of the hydrocarbon chain (Fig. 1). The higher content of DHA in the experimental group compared to the control group is the result of the metabolism of ALA and the ingested diet. In the case of EPA, it is formed by biosynthesis by enzymatic interconversion of ALA, which occurs (to a very limited degree) in the human body. At this stage, there are several steps of the fatty acid chain elongation, i.e., the step of lengthening the carbon chain of two carbon atoms, and the desaturation step, in which the fatty acid molecule produces next unsaturated bonds. These transformations are catalyzed by enzymes Δ6 desaturase and Δ5 and elongase (Fig. 1). The highest content of AA (C20: 4) was observed for the control group. This acid is biosynthesized in vivo from LA. The highest content of LA was detected in the control group without supplementation of the diet; therefore, the content of AA was significantly higher in the control group compared to the experimental group. Gamma-linolenic acid (GLA, C18:3) was determined only for the control group at the level of 0.05 %. GLA is produced from linoleic acid. The mixture of such enriched feed added to hens' diet forced the reduction in the content of monounsaturated fatty acids in raw egg.

Feeding hens with enriched feed had a significant effect on the ratio of ω-6/ω-3 of fatty acids, significantly reducing the value of 25.72 % for the test group and below 4 % for other groups (Table 5). Fat in eggs of quails fed with basic diet contained significantly less essential fatty acids from the omega-3 group.

Factor analysis has been performed in the range of all investigated fatty acids (20 parameters). Three repetitions of each group were used as cases. Three factors with eigenvalue bigger than 1 explained 97 % of total variance of data and grouped the cases into four previously defined classes (Fig. 4).

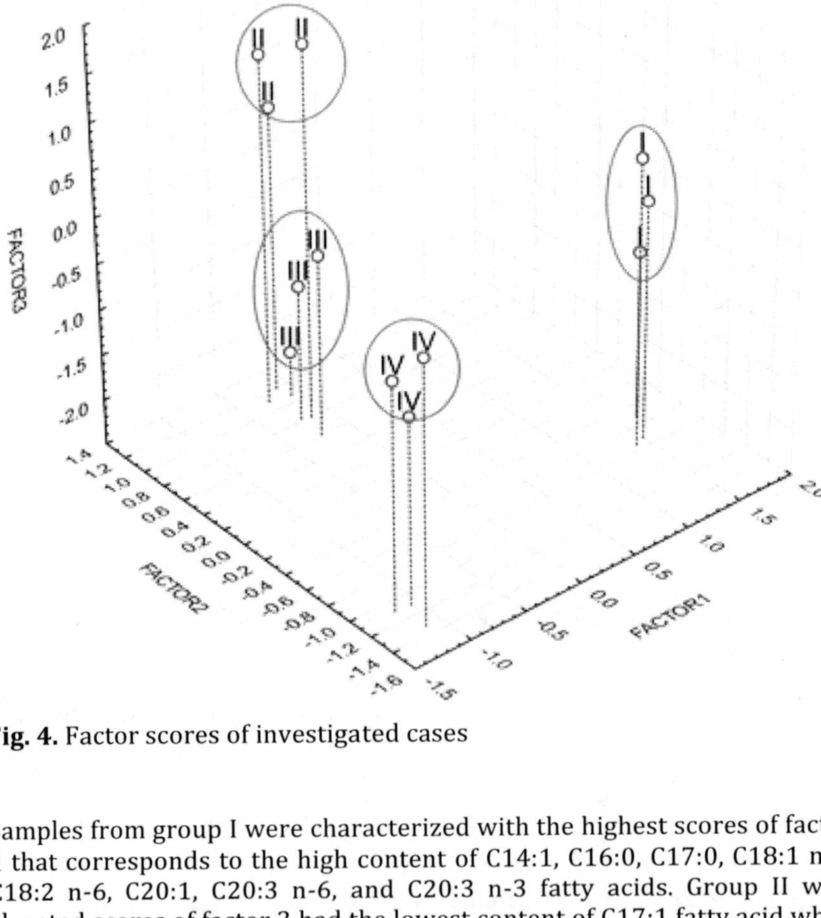

Fig. 4. Factor scores of investigated cases

Samples from group I were characterized with the highest scores of factor 1 that corresponds to the high content of C14:1, C16:0, C17:0, C18:1 n-9, C18:2 n-6, C20:1, C20:3 n-6, and C20:3 n-3 fatty acids. Group II with elevated scores of factor 3 had the lowest content of C17:1 fatty acid while distinguishing between cases from groups III and IV was possible through scores of factor 2 positively correlated with amounts of C20:5 n-3, C22:5 n-6, and C22:6 n-3 fatty acids.

The increase in the amount of linseed oil and the reduction in marine fish oil in the diet cause a significant decrease in saturated fatty acids and an increase in essential fatty acids of the n-3 family in egg yolk. Many authors conclude that hens have the ability to synthesize EPA and DHA from linoleic acid if they receive it in sufficient amounts in the diet (Kralik et al. 2008). Literature reports that egg yolk is a rich source of EFAs of the n-3 and n-6 acid family, the amount of which depends mainly on the genotype and age of laying hens (Polat et al. 2013). Bean and Leeson show that there is no effect of hens' origin on the level of ALA (C18:3) in the egg yolk (Bean and Leeson 2003). The recommended intake of omega-3 is 0.5–2 % of the total energy supplied. According to the recommendation of the American Heart Association, adults should eat meals composed of fish that are high in fat, at least twice a week; for patients with coronary heart disease, a daily intake 1 g of EPA and DHA is recommended (Block and Pearson 2006). According to the recommendations of the International Society for the Study of Fatty Acids and Lipids, the amount of energy from the intake of LA should be 2 %, and in the case of ALA it should be 0.7 % (Sanders 2000). The results of several studies suggest that the value of the omega-6/omega-3 ratio should be low, and the approximately 4–5:1 ratio of omega-6 to omega-3 fatty acids is considered as most favorable. In addition, it is believed that the high ratio of omega-6 to omega-3 acids in the diet promotes the development of cardiovascular diseases, cancer, inflammatory diseases, and autoimmune diseases (Simopoulos 2008).

CONCLUSION

The GC-MS methodology is a powerful tool for exploring differences at the metabolite level. Furthermore, we developed a sensitive and reliable HILIC-MS/MS method for the simultaneous qualitative analysis of 15 PC and 8 PE species, containing polyunsaturated fatty acids in two groups of egg yolk. Additionally, a new stationary bonded phase was applied, which demonstrated good separation of phospholipid classes in a short time. The results indicate the possibility of enriching quail eggs with polyunsaturated fatty acids, in particular EPA and DHA, by using feed containing selenium yeast, algae, and flaxseed for feeding poultry. The explanation of the mechanism of polyunsaturated fatty acids allows properly balanced diet in order to achieve good health and planned prophylactic effect. When using feed supplementation, an increase in n-3 PUFA and a reduction in ω-6/ω-3 ratio in egg yolk lipids were observed.

ACKNOWLEDGMENTS

This work was supported by the National Science Center (Cracow, Poland), no. NCN 2013/11/N/ST4/01838 (for period 2014–2016), NCBiR no. BIOSTRATEG-II/389/2015 (for period 2016–2019), Faculty of Chemistry NCU grant for young scientists nos. 2084-Ch and 2354-Ch, budget of Kuyavia and Pomerania and the European Regional Development Fund under the ROP for the years 2007–2013, project no.

RPKP.05.04.00-04-003/13, and "Krok w przyszłość" (Step into the Future) Sectoral Operational Program—Human Resources, 2015–2016.

REFERENCES

1. Ahn DU, Lee SH, Singam H, Lee EJ, Kim JC (2006) Sequential separation of main components from chicken egg yolk. Food Sci Biotechnol 15:189–195
2. Anesi A, Guella G (2015) A fast liquid chromatography-mass spectrometry methodology for membrane lipid profiling through hydrophilic interaction liquid chromatography. J Chromatogr A 1384:44–52
3. Bean LD, Leeson S (2003) Long-term effects of feeding flaxseed on performance and egg fatty acid composition of brown and white hens. Poult Sci 82:388–394
4. Behr SR (1994) Effects of fish oil and vegetable oil formula on platelet aggregation and neutrophils. Am J Clin Nutr 5:717
5. Block RC, Pearson TA (2006) The cardiovascular implications of omega-3 fatty acids. Cardiol J 13:557–569
6. Bocian S, Dziubakiewicz E, Buszewski B (2015) Influence of the charge distribution on the stationary phases zeta potential. J Sep Sci 38:2625–2629
7. Bojarowicz H, Wozniak B (2008) Wielonienasycone kwasy tłuszczowe oraz ich wpływ na skórę. Prob Hig Epidemiol 89(4):471–475
8. Brugger B, Erben G, Sandhoff R, Wieland FT, Lehmann WD (1997) Quantitative analysis of biological membrane lipids at the low picomole level by nano-electrospray ionization tandem mass spectrometry. Proc Natl AcadSci USA 94:2339–2344
9. Buszewski B, Noga S (2012) Hydrophilic interaction liquid chromatography (HILIC)—a powerful separation technique. Anal Bioanal Chem 402(1):231–247
10. Buszewski B, Jaroniec M, Gilpin RK (1994) Influence of eluent composition on retention and selectivity of alkylamide phases under reversed-phase conditions. J Chromatogr A 668:293–299
11. Christie WW (1984) In: K. MH (ed) Handbook of chromatography, vol. 1. CRC Press, Boca Raton, pp. 33–46
12. Fetterman JW, Zdanowicz MM (2009) Therapeutic potential of n-3 polyunsaturated fatty acids in disease. Am J Health Syst Pharm 66(13):1169–1179
13. Flachs P, Rossmeisl M, Bryhn M, Kopecky J (2009) Cellular and molecular effects of n−3 polyunsaturated fatty acids on adipose tissue biology and metabolism. Clin Sci 116:1–16
14. Frega NG, Pacetti D, Boselli E (2012) Characterization of phospholipid molecular species by means of HPLC-tandem mass spectrometry—applications and principles. INTECH 27:637–672
15. Friedman A, Moe S (2006) Review of the effects of omega-3 supplementation in dialysis patients. Clin J Am Soc Nephrol 1(2):182–192
16. Gladkowski W, Kiełbowicz G, Chojnacka A, Bobak Ł, Spychaj R, Dobrzański Z, Trziszka T, Wawrzeńczyk C (2014) The effect of feed supplementation with dietary sources of n-3 polyunsaturated fatty acids, flaxseed and algae Schizochytrium sp., on their incorporation into lipid fractions of Japanese quail eggs. Int J Food Sci Tech 49(8):1876–1885
17. Gladkowski W, Kiełbowicz G, Chojnacka A, Gil M, Trziszka T, Dobrzański Z, Wawrzeńczyk C (2011) Fatty acid composition of egg yolk phospholipids fractions following feed supplementation of Lohmann Brown hens humic-fat preparations. Food Chem 126:1013–1018
18. Hansen HS (1994) New biological roles of omega 6 and omega 3 fatty acids. Nutr Rev 5:162

19. Hibbeln JR, Salem M (1995) Dietary PUFAs and depression. Am J Clin Butr 1:42
20. Holčapek M, Cifkova E, Cervena B, Lisa M, Vostalova J, Galuszka J (2015) Determination of nonpolar and polar lipid classes in human plasma, erythrocytes and plasma lipoprotein fractions using ultrahigh-performance liquid chromatography-mass spectrometry. J Chromatogr A 1377:85–91
21. Kralik G, Skrtic Z, Suchy P, Strakova E, Gajcevic Z (2008) Feeding fish oil to laying hens to increase the n-3 PUFA of egg yolk. Acta Vet Brno 77:561–568
22. Lisa M, Cifkova E, Holčapek M (2011) Lipidomic profiling of biological tissues using off-line two-dimensional high-performance liquid chromatography–mass spectrometry. J Chromatogr A 1218:5146–5156
23. Namiki M (2007) Nutraceutical functions of sesame: a review. Crit Rev Food Sci Nutr 47:651–673
24. Newton JS (1996) Long chain fatty acids in health and nutrition. J Food Lipids 3:233–249
25. Pari S, Nie B, Arnold RD, Cummings BS (2016) Extraction, chromatographic and mass spectrometric methods for lipid analysis. Biomed Chromatogr 30:695–709
26. Perona JS, Ruiz-Gutierrez V (2003) Simultaneous determination of molecular species of monoacylglycerols, diacylglycerols and triacylglycerols in human very low density lipoproteins by reversed phase liquid chromatography. J Chromatogr B 785(1):89–99
27. Polat ES, Citil OB, Garip M (2013) Fatty acid composition of yolk of nine poultry species kept in their natural environment. Anim Sci Pap Rep 31:363–368
28. Sanders TAB (2000) Polyunsaturated fatty acids in the food chain in the food chain in Europe. Am J Clin Nutr 71:176–178
29. Schwalbe-Herrmann M, Willmann J, Leibfritz D (2010) Separation of phospholipid classes by hydrophilic interaction chromatography detected by electrospray ionization mass spectrometry. J Chromatogr A 1217:5179–5183
30. Simopoulos AP (2008) The importance of the omega-6/omega-3 fatty acid ratio in cardiovascular disease and other chronic diseases. Exp Biol Med 233:674–688
31. Viidanoja J (2015) Analysis of phospholipids in bio-oils and fats by hydrophilic interaction liquid chromatography–tandem mass spectrometry. J Chromatogr B 1001:140–149
32. Whelan J, Rust C (2006) Innovative dietary sources of n-3 fatty acids. Ann. Rev Nutr 26:75–103
33. Zhao J, Jiang Z-T, G-R L, Tan J, Li R (2010) Determination of phosphatidylcholine in soybean lecithin samples by high performance liquid chromatography on titania. Anal Chem 2:1779–1783

CHAPTER 20

A Simple and Robust Protocol for fast RP-HPLC Determination of Salicylates in Foods

Michał Szkop[1,2] & Urszula Szkop[3] & Paulina Kęszycka[4] & Danuta Gajewska[4]

[1].*Department of Biochemistry, Faculty of Agriculture and BiologyWarsaw University of Life Sciences – SGGWWarsawPoland*

[2].*Department of BiochemistryNencki Institute of Experimental BiologyWarsawPoland*

[3].*Autonomous Department of Microbial Biology, Faculty of Agriculture and DiologyWarsaw University of Life Sciences – SGGWWarsawPoland*

[4].*Department of Dietetics, Faculty of Human Nutrition and Consumer SciencesWarsaw University of Life Sciences – SGGWWarsawPoland*

ABSTRACT

In this methodological report, we present a simple, versatile, and reliable procedure for quantitative determination of free and conjugated forms of salicylic acid (SA) in various food products using reversed phase high-performance liquid chromatography (RP-HPLC) with fluorescence detection. The presented sample preparation protocol is considerably simplified in comparison to procedures applied previously and is based on three simple and fast extraction steps providing a supreme tool for large-scale routine assays. The limits of detection were approximately 0.021 μg g^{-1} of dried weight for spices, lyophilized fruits, or vegetables, and

0.001 μg ml^{-1} for beverages. The recoveries of the spiked SA were in the range of 87.6 to 96.6 % for all studied products. Applicability of the method was verified by the analysis of salicylate content in a wide range of products including spices (curry, oregano, red pepper), beverages (beer, brewed tea, milk, wine), lyophilized fruits (apricot, strawberry, watermelon), and vegetables (cucumber, tomato).

Keywords: Salicylic acid Salicylates Foods RP-HPLC

INTRODUCTION

Salicylates are a group of chemical compounds comprising free salicylic acid (SA) and its various derivatives, mainly glycosides and esters. They occur naturally in many food products, including fruits, vegetables, beverages, herbs, and spices (Swain et al. 1985, Scotter et al., 2007, Wood et al. 2011). Plant salicylates act as protection agents against harmful insects, fungi, and bacteria. Synthetically produced salicylates are found in over-the-counter medications such as aspirin, as well as numerous prescribed medications, cosmetics, and preservatives.

Several studies indicate that increased intake of salicylates from natural sources may be beneficial (Duthie and Wood 2011). However, there is a group of adults and children for which even a small dose of salicylates can cause serious health problems. Consumption of salicylates by these patients may cause the symptoms such as asthma, nasal polyps, urticaria, or angioedma. Patients with salicylates hypersensitivity also experienced abdominal pain, diarrhea, headache, and constitutional symptoms (Swain 1988, Morwood et al. 2005). One of the factors which may beneficially reduce the symptoms in salicylate-sensitive individuals is a proper, low-salicylate diet (Swain 1988, Cunningham 2010). For this purpose, the patients, as well as dietitians planning salicylate-restricted diets, should obtain the detailed information regarding salicylate content in foods. Unfortunately, these data are rather incomplete and often contradictory (Swain 1988, Janssen et al. 1996, Wood et al. 2011), which might be associated with the lack of the relatively cheap, fast, and simple method that allows the routine determination of salicylates in a wide range of food products. Most of the existing methods are dedicated to analyze cosmetics, pharmaceuticals, plasma, or urine (Mikami et al. 2002, Torriero et al. 2004, Vree et al. 1994, Kees and Jehnich 1996, Salinas et al. 1990), where due to the uncomplicated matrix with a low content of interfering substances and rather large content of an analyte, essential sample preparation process is not required. In turn, only a limited number of the protocols have been developed for the determination of free SA or its derivatives in foods. Moreover, these protocols included complicated and cumbersome multi-step sample processing (Swain et al. 1985, Venema et al. 1996), required rather sophisticated and expensive equipment (Scotter et al. 2007, Gruz et al. 2008), or were limited to one group of food products (Paterson et al. 2006, Irakli et al. 2012).

Therefore, the purpose of this study was to develop a simple, versatile, and fast method for the determination of salicylates in a large number of samples from various food products. Taking into consideration that SA is present in foods either in the free or bound form, and presumably the physiological effects of these two forms may be different (Hare et al. 2003), it was aimed to elaborate the method in such a manner that it allows the determination of free SA content and the overall amount of salicylates under possibly uniform sample preparation conditions. Especially, that up to date two separate protocols utilizing different extraction conditions were applied for the determination of free SA and free + bound SA content in foods (Venema et al. 1996, Irakli et al. 2012). In our opinion, such an approach may lead to some data inconsistencies due to possible differences in the free SA extraction levels when two different solvents are used. It is worth to note that most of the recently developed protocols enhanced only the methodology of free SA content determination (Scotter et al. 2007, Gruz et al. 2008, de Souza et al. 2014).

MATERIALS AND METHODS

Reagents

Salicylic acid was purchased from Sigma-Aldrich Inc. (St. Louis, MO, USA). HPLC-grade acetonitrile and ethyl acetate were purchased from Roth (Karlsruhe, Germany). Potassium dihydrogen phosphate and dipotassium hydrogen phosphate were purchased from POCH SA (Gliwice, Poland). All aqueous solution were prepared using ultra-pure Milli-Q water. HPLC eluents were filtrated prior use.

Initial Processing of Food Products

Fresh fruits and vegetables were freeze-dried with a freeze dryer Labconco 2.5 at a temperature of −40 °C and under pressure of 0.100 mBa. The freeze-dried samples were pulverized to a fine powder using a laboratory mill A-11. The pulverized fruits and vegetables were placed in plastic tubes and stored at −80 °C to prevent the loss of salicylates. Oregano was pulverized to a fine powder with mortar and pestle before extraction. Tea was prepared as infusion (2 g of tea brewed for 2 min in 200 mL of boiling water). Curry, red pepper, beer, milk, and wine were directly subjected for extraction.

Sample Preparation Procedure

Sixty milligram weights of dry food products (i.e., spices, lyophilized fruits and vegetables) were placed in 2.0-mL eppendorf tubes, and to each tube, 1.5 mL of 0.4 M K_2HPO_4 was added. For beverages (i.e., beer, brewed tea, milk, or wine), 1 mL of each product was mixed with 0.5 mL of 1.2 M K_2HPO_4. All the samples were then vigorously vortexed until fine mixture was obtained (approximately 30–180 s depending on product) and subsequently incubated for 15 min at 70 °C. After incubation, the samples were additionally vortexed for approximately 30 s, and centrifuged at 16,000×*g* for 10 min. Subsequently, 1 mL of the supernatants were taken up and placed in the 2-mL safe-lock eppendorf tubes containing 150 μL of 10 M HCl. Obtained acidified supernatants were then either directly extracted with ethyl acetate (*free SA determination*) or subjected to hydrolysis by incubation at 95 °C for 90 min and cooled down to a room temperature prior to ethyl acetate extraction (*free + bound SA determination*). Ethyl acetate extractions were performed as follows: 850 μL portions of ethyl acetate were added to the acidified supernatants, the mixtures were then extensively vortexed for 60 s and centrifuged at 16,000×*g* for 10 min. Next, 600-μL portions of the upper organic phase were removed and placed in the 2-mL safe-lock eppendorf tubes containing 600 μL of 0.2 M phosphate buffer (pH 7.8; prepared by dissolving 1.25 g of KH_2PO_4 and 15.82 g K_2HPO_4 in 0.5 L of H_2O). The samples were then vigorously vortexed for about 60 s and centrifuged at 16,000×*g* for 1 min. Finally, 200 μL portions of the lower aqueous phase were removed and placed either in HPLC vials for direct analysis, or in a new eppendorf tube and stored at −80 °C. Schematic representation of the sample preparation process is shown in Fig. 1.

Analytical Equipment and Chromatographic Conditions

The HPLC system consisted of a binary pump (Model 1525, Waters Corporation, Milford, MA, USA), a fluorimetric detector (Model 474, Waters), an autosampler (Model 717plus, Waters), and a personal computer with Breeze data acquisition and integration software (Waters). Chromatographic separations were carried out at room temperature on a C8 column (Symmetry 4.6 × 150 mm, 5 μm, Waters) guarded by a C8 precolumn (Symmetry 3.9 × 20 mm, 5 μm, Waters) using a linear gradient elution. Sixty millimolar phosphate buffer pH 6.8 (prepared by dissolving 4.08 g of KH_2PO_4 and 5.22 g of K_2HPO_4 in 1 L of H_2O) and 80 %:20 % (*v*/*v*) acetonitrile:H_2O were used as eluent A and eluent B, respectively. The flow rate of the mobile phase was 1 mL min^{-1}. The mobile phase started with eluent A: eluent B at 90 %:10 %, and varied to 70 %:30 %, 0 %:100 % and 90 %:10 % in 10, 13, and 22 min, respectively. The total run time was 27 min. The fluorimetric detector was set to excitation and emission wavelengths of 295 and 405 nm, respectively. The injection volumes were 20 µL. The content of SA in foods was quantified based on the prepared calibration curve.

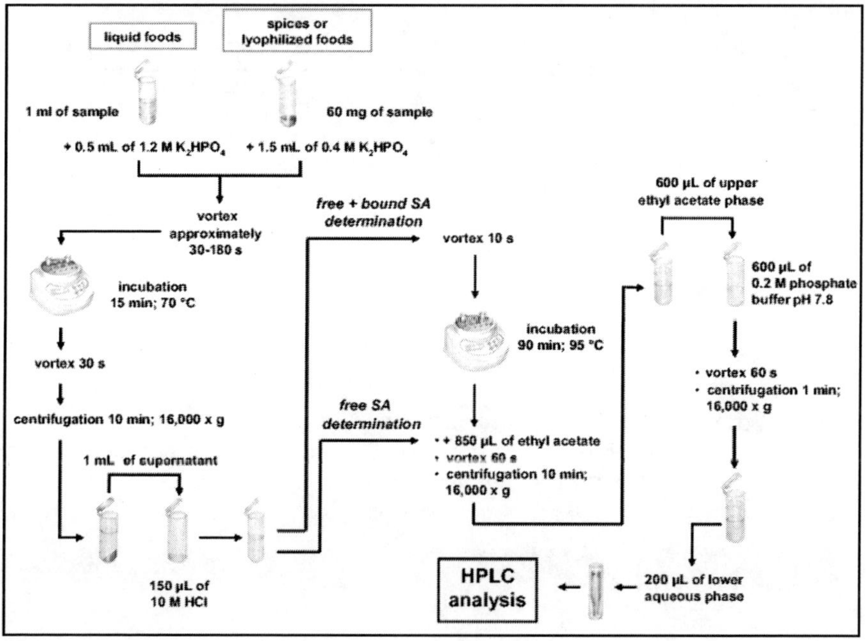

Fig. 1. Schematic representation of the sample preparation procedure

Calibration Curve, Limits of Detection, and Quantification

A set (n = 14) of working solutions, containing 3.5, 10, 25, 37.5, 50, 62.5, 125, 250, 500, 1250, 2500, 5000, 7500, and 10,000 ng mL^{-1} of SA, were prepared in triplicate in 0.2 M phosphate buffer (pH 7.8) and analyzed. The injection volumes were 20 µL. Calibration curve was prepared by least squares linear regression analysis of the peak area versus amount of the injected SA. The limits of SA detection (LOD) and quantification (LOQ) were determined from chromatograms based on signal-to-noise ratio of 3 and 10, respectively. Using this approach, the lowest detectable and quantifiable amounts of the SA in the injection volume of 20 μL were estimated to be 0.02 and 0.07 ng, respectively. These values were subsequently recalculated and expressed in micrograms per gram of the dry weight for spices, lyophilized fruits and vegetables, and in micrograms per milliliter for beverages. The recalculations were based on the following formulas:

$$LOD, LOQ = \left[\frac{\mu g}{\text{g of dry weight}}\right] = \left[\frac{ng \times 30^{a} \times 850^{b} \times 1.5^{b} \times 1000^{d}}{600^{b} \times 1000^{c} \times 60^{d}}\right] = ng \times 1.06$$

$$LOD, LOQ = \left[\frac{\mu g}{ml}\right] = \left[\frac{ng \times 30^{a} \times 850^{b} \times 1.5^{b}}{600^{b} \times 1000^{c}}\right] = ng \times 0.06$$

where the individual coefficients in the equations are related to the: [a] injection volume of 20 μL out of 600 μL, [b] volume correction factor for extraction steps, [c] units conversion from ng to μg, and [d] initial sample weight.

Recovery Studies

Recovery analysis was carried out by spiking each studied food product with SA standard at two different levels corresponding to 100 % of *free SA* and 100 % of *free + bound SA* amounts found in the same non-spiked food sample. For milk sample, where *free SA* and *free + bound SA* content was below quantification limit, the fortification level was 20 ng. Subsequently, spiked samples were processed in the identical way as non-spiked samples according to the sample preparation procedure described above. Recovery (%) was calculated by comparison of SA content in spiked and non-spiked samples.

Precision of the Method

The precision of the method was evaluated in terms of intra-day and inter-day variability. Measurements were performed on curry, brewed tea, tomato and strawberry samples. In each of these products, *free SA* and *free + bound SA* content was analyzed successively four times per day during 3 consecutive days. Intra-day and inter-day precisions were expressed as the relative standard deviations (RSD) of results from repeated measurements.

RESULTS AND DISCUSSION

In order to substantially simplify and unify the methodology of salicylate determination in foods, we have developed an efficient and versatile protocol utilizing RP-HPLC with fluorescence detection. The established sample preparation procedure is characterized by uniform extraction conditions for *free SA* and *free + bound SA* determination in both lyophilized and liquid foods, lack of sample drying step and high compactness, i.e., all sample preparation steps can be accomplished in the eppendorf tubes in a limited time (see Fig. 1). The sample preparation procedure is based on three simple extraction steps. In the first stage, SA and its derivatives are extracted from foods under optimized conditions using relatively concentrated (0.4 M) K_2HPO_4. We have found that the salicylate extraction yield increased together with increasing ionic strength and pH of the extraction buffer, although very alkaline buffers could not be applied due to hydrolysis of SA derivatives (data not shown). 0.4 M K_2HPO_4 solution was characterized by sufficiently high pH and ionic strength to exhaustively extract salicylates, buffered against pH changes toward the acidic direction, and also prevented potential pH shift to very alkaline region (phosphate pK_{a3} is ≈12.5). Two further steps of sample preparation, namely ethyl acetate extraction and back extraction to the phosphate buffer, took advantage of the physicochemical properties of SA. pKa of SA carboxyl group is about 2.97, and therefore by varying the pH of the solution, SA partition coefficient between aqueous and organic phase changes significantly. In our opinion, this extraction-re-extraction steps serve an excellent alternative to the commonly applied laborious and time consuming sample drying step (Swain et al. 1985, Venema et al. 1996, Scotter et al. 2007). Especially, that due to the volatility of SA, additional precautions are necessary during sample drying (Verberne et al. 2002). Moreover, applied back extraction buffer possesses not only sufficiently high concentration and pH to thoroughly re-extract SA from ethyl acetate layer, but was also found to be a suitable injection solvent. The developed sample preparation procedure allowed to obtain adequate samples purity with low level of interferences for all tested food products comprising spices (curry, oregano, red pepper), beverages (beer, brewed tea, milk, wine), lyophilized fruits (apricot, strawberry, watermelon), and vegetables (cucumber, tomato) (Fig. 2a, b). Under established and optimized chromatographic conditions, the retention time of SA was approximately 8.1 min. The dependence of the detector response on the amount of injected SA was found to be linear in the range from 0.07 to 200 ng. The resulting calibration curve ($y = 1817.1x - 168.5$) was characterized by a high determination coefficient ($R^2 = 0.9999$). Precision of the method was assessed by determining intra-day and inter-day variability of *free SA* and *free + bound SA* content in four selected food products representing lyophilized fruits and vegetables (strawberry, tomato), spices (curry), and beverages (brewed tea). The RSD values of *free SA* and *free + bound SA* content for both intra-day and inter-day precisions were between 2.99 and 5.71 (Table 1). These results indicate that the method is characterized by a satisfactory precision, irrespective of matrix type in which salicylate content is determined. The recovery rates of SA, analyzed for all 12 studied products, were in the range of 88.2 to 95.8 % for *free SA* determination method and in the range of 87.6 to 96.6 % for *free + bound SA* determination method (Table 2). The LOD values were approximately 0.021 $\mu g\ g^{-1}$ of dried weight for spices, lyophilized fruits and vegetables, and 0.001 $\mu g\ ml^{-1}$ for beverages, while the LOQ values were approximately 0.074 $\mu g\ g^{-1}$ of dried weight for spices, lyophilized fruits, and vegetables, and 0.004 $\mu g\ ml^{-1}$ for beverages (Table 2). The obtained recovery ranges and LOD values are similar with those obtained by Venema et al. (1996). The LOQ values of the method were low enough to allow quantitative determination of salicylates content in 11 among 12 tested products. The amounts of *free SA* ranged from 3.34 to 7.69 $\mu g\ g^{-1}$ in spices, 0.31 to 4.37 $\mu g\ g^{-1}$ in lyophilized fruits and vegetables, and 0.02 to 0.68 $\mu g\ ml^{-1}$ in beverages, whereas *free + bound SA* content ranged from 7.72 to 35.61 $\mu g\ g^{-1}$ in spices, 1.18 to 34.04 $\mu g\ g^{-1}$ in lyophilized fruits and vegetables, and 0.05 to 0.70 $\mu g\ ml^{-1}$ in beverages (Table 2). The determined values of salicylate content in the tested food products are, where comparison is possible, in agreement with the data available in the literature (Venema et al. 1996, Wood et al. 2011). The remarkable differences in the *free SA* and *free + bound SA* content observed in the case of strawberry, cucumber, and oregano (approximately 50, 10, and 5-fold increase of SA content after acidic hydrolysis, respectively), together with obtained recovery rates for SA, confirm that applied initial sample extraction step allows efficient extraction of both *free SA* and its various derivatives. For comparison, Venema et al. (1996) reported about 20, 4, and 2-fold increase of SA content in these products after alkaline hydrolysis of the samples.

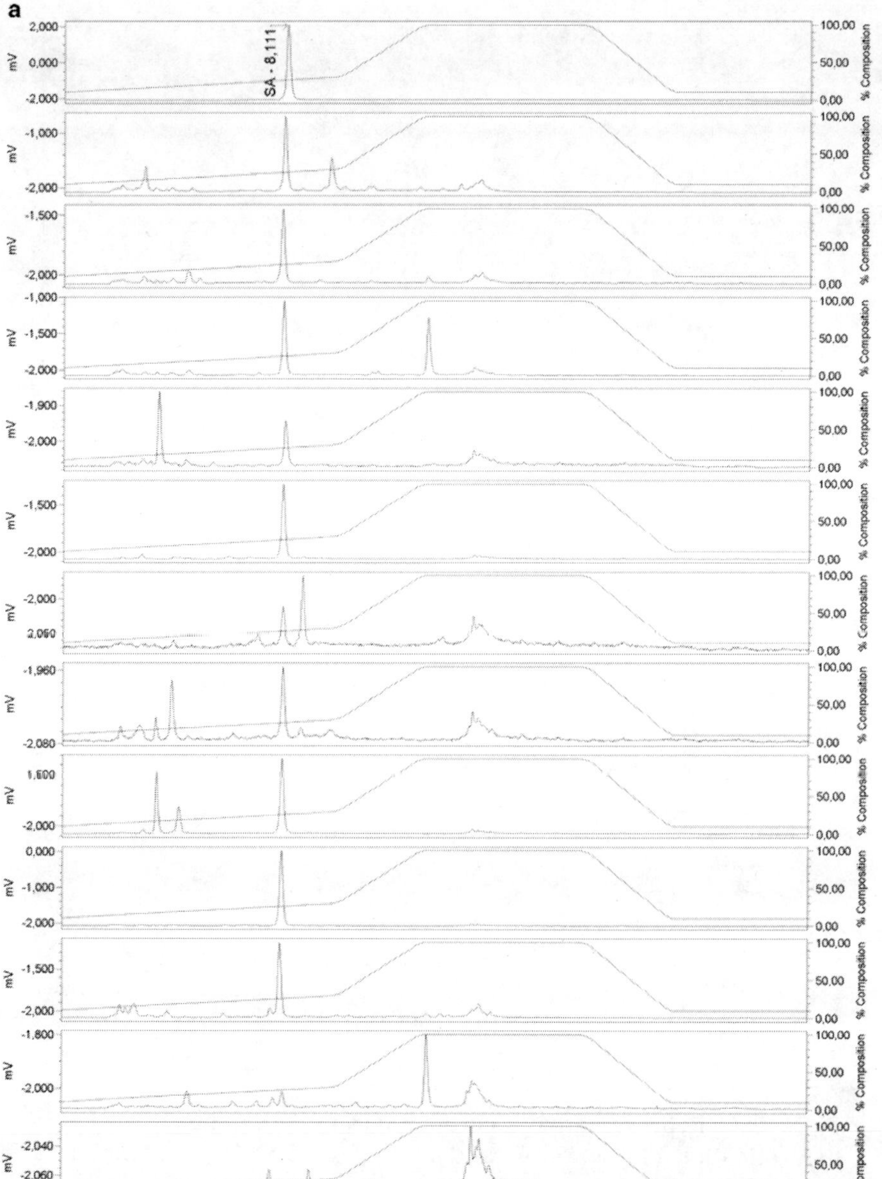
a
SA - 8,111
mV
% Composition
Minutes
2,00
4,00
6,00
8,00
10,00
12,00
14,00
16,00
18,00
20,00
22,00
24,00
26,00
100,00
50,00
0,00

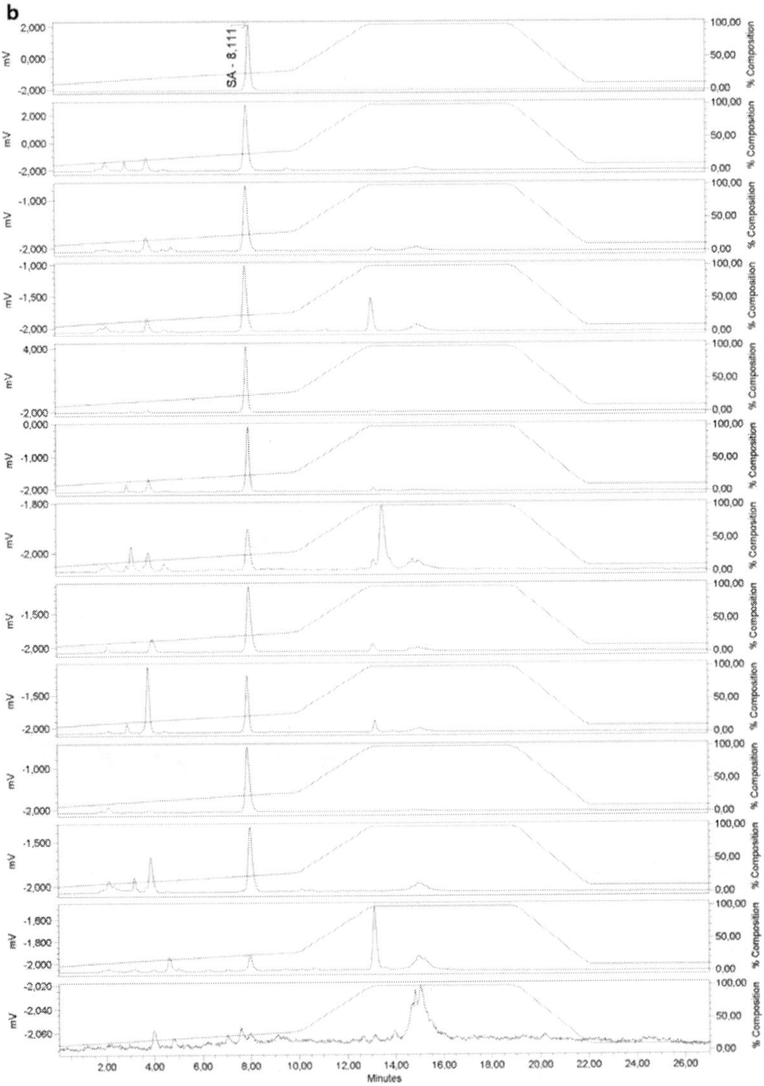

Fig. 2. Representative chromatograms obtained by RP-HPLC determination of *free SA* (**A**) and *free + bound SA* (**B**) content in various foods. Chromatogram on the top shows SA standard (1.25 μg ml^{-1}) analysis. Chromatograms from *top* to *bottom* show the analysis of samples prepared from oregano, curry, red pepper, strawberry, watermelon, apricot, cucumber, tomato, brewed tea, wine, beer, and milk. *Solid line* represents the gradient profile. All chromatograms were normalized to the highest peak

Table 1. Precision of the method in terms of intra-day and inter-day variability

Product	Intra-day variation				Inter-day variation			
	Free SA determination		Free + bound SA determination		Free SA determination		Free + bound SA determination	
	Mean ± SD	RSD (%)	Mean ± SD	RSD (%)	Mean ± SD	RSD (%)	Mean ± SD	RSD (%)
Curry	3.29 ± 0.14	4.26	9.73 ± 0.39	4.01	3.33 ± 0.15	4.50	9.72 ± 0.38	3.91
Strawberry	0.71 ± 0.03	4.23	33.82 ± 1.41	4.17	0.73 ± 0.04	5.48	33.47 ± 1.56	4.66
Tomato	3.79 ± 0.17	4.49	4.74 ± 0.15	3.16	3.80 ± 0.16	4.21	4.69 ± 0.15	3.20
Brewed Tea	0.67 ± 0.02	2.99	0.70 ± 0.04	5.71	0.69 ± 0.03	4.35	0.72 ± 0.04	5.56

SA content is expressed in μg g^{-1} of dry weight for curry, strawberry, and tomato, and in μg ml^{-1} for brewed tea. Intra-day and inter-day variations are expressed as the relative standard deviations (RSD) of results obtained on day 1 and over 3 days, respectively

Table 2. Recovery studies and the salicylates content in the tested food products

Product	SA content[a]		Recovery studies[b]		Limits of detection and quantification[c]	
	Free	Free + bound	Free SA determination	Free + bound SA determination	LOD	LOQ
Oregano	7.69 ± 0.22	35.61 ± 1.37	91.4 ± 3.18	89.1 ± 2.55	0.021	0.074
Curry	3.34 ± 0.10	9.78 ± 0.20	93.8 ± 3.65	91.7 ± 1.60		
Red pepper	5.47 ± 0.17	7.72 ± 0.18	95.3 ± 2.79	88.8 ± 3.52		
Strawberry	0.73 ± 0.03	34.04 ± 0.99	95.8 ± 2.47	94.4 ± 3.54		
Watermelon	4.37 ± 0.12	10.62 ± 0.14	92.7 ± 3.45	93.1 ± 3.16		
Apricot	0.31 ± 0.01	1.18 ± 0.05	94.2 ± 4.07	87.6 ± 4.56		
Cucumber	0.65 ± 0.02	6.71 ± 0.20	88.2 ± 2.76	91.7 ± 3.39		
Tomato	3.94 ± 0.11	4.67 ± 0.08	92.1 ± 3.38	96.6 ± 4.12		
Brewed Tea	0.68 ± 0.01	0.70 ± 0.02	93.3 ± 2.19	91.0 ± 3.37	0.001	0.004
Wine	0.29 ± 0.01	0.32 ± 0.01	94.1 ± 3.18	89.1 ± 2.14		
Beer	0.02 ± 0.00	0.05 ± 0.00	89.6 ± 3.52	91.5 ± 2.98		
Milk	<LOQ	<LOQ	90.2 ± 2.93	88.0 ± 4.18		

[a] SA content is expressed in μg g^{-1} of dry weight for spices and lyophilized foods, and in μg ml^{-1} for beverages. The results are the mean ± SD of four replicates

[b] The results are the mean ± SD of four replicates

[c] LOD and LOQ values are expressed in μg g^{-1} of dry weight for spices and lyophilized foods, and in μg ml^{-1} for beverages

All these results, taken together, indicate that the presented method is characterized by an adequate accuracy and precision for fast and reliable determination of salicylates in spices, beverages, lyophilized fruits or vegetables, and presumably also in the other food items that can be freeze-dried. The use of the lyophilized products, which are much easier to handle than the fresh products, leads to a considerable unification of the assay and facilitates the sample preparation for HPLC analyses. Therefore, the developed method should be a versatile tool for routine assays.

ACKNOWLEDGMENTS

We thank Dr. E. Hallmann for help during initial processing of food products.

CONFLICT OF INTEREST

Michał Szkop declare that he has no conflict of interest. Urszula Szkop declare that she has no conflict of interest. Paulina Kęszycka declare that she has no conflict of interest. Danuta Gajewska declare that she has no conflict of interest.

REFERENCES

1. Cunningham E (2010) Are there foods that should be avoided if a patient is sensitive to salicylates? J Am Diet Assoc 110(6):976
2. de Souza GCS, da Silva PAB, Leoterio DMDS, Paim APS, Lavorante AF (2014) A multicommuted flow system for fast screening/sequential spectrophotometric determination of dichromate, salicylic acid, hydrogen peroxide and starch in milk samples. Food Control 46:127–135
3. Duthie GG, Wood AD (2011) Natural salicylates: foods, functions and disease prevention. Food Funct 2:515–520
4. Gruz J, Novák O, Strnad M (2008) Rapid analysis of phenolic acids in beverages by UPLC–MS/MS. Food Chem 111(3):789–794
5. Hare LG, Woodside JV, Young IS (2003) Dietary salicylates. J Clin Pathol 56(9):649–650
6. Irakli MN, Samanidou VF, Biliaderis CG, Papadoyannis IN (2012) Development and validation of an HPLC-method for determination of free and bound phenolic acids in cereals after solid-phase extraction. Food Chem 134(3):1624–1632
7. Janssen PLTMK, Hollman PCH, Venema DP, van Staveren WA, Katan MB (1996) Salicylates in foods. Nutr Rev 54(11):357–359
8. Kees F, Jehnich D (1996) Simultaneous determination of acetylsalicylic acid and salicylic acid in human plasma by high-performance liquid chromatography. J Chromatogr B Biomed Appl 677(1):172–177
9. Mikami E, Goto T, Ohno T, Matsumoto H, Nishida M (2002) Simultaneous analysis of dehydroacetic acid, benzoic acid, sorbic acid and salicylic acid in cosmetic products by solid-phase extraction and high-performance liquid chromatography. J Pharm Biomed Anal 28(2):261–267
10. Morwood K, Gillis D, Smith W, Kette F (2005) Aspirin-sensitive asthma. Intern Med J 35:240–246
11. Paterson JR, Srivastava R, Baxter GJ, Graham AB, Lawrence JR (2006) Salicylic acid content of spices and its implications. J Agric Food Chem 54(8):2891–2896
12. Salinas F, Muñoz de la Peña A, Durán-Merás I, Soledad Durán M (1990) Determination of salicylic acid and its metabolites in urine by derivative synchronous spectrofluorimetry. Analyst 115(7):1007–1011

13. Scotter MJ, Roberts DP, Wilson LA, Howard FA, Davis J, Mansell N (2007) Free salicylic acid and acetyl salicylic acid content of foods using gas chromatography–mass spectrometry. Food Chem 105(1):273–279
14. Swain AR, Dutton SP, Truswell AS (1985) Salicylates in foods. J Am Diet Assoc 85:950–96
15. Swain, AR (1988) The role of natural salicylates in food intolerance. PhD thesis. Sydney Australia: University of Sydney
16. Torriero AA, Luco JM, Sereno L, Raba J (2004) Voltammetric determination of salicylic acid in pharmaceuticals formulations of acetylsalicylic acid. Talanta 62(2):247–254
17. Wood A, Baxter G, Thies F, Kyle J, Duthie G (2011) A systematic review of salicylates in foods: estimated daily intake of a Scottish population. Mol Nutr Food Res 55:7–14
18. Venema DP, Hollman PC, Janssen KP, Katan MB (1996) Determination of acetylsalicylic acid and salicylic acid in foods, using HPLC with fluorescence detection. J Agric Food Chem 44(7):1762–1767
19. Verberne MC, Brouwer N, Delbianco F, Linthorst HJ, Bol JF, Verpoorte R (2002) Method for the extraction of the volatile compound salicylic acid from tobacco leaf material. Phytochem Anal 13(1):45–50
20. Vree TB, van Ewijk-Beneken Kolmer EW, Verwey-van Wissen CP, Hekster YA (1994) Direct gradient reversed-phase high-performance liquid chromatographic determination of salicylic acid, with the corresponding glycine and glucuronide conjugates in human plasma and urine. J Chromatogr 652(2):161–170

Index